Plant Conservation

PEOPLE AND PLANTS CONSERVATION SERIES

Series Editor
Martin Walters

Series Originator
Alan Hamilton

People and Plants is a joint initiative of WWF,
the United Nations Educational, Scientific
and Cultural Organization (UNESCO)
and the Royal Botanic Gardens, Kew

AVAILABLE

Applied Ethnobotany: People, Wild Plant Use and Conservation
Anthony B. Cunningham

Biodiversity and Traditional Knowledge: Equitable Partnerships in Practice
Sarah A. Laird (ed)

Carving Out a Future:
Forests, Livelihoods and the International Woodcarving Trade
Anthony Cunningham, Brian Belcher and Bruce Campbell

Ethnobotany: A Methods Manual
Gary J. Martin

People, Plants and Protected Areas: A Guide to In Situ *Management*
John Tuxill and Gary Paul Nabhan

Plant Conservation
Alan Hamilton and Patrick Hamilton

Plant Invaders: The Threat to Natural Ecosystems
Quentin C. B. Cronk and Janice L. Fuller

Tapping the Green Market:
Certification and Management of Non-Timber Forest Products
Patricia Shanley, Alan R. Pierce, Sarah A. Laird and Abraham Guillén (eds)

Uncovering the Hidden Harvest:
Valuation Methods for Woodland and Forest Resources
Bruce M. Campbell and Martin K. Luckert (eds)

FORTHCOMING

Plant Identification:
Creating User-Friendly Field Guides for Biodiversity Management
William Hawthorne and Anna Lawrence

Plant Conservation

An Ecosystem Approach

Alan Hamilton and Patrick Hamilton

WWF

UNESCO

ROYAL BOTANIC GARDENS KEW

EARTHSCAN

London • Sterling, VA

First published by Earthscan in the UK and USA in 2006

ISBN-10: 1-84407-083-2 paperback
1-84407-082-4 hardback

ISBN-13: 978-1-84407-083-1 paperback
978-1-84407-082-4 hardback

Typesetting by MapSet Ltd, Gateshead, UK
Cover design by Yvonne Booth
Diagrams drawn by Martina Höft

Printed and bound in the UK by Bath Press

For a full list of publications please contact:

Earthscan
8–12 Camden High Street
London, NW1 0JH, UK
Tel: +44 (0)20 7387 8558
Fax: +44 (0)20 7387 8998
Email: earthinfo@earthscan.co.uk
Web: **www.earthscan.co.uk**

22883 Quicksilver Drive, Sterling, VA 20166-2012, USA

Earthscan is an imprint of James and James (Science Publishers) Ltd and publishes in
association with the International Institute for Environment and Development

A catalogue record for this book is available from the British Library

Library of Congress Cataloging-in-Publication Data

Hamilton, A. C. (Alan Charles), 1945-
  Plant conservation / Alan Hamilton and Patrick Hamilton.
      p. cm.
  Includes bibliographical references and index.
  ISBN 1-84407-082-4   ISBN 1-84407-083-2
  1. Plant conservation. I. Hamilton, Patrick. II. Title.
  QK86.A1H36 2005
  333.95'316—dc22

                                                                    2005019429

Contents

List of figures, tables and boxes

FIGURES

TABLES

BOXES

The People and Plants Initiative

This is one of the last books to be produced as a product of the People and Plants Initiative (1992–2005), a partnership programme of WWF, UNESCO and the Royal Botanic Gardens, Kew. People and Plants was designed to increase capacity for community-based plant conservation internationally. Rooted in a dedicated core team, People and Plants has mounted integrated projects of capacity-building in applied ethnobotany in several countries and regions, most notably Bolivia, Kenya, Malaysia, Mexico, Nepal, Pakistan, the South Pacific, Southeast Asia (especially Sabah), Tanzania, Uganda and Zimbabwe. There have been further activities in Bangladesh, Bhutan, Cameroon, China, the Dominican Republic, Ethiopia, India, Indonesia, Mozambique, the Philippines, Rwanda and elsewhere. The present book draws on many field examples of the People and Plants Initiative. However, the subject matter is not limited to applied ethnobotany, extending into other aspects of plant conservation.

Apart from its books, several other series of publications have been produced under the People and Plants Initiative. A full list can be viewed at:

www.peopleandplants.org

People and Plants assisted in the founding of a number of new conservation groups. One of these, People and Plants International (*not* to be confused with its forerunner, the People and Plants Initiative), is a new international non-profit group, registered in New York, dedicated to the pursuit of conservation and sustainable management of natural resources. Carrying forward the work started by the People and Plants Initiative, People and Plants International has now embarked on its mission of facilitating organizations to obtain the resources they need for sustainable resource management, training and local conservation work. People and Plants International brings together an invited network of some of the most experienced global practitioners in applied ethnobotany. Please view the website above for further details.

Alan Hamilton
Plant Conservation and Livelihoods Programme Manager
Plantlife International
Formerly WWF Programme Coordinator of the People and Plants Initiative

Foreword

The 14-year-long People and Plants Initiative (PPI) of the World Wide Fund for Nature (WWF), the United Nations Educational, Social and Cultural Organization (UNESCO) and the Royal Botanic Gardens, Kew (RBG) that ran from 1992 to 2005 has done much to integrate applied ethnobotany within the conservation world. It is therefore highly appropriate that the coordinator of the programme, Alan Hamilton, should (together with his son) produce a book that brings together many of the aspects addressed and the lessons learned from the programme. This book, which is a practical manual for any conservationist, goes far beyond the topic of applied ethnobotany. It skilfully weaves together in a logical and practical way much of the information gathered by the PPI; but it also includes and addresses the major aspects of contemporary plant conservation. It is full of useful examples from many different developing countries where biodiversity is now being managed and conserved through the initiatives of the PPI. We read about the management of resources in Ayubia National Park in Pakistan, the conservation of the medicinal plants by the *amchis* in Nepal, the control of wood for carving in Kenya and many other examples of practical resource conservation in such places as Malaysia, Uganda and Zimbabwe.

It is clearly outlined why plant conservation is so important and why we need the many methods, approaches and techniques described here. The threats to plants are numerous, varying from habitat destruction and loss, through invasive species of plants and animals, to the threats imposed by climate change. The many services to the environment and the many uses for plants listed here demonstrate how important it is to conserve and to use sustainably as many species as possible. Here we have a balanced approach from theoretical and legal matters to the practical, from the field to the laboratory, and about both *in situ* and *ex situ* methods. This is not just a book to read; it is one that calls every reader to action either locally or on a more global scale.

The clear message that comes through this book is the importance of involving local people if conservation is to be successful. When they are involved, sustainable uses of the resources can be achieved. It is no good just setting up exclusive areas where the locals have no say in matters. The 14 years of the People and Plants Initiative and its series of publications have produced much useful information and it is good to summarize and evaluate it in a single volume. This will prove a useful tool for those people who are putting the recently drafted Global Strategy for Plant Conservation into action and especially those who are working with local people. However, the extensive examples show that everyone can get involved in some way in plant conservation, from the schoolchild to the professional plant conservationist. Here are many useful suggestions and resources for the practical conservationist. It is to be hoped that the ethnobotanical approach to conservation espoused here increases rather than decreases in spite of the conclusion of the People and Plants Initiative.

Professor Sir Ghillean Prance
Former Director, Royal Botanic Gardens, Kew
Scientific Director, The Eden Project, Cornwall
Lyme Regis, May 2005

Introduction

This book is an exploration of how to save the world's plants. The evidence suggests that we live at a special time in relation to plant conservation because right now the world is on track to lose a substantial part of its flora (Chapter 1). Plants are not an obscure group of organisms. Whether wild or cultivated, plants constitute a basic feature of the land, contributing to its overall interest and diversity. Plants are involved in many ecological processes and are intimately tied to human economies. How plants are managed will fundamentally determine the long-term success of humanity in terms of both conservation and development.

The types of plants and vegetation found at any locality on the world's surface are major influences over the types of animals that exist, the local micro-climate, supplies of water yielded and erosion of the soil. Products from locally growing plants, both cultivated and wild, play essential roles in the lives of many rural people in developing countries. The rights of people to use local wild plant resources and the conditions attached to this use can have major implications for how habitats are managed (Chapter 10). Those interested in saving charismatic animals, such as gorilla and snow leopard, should note that it is hardly an exaggeration to say that *if you want to save the animals, work on the plants* (see the case studies in Chapters 9 and 11).

Progress in plant conservation will depend to a great degree upon the thoughtfulness, drive and influence of those people who have a special knowledge of, and concern for, plants (Chapter 3). Because of this, scientifically trained botanists have a particular role to play in plant conservation. A fundamental problem in plant conservation today is that in many countries fewer students are being trained in 'whole plant' botany – the sort of botany that is useful for conservation and sustainable development. At the same time, folk traditions of the uses and methods of managing plants are in decline. It is an irony that, just when so much concern is being expressed about the environment, there is a growing shortage of people – whether scientifically-trained botanists or traditional experts – who are knowledgeable about plants.

This book has been greatly influenced by the experience of Alan Hamilton as the World Wide Fund for Nature (WWF) coordinator of the People and Plants Initiative. This was a programme of WWF, the United Nations Educational, Social and Cultural Organization (UNESCO) and the Royal Botanic Gardens, Kew (RBG) that spanned the period from 1992 to 2005 and was designed to increase the capacity for work internationally in applied ethnobotany. Despite its esoteric-sounding name, applied ethnobotany is arguably the basic discipline needed for plant conservation (Chapter 12). This is because *ethnobotany* deals with the relationships between people and plants, while *applied* ethnobotany is the branch of ethnobotany concerned with its applications to conservation and sustainable development. Placing relationships between people and plants at the centre of attention helps to orientate plant conservation activities, including the identification of key tasks required for success (see Chapter 1).

This work contains accounts of many aspects of plant conservation, including types of threats (Chapter 2), the roles of learning and research (Chapter 4), methods of managing plants (Chapter 6), approaches to *in situ* plant conservation (Chapter 11), working with communities (Chapter 12), *ex situ* conservation (Chapter 13) and working on trade systems (Chapter 14). Background material is provided on plant types and their lives (Chapter 5), plant geography (Chapter 7) and the selection of conservation priorities (Chapter 9). Plant conservation is a young discipline with many uncertainties about how best to proceed. The authors hope that this book will contribute in a small way to advancing the subject and stimulate more people to play their parts in this, one of the most vital challenges of modern times.

People and Plants Partners

WWF

The World Wide Fund for Nature (WWF), founded in 1961, is the world's largest private nature conservation organization. It consists of 29 national organizations and associates and works in more than 100 countries. The coordinating headquarters are in Gland, Switzerland. The WWF mission is to conserve biodiversity, to ensure that the use of renewable natural resources is sustainable and to promote actions to reduce pollution and wasteful consumption.

UNESCO

The United Nations Educational, Scientific and Cultural Organization (UNESCO) is the only UN agency with a mandate spanning the fields of science (including social sciences), education, culture and communication. UNESCO has over 40 years of experience in testing interdisciplinary approaches to solving environmental and development problems in programmes such as that on Man and the Biosphere (MAB). An international network of biosphere reserves provides sites for conservation of biological diversity, long-term ecological research and demonstrating approaches to the sustainable use of natural resources.

ROYAL BOTANIC GARDENS, KEW

The Royal Botanic Gardens (RBG), Kew, has 150 professional staff and associated researchers, and works with partners in over 42 countries. Research focuses on taxonomy, preparation of floras, economic botany, plant biochemistry and many other specialized fields. The Royal Botanic Gardens has one of the largest herbaria in the world and an excellent botanic library.

During its last and most recent phase (2001–2005), the People and Plants Initiative was financially supported by the Darwin Initiative and the Department for International Development (DFID), UK. Plantlife International has provided essential moral support, without which this book might not have been completed. Its expert staff have contributed much useful advice and information.

PLANTLIFE

DISCLAIMER

While the organizations concerned with the production of this book firmly believe that its subject is of great importance for conservation, they are not responsible for the detailed opinions expressed.

Plant conservation websites

CONSERVATION ORGANIZATIONS, INITIATIVES, ISSUES

ARC: www.arcworld.org (Alliance of Religions and Conservation)

BGCI: www.bgci.org.uk/home/index.html (Botanic Gardens Conservation International; links to botanic gardens)

CBD: www.biodiv.org (Convention on Biological Diversity; includes Global Strategy for Plant Conservation)

CGIAR: www.cgiar.org (Consultative Group on International Agricultural Research)

CITES: www.cites.org

Climate change: www.ipcc.ch (Intergovernmental Panel on Climate Change; includes assessment of carbon sequestration)

DELTA: delta-intkey.com/ (Descriptive Language for Taxonomy)

FAO: www.fao.org (Food and Agriculture Organization of the UN; links to other information on plant genetic resources)

Forest Stewardship Council: www.fsc.org

Global Trees Campaign: www.globaltrees.org

ILDIS: www.ildis.org (International Legume Database and Information Service)

IUCN: www.issg.org (Invasive Species Specialist Group)

IUCN: www.iucn.org/themes/ssc/plants/plantshome.html (Plant Conservation site)

IUCN: www.redlist.org (Red List; links to many plant conservation sites)

IUCN: www.iucn.org/themes/ssc/programs/sisindex.htm (Species Information Service)

IUCN: www.iucn.org/themes/wcpa/wcpa/protectedareas.htm (World Commission on Protected areas)

People and Plants International: www.peopleandplants.org

Plant Talk: www.plant-talk.org (plant conservation magazine)

Plantlife International: www.plantlife.org.uk

Rainforest Alliance: www.rainforest-alliance.org/

TRAFFIC: www.traffic.org (conservation and international plant trade)

UNESCO: www.unesco.org/mab/ (Man and Biosphere)

UNESCO: whc.unesco.org/ (World Heritage Convention)

WCMC: www.wcmc.org.uk (World Conservation Monitoring Centre)

WHO: www.who.int/medicines/library/trm/medicinalplants/agricultural.shtml (World Health Organization, Good Agricultural and Collection Standards for Medicinal Plants)

WWF: www.wwf-uk.org/core/wildlife/fs_0000000029.asp (plant conservation site)

REGIONAL AND NATIONAL SITES

Africa: www.prota.org (Plant Resources of Tropical Africa)

Australia (Western): http:www.calm.wa.gov.au (includes FloraBase)

Australian Network for Plant Conservation: www.anbg.gov.au/anpc

Caribbean: http://funredes.org/endacaribe/tramil.html (TRAMIL, medicinal plants)

Colombia: www.humboldt.org.co (Humboldt Institute)

Ethiopia: www.telecom.net.et/~ibcr/ (Institute of Biodiversity Conservation and Research)

Europe: www.plantaeuropa.org (Planta Europa, European Plant Conservation Strategy)

Guyanas: www.mnh.si.edu/biodiversity/bdg/medicinal (medicinal plants)

India: www.frlht-india.org (Foundation for the Revitalisation of Local Health Traditions)

South Africa: www.dwaf.gov.za/wfw (Department of Water Affairs and Forestry, Working for Water Programme; links to invasive species sites)

UK: www.nccpg.com (The National Council for the Conservation of Plants and Gardens)

UK: www.rbgkew.org.uk (Royal Botanic Gardens, Kew)

USA: www.centerforplantconservation.org (Center for Plant Conservation; US threatened species)

USA: www. plantsavers.org (United Plant Savers; includes wildcrafting guidelines)

Acknowledgements

The authors are very grateful to a number of people who kindly provided information for this book or checked specific passages. We wish to express our particular gratitude to Pierre Binggeli for providing Figure 10.4 (monastery forest in Ethiopia); Irwin M. Brodo, Elizabeth Haworth and Michael Wynne for information on species numbers (Box 1.1); Anthony Cunningham for some corrections to the section on 'Assessing sustainability of resource supply' in Chapter 11; Craig Hilton-Taylor for information on the World Conservation Union (IUCN) Red List categories in Chapter 9; Robert Höft for information on the Convention on Biological Diversity (CBD) in Chapter 3; Nicola Hutchinson for specifics on UK legislation in Chapter 10; Neville Marchant for details of the Western Australia plant network in Chapter 3; Pei Shengji for information on plant conservation in China and general orientation, particularly regarding Chapters 5 and 7; Elizabeth Radford of Plantlife International for information in Chapter 9 on Important Plant Areas; Dawn Robinson for details of land tenure in Mexico in Chapter 10; Alison Rosser for information on the Convention on the International Trade in Endangered Species of Wild Fauna and Flora (CITES) in Chapters 3 and 14; Susanne Schmitt for information on plant trade in Chapter 14; Suzanne Sharrock and Peter and Diane Wyse Jackson for details of botanic gardens in Chapters 3 and 13; and Tim Wilkins for data on managing plants in Box 6.1.

Special thanks are due to Martina Höft for drawing many of the figures, and to Jonathan Sinclair Wilson, publishing director of Earthscan, and Martin Walters, series editor of the People and Plant Conservation Series, for their interest in the production of this book, as well as their practical assistance. The authors further appreciate the care with which Hamish Ironside of Earthscan and copy-editor Andrea Service have edited and prepared this book for publication.

Alan Hamilton has had the privilege to learn from many knowledgeable and inspiring colleagues during the course of the People and Plants Initiative. They are far too numerous to mention all here – and apologies for those omitted – but special thanks are due to: Rabia Afza; Agnes Lee Agama; Habib Ahmad; Zakiya Aloyce; Reza Asmi; Yildiz Aumeeruddy-Thomas; Abdullah Ayaz; Michael Berjak; Remegius Bukenya-Ziraba; Dominic Byarugaba; Javier Caballero; June Cale; Bruce Campbell; C. L. Chan; Sudipto Chatterjee; Simon Choge; Pam Coghlan; Quentin Cronk; Anthony Cunningham; Geoff Davison; Olga Lucia Sanabria Diago; Saskia Flipsen; Steve Gartlan; Suresh Ghimire; Gabriella Graz; Jeremy Grimsdell; Chandra Gurung; Kala Gurung; Ali Habib;

Malcolm Hadley; Debbie Heaney; Mary Lou Higgins; Inés Hinojosa; Alison Hoare; Martina and Robert Höft; Hu Huabin; Rusaslina Idrus; Asma Jabeen; Shahril Kamarulzaman; Sam Kanyamibwa; Claudia Karez; Peris Kariuki; Wolfgang Kathe; Mikaail Kavanagh; Fassil Kebebew; John Kessy; Ashiq Ahmad Khan; Mir Ajab Khan; Khasbagen; Sonia Lagos Witte; Sarah Laird; Yeshi Choden Lama; Danna Leaman; Anna Lewington; Liu Wenjiang; David Maingi; Gary Martin; Khan Bahadar Marwat; Patrick Maundu; Fonki Mbenkum; James Morley; Patrick Mucunguzi; Jackson Mutebi; Hermann Mwageni; Gary Nabhan; Jamili Nais; Gezahegn Negussie; Mark Nesbitt; Peter Newbourne; Deborah Nicholls; Sally Nicholson; Joseph Obua; Raymond Obunga; Patrick Omari; Arthur Pei; Pei Shengji; Balu Perumal; Chuck Peters; Michel Pimbert; Ghillean Prance; Hew Prendegast; Ruben Prieto; Yuriko Prod'hom; Louis Putzel; Peter Ramshaw; Ajay Rastogi; Sarah Safdar; Silvia Salas; Paul Siegel; Patricia Shanley; Mingma Norbu Sherpa; Zabta Shinwari; Krishna Shrestha; Tirtha Shrestha; Stella Simiyu; Adrian Stratta; Peter Sumbi; Kesaia Tabunakawai; Chusie Trisonthi; John Tuxill; Justine Vaz; Martin Walters; Dominic White; Rob Wild; Chris Wilde; Robert Wilkinson; Jonathan Sinclair Wilson; Sejal Worah; Yang Zhiwei; and Wendy Yap.

Apart from those involved in the People and Plants Initiative, others who have contributed to Alan Hamilton's education in plant conservation include Lyn de Alwis; Sema Atay; Jonathan Baranga; Pierre Binggeli; Tom Butynski; Andy Byfield; Henrique Costa Neves; Michael Daniel; Steve Davis; Asha de Silva; Gunaratne de Silva; Desalegn Desissa; Ehsan Dulloo; Fang Zhen-dong; Danielle Florens; Tara Gandhi; David Given; He Shanan; Inga Hedberg; Olov Hedberg; Vernon Heywood; Susanne Honnef; Peter Howard; Kim Howell; Tony Katende; Jon Lovett; Quentin Luke; Maati Maatta; Olga Herrera MacBryde; K. M. Matthew; Israel Mwasha; Mianda-Bungi Ndjele; Matti Nummelin; Hannington Oryem-Origa; Nerimen Özhatay; Dimitrios Phitos; Alan Pierce; Nat Quansah; Raimundo Quintal; Ann Robertson; Christopher Ruffo; Wendy Strahm; Don Sumithraarachchi; Eleni Svoronou; Hugh Synge; and Clive Wicks. Alan Hamilton would also like to give a special word of thanks to his former colleagues in the now defunct Plants Conservation Unit of WWF International (later in WWF-UK), Ros Coles and Susanne Schmitt. We did try hard for plant conservation!

Alan Hamilton and Patrick Hamilton would like to thank Naomi Hamilton and Susan Manby for their encouragement during the writing of this book. Patrick Hamilton is grateful to Linda and Chloe Sadlier for their continued support.

1

Perspectives on Plant Conservation

PLANTS AND CONSERVATION

Plants are fundamental components of many ecosystems on Earth, forming their productive bases and physical structures, and producing resources which support a diversity of other organisms. In total, there are around 300,000 species of plants, excluding lichens and fungi (see Box 1.1). Plants provide food and other materials essential for human livelihoods, and are involved in many ecological processes that benefit people. The plant cover of the land can significantly influence climate, water supplies and the stability of the soil.

Over the last 500 years – and especially during the last 50 – relationships between people and their planet, including its plants, have been transformed. Before then, the great majority of people lived in rural settings, and used and depended extensively upon locally growing plants, both cultivated and wild. Subsequently, people worldwide have increasingly become part of ever more extensive economic, cultural and social systems. As this has been happening, human pressure on the Earth's resources has become greatly magnified through a dramatic rise in the size of the human population, a considerable proportion of which has come to enjoy levels of use of natural resources that were once the preserve of kings.

As human numbers and pressure on resources have increased, so conservation problems have developed on larger scales. Human populations have long caused local extinctions of organisms (better documented for animals than plants), suffered periodic shortages of natural resources (for example, as demonstrated by recurrent famines) and been responsible for local environmental degradation (such as deforestation and soil erosion). What has now happened is that people are endangering species, overexploiting natural resources and causing pollution on a global scale. The purpose of this book is to explore the steps that can be taken to improve matters with respect to plants.

The term conservation has both active and passive meanings. In an *active* sense, conservation is the taking of actions to ensure that things that are valued stand a better chance of persisting into the future. There are various features

---

BOX 1.1 ESTIMATED NUMBER OF SPECIES IN THE MAIN GROUPS OF PLANTS (INCLUDING FUNGI AND LICHENS)

Fungi: 1,500,000 species, of which approximately 80,000 are more conspicuous, mushroom types of fungi;

Lichens: about 20,000 species; these include more conspicuous macrolichens (larger lichens);

Brown algae: about 1500 species, the great majority of which are marine (brown seaweeds);

Red algae: about 4000–6000 species, the majority being marine (red seaweeds);

Green algae: about 30,000 species of which most are freshwater and roughly 1000 are marine (green seaweeds);

Bryophytes (mosses, liverworts and hornworts): 15,000 species;

Vascular (higher) plants: 272,655 species;

  Pteridophytes (ferns and allies): 13,025 species;

  Seed plants (spermatophyta): 259,630 species;[*]

    Gymnosperms (vascular plants carrying naked seeds): 980 species;

    Flowering plants (angiosperms; vascular plants with seeds enclosed within an ovary): 258,650 species;

      Dicotyledons (the largest group of flowering plants): 199,350 species; and

      Monocotyledons (flowering plants with only one cotyledon, or seed leaf, as opposed to dicotyledons, which normally have two): 59,300 species.

These are the groups commonly recognized as plants, according to traditional botanical thinking. Nevertheless, they are not all natural groups or even 'plants' in an evolutionary sense (see Figure 8.2), and there is considerable uncertainty about the number of species in all groups because of inadequate primary exploration and confused taxonomy.

Note: [*] Other estimates for the number of species of seed plants are 223,300 (Scotland and Wortley, 2003), 300,000–320,000 (Prance et al, 2000), 421,968 (Bramwell, 2002) and 422,127 (Govaerts, 2001).

Sources: Galloway, 1992; Hawksworth et al, 1995; Norton et al, 1996; Graham and Wilcox, 2000; Hallingbäck and Hodgetts, 2000; Hawksworth, 2001; Porembski and Barthlott, 2000; Baillie et al, 2004; Elizabeth Haworth, personal communication, 2005

---

of the world of plants than can be subjects of conservation attention, including plant species, the genetic diversity found within plant species, plant resources of various sorts and types of vegetation. Conservation in the active sense can be closely related to restoration, which involves extending efforts beyond just trying to protect those aspects of the plant world that are of interest, to try to enhance their conservation worth.

In a *passive* sense, the term conservation refers to actions beneficial for plant conservation carried out by people with conservation not, or only partly, in mind, as demonstrated by some traditional conservation practices (see Chapter 10). A major aim in *active* plant conservation is to institutionalize the everyday activities of people in order to favour plant conservation in this passive way. In brief, conservation should be promoted as a culture (Long Chunlin and Pei Shengji, 2003). Plant conservation should not just be a crisis discipline, but also an aspiration in terms of how people normally behave.

CONSERVATION CONCERNS RELATING TO PLANT DIVERSITY

One of the deepest concerns in plant conservation is the large number of species under threat of extinction. Preventing the extinction of species is a top conservation priority because, once lost, species are gone forever. From admittedly very poor quality data, it seems that we still have time to save many species of plants, provided that we act now in a systematic and concerted way. 'Only' 380 species of higher plants are known to have become recently extinct (Walter and Gillett, 1997). Although certainly an underestimate, this relatively low figure (compared with the total number of plant species on Earth – see Box 1.1) gives a measure of reassurance. It appears that the current extinction episode for plants is just starting, rather than already well under way, as it is with larger animals. The massive number of extinctions of large mammals that has already occurred during the current human-induced extinction episode emphasizes how foolish we would be not to take alarm bells rung for plants extremely seriously. Seventy-three to 86 per cent of all species of larger mammals in Australia and the Americas have already been exterminated (Barnosky, 1989). The present wave of extinctions of animals and plants (already extinct or predicted) is very exceptional by the standards of geological time. The modern extinction rate is estimated to be 100 to 1000 times faster than the average over the last 570 million years (Myers, 1996; Novacek and Cleveland, 2001).

One way of assessing the number of plant species under threat of extinction is through species-by-species estimates. This is only possible with any degree of reliability for plant groups or places that are better known. There are extensive parts of the world, especially in the tropics and subtropics, which cannot be assessed properly using this method. However, other approaches, such as calculating extinction rates using biogeographic theory, suggest that the problem is universal (Lovejoy, 2002). Table 1.1 shows estimates for the number of threatened species for some cases that are considered to be reasonably accurate. It also includes less reliable estimates made for the world flora as a whole and for world trees. Estimates for the proportion of species under threat of extinction in this sample range from 9 to 52 per cent, which is probably a fair reflection of reality because there is no doubt that certain plant groups and certain places are much more threatened than others. For example, the high vulnerability of cycads as a group (52 per cent of species threatened) reflects the fact that these are slow-growing plants, have few growing points (and are thus vulnerable to damage) and are often the targets of plant collectors (see Figure 1.1). Craig Hilton-Taylor, who is the Red List Programme

Table 1.1 *Estimates of numbers of threatened plant species for various plant groups and areas*

Plant group or part of world	Number of species in flora	Number of threatened species	Percentage of flora threatened	IUCN Red List categories and criteria used	Comments
Global flora	270,000	33,798	> 13	Pre-1994	The figure of 33,798 includes 380 species classified as Extinct in the Wild, 371 species as Extinct in the Wild or Endangered, and 14,504 species as Rare*. The figure given for the percentage of flora threatened is a minimum estimate, especially because of very incomplete coverage of the tropics (Bramwell, 2002). Total size of world flora from Baillie et al, 2004, rounded to nearest 1000.
Global tree flora	80,000–100,000	8753	9–11	Expert assessments, starting with those in Walter and Gillett, 1997	The figure of 8753 includes 77 species classified as Extinct and 18 species classified as Extinct in the Wild.
Global conifers	618	153	25	1994 and 2001	Comprehensively reviewed.
Global cycads	288	151	52	2001	Comprehensively reviewed. The figures exclude two species that are extinct in the wild but survive in cultivation.
Australia	15,638	2245	14	Pre-1994	Number of threatened species based on adding together assessments for individual states or territories.
Ecuador (excluding Galapagos)	15,492 (of which 4011 are endemic)	2884	>19 (72% of endemic species)	2001	This analysis is for endemic species only; if non-endemics were included, the number of species threatened would be higher.
Mauritius (island)	750	294	39	Pre-1994	
Socotra	825 (of which 306 are endemic)	149	>18 (49% of endemic species)	2001	This analysis is for endemic species only; if non-endemics were included, the number of species threatened would be higher.
South Africa	23,420	2215	9	Pre-1994	Based on the conservation status of species over their entire ranges (not just in South Africa).
US	16,108	4669	29	Pre-1994	

Notes: Estimates for countries and regions are often based entirely or largely upon endemic species and are therefore likely to be underestimates. For information on the IUCN Red List categories and criteria, see Chapter 9. Figures refer to vascular plants, unless otherwise stated.
* All of these terms are defined categories of conservation status.

Sources: Table originally produced for this book using the following publications: Walter and Gillett, 1997 (global flora, Australia, Mauritius, South Africa, US); Oldfield et al, 1998 (global tree flora); Valencia et al, 2000 (Ecuador); Baillie et al, 2004 (global conifers, global cycads, Socotra). See also Farjon and Page, 1999 (global conifers) and Donaldson, 2003 (global cycads).

Figure 1.1 Macrozamia riedlei, *a cycad*

Note: Cycads as a group are particularly endangered because they are slow growing, have few growing points and are often the targets of collectors. However, this particular species, from south-west Australia, is not considered to be at risk.
Source: Alan Hamilton

Officer of the World Conservation Union (IUCN) and who is in one of the best positions internationally to make an informed judgement, thinks that about 10 to 20 per cent of vascular plant species are threatened (Hilton-Taylor, personal communication, 2005).

One reason for concern about the loss of plant species is that this will result in the loss of other species as well. This worry stems from a recognition of the key position of plants in ecological systems – as primary producers of organic matter and providers of physical spaces – and also because of the close dependency of some organisms upon only one or a few particular species of plants. A number of species of fungi will likely accompany each species of higher plant into extinction. There are around five to ten times as many species of fungi as plants, many associated with single species of plants (Hawksworth, 2001). With regard to animals, there are some pollinators that are closely tied to particular species or groups of plants and will become extinct or at least more endangered should these plant species disappear. Every species of fig *Ficus* has its own species of wasp, responsible for its pollination. The average number of animals that will go extinct with each species of plant is unknown, but will be greatly dependent upon how many species of arthropods live in tropical forest and upon their degrees of reliance upon particular types of plants – matters that are still poorly understood. Arthropods living on tropical forest trees are believed to form the single most numerous category of living things on Earth that can be seen with the naked eye (Erwin, 1991; Stork, 1993; Kitching et al, 1997).

Plant species cannot survive without habitats. Habitats differ in the numbers and uniqueness of the plant species they contain, and are under varying degrees of pressure. It follows that there are certain floras and habitats which are of particular concern from the point of view of saving plant species. Box 1.2 provides some examples. Special emphasis is placed in this box on **tropical forest** to draw attention to the particular conservation importance of this habitat type, which is especially rich in plant diversity. Tropical forest is also of exceptional value for moderating global and regional climates (see the section on 'Conservation concerns relating to ecosystem services'). Attention is further drawn to the exceptional and often highly threatened floras of **oceanic islands**, which, among other problems, are often under severe attack from invasive animals and plants. Places with unusual geology also merit special mention because of their sometimes unusual associated floras. **Limestone and ultramafic** rocks are the most widely distributed types of rocks that often carry distinctive species.

Another aspect of plant diversity about which there is conservation concern is loss of genetic diversity within particular populations of species or, indeed, within species as a whole. The existence of broad genetic diversity increases the abilities of species or populations to survive (see Chapter 2) and also ensures that more genes and strains of plants are available for potential human benefit. The problems that can result from having a narrow genetic base can be illustrated by the cases of the small-leaved elm and the potato. Up until the late 1960s, the small-leaved elm *Ulmus procera* was one of the commonest large trees in the English countryside. However, today virtually no large trees remain thanks to the ravages of Dutch elm disease, caused by the fungus *Ophiostoma ulmi*, carried from tree to tree by bark beetles. Although *U. procera* is popularly regarded as a 'wild' tree in England, DNA evidence has revealed that the entire English *U. procera* population is descended from a single clone introduced during Roman times (AD 43–410). It is suspected that the vulnerability of *U. procera* in England is related to its low genetic diversity. The varieties of the potato *Solanum tuberosum* that were grown in Ireland during the mid-1800s lacked resistance to potato blight, caused by the fungus-like organism *Phytophthora infestans*. Attacks of blight, starting in 1845, led to mass starvation and the emigration of millions of people. The trait of resistance to blight is found in several wild potato species and can be introduced into cultivated potatoes to confer resistance (Hawkes et al, 2001).

It is certain that loss of genetic diversity within species is a widespread phenomenon in the plant kingdom today because populations of many species are being eradicated or reduced in size. However, it has been little measured. The usefulness of access to relevant genetic diversity for plant breeding and the extent to which this diversity has declined are best known for crops. The incorporation of genes from traditional landraces of crops and wild crop relatives into modern crop varieties has been responsible for dramatic increases

BOX 1.2 EXAMPLES OF CONSERVATION CONCERNS RELATING
TO PARTICULAR FLORAS AND HABITATS

Tropical forest

- The area of tropical forest is declining at a rate of 1 to 4 per cent per year, with higher rates in some areas than others – for example, in West Africa compared with central Africa.
- Lowland forest is generally being lost at a much higher rate than montane forest – for example, in east Kalimantan (Indonesia), where the amount of reduction during 1984 to 2000 for lowland forest types was 23 per cent for alluvial forest, 38 per cent for forest on sandy terraces and 58 per cent for forest on plains, compared with less than 1 per cent for forest on hills and mountains.
- Tropical dry forests are regarded as one of the world's most threatened and least protected habitat types.
- Cloud forests are widely threatened around the world.
- The floristically distinctive coastal and Eastern Arc forests of Eastern Africa are rich in endemic species, small and highly threatened.
- Some countries that were once extensively forested now have very little forest – for example, only 2.5 per cent of Ethiopia is still forest covered.
- Hilltops in tropical forest can be floristically distinct and especially threatened – for example, through construction of communication masts.

Oceanic islands

- The percentage of rare or endangered plant taxa that are endemic to various medium-sized islands or island groups (for example, the Canary Islands, Juan Fernández, Madeira, Mauritius and Socotra) ranges from 55 to 97 per cent.
- 150 (66 per cent) of the 229 plant taxa only known from the Galapagos Islands are rare or threatened.
- The native flora of Hawaii consists of 1100 native species, of which 90 per cent are endemic. Out of these, 100 are already extinct in the wild, 12 are known only from single wild plants and 180 have fewer than 100 wild individuals remaining; altogether, 60 per cent of the endemic species elicit some conservation concern.

Other habitat types

- Limestone outcrops in Peninsular Malaysia have 130 forest plant species found nowhere else on Earth, some confined to single sites. These hills are threatened by quarrying, land clearance for agriculture and other developments.
- Coastal ecosystems can be especially threatened – for example, sand dunes around the Mediterranean.
- Today, tropical thorn forest in Pakistan covers only 2 per cent of its documented historical extent. Its area in Lahore District has shrunk from 1813km^2 in 1940 to 2km^2 in 1990.

> - In Turkey, 87 per cent of peatland sites have been destroyed, with a reduction in total area from 24,000ha to 3000ha.
> - In England and Wales, 97 per cent of enclosed unimproved grassland was lost between 1932 and 1984.
> - Only 17 per cent of the dry lowland heathland present in the UK in 1800 still remains.
>
> *Sources:* Davis et al, 1986; Kiew, 1991; Khan, 1994b; Byfield and Özhatay, 1997; Green et al, 1997; Burgess et al, 1998; Meilleur, 1999; Jepson et al, 2001; Northwood et al, 2001; Wilkie and Laporte, 2001; Critchley, 2003; Peters et al, 2003; Basset et al, 2004

in crop yields over the last decades (Phillips and Meilleur, 1998; Hawkes et al, 2001). In India, for example, such transfers of genes resulted, during the 1960s, in the country becoming a net exporter of food, rather than an importer. With regard to decline in genetic diversity, there are many reports of loss of traditional varieties of crops over recent decades, ironically with displacement by modern crop varieties (benefiting from genes from traditional varieties) being one of the major causes of their extinction. Thousands of landraces of rice were cultivated in the Philippines until recently; but today just two Green Revolution varieties account for 98 per cent of the entire cultivated area. In China, nearly 10,000 wheat varieties were in use in 1949; but, by the 1970s, only 1000 could still be found. In the US, between 81 and 95 per cent of all apple, cabbage, field maize, pea and tomato varieties documented in crop registers before 1904 appear to have vanished completely and are not even represented in *ex situ* genebanks (Cameron, 1994; Hawkes et al, 2001; Tuxill and Nabhan, 2001).

The European wine industry presents an example of the usefulness of retaining plant diversity for crop production. This industry, which is based on the European vine *Vitis vinifera*, was almost completely annihilated by the grape louse *Phylloxera vitifoliae* following the arrival of this pest from North America (first recorded in Europe in 1863). What saved the European industry was the import of new types of disease-resistant rootstocks for grafting. These rootstocks are of native American species of vines such as *Vitis labrusca*.

CONSERVATION CONCERNS RELATING TO PLANT RESOURCES

A plant resource can be defined as a plant or part of a plant that is of actual or potential value to people. There are many types of values, utilitarian and otherwise (see Chapter 7). Because the definition is so broad, it is possible to regard all plant species and all plant genetic diversity as resources. From the perspective of material use, perhaps around 20 per cent or more of global plant species have already been used by people for various purposes (circa 50,000 species

---

**BOX 1.3 SOME MATERIAL USES OF PLANTS AND TYPES OF
PLANT-BASED PRODUCTS**

Material uses of plants and types of plant-based products include the following:

- *ornamental uses* of plants or parts of plants, including for horticulture (for example, bulbs, potted plants and cut flowers);
- uses of *wood 'in the round' or split*, including for house construction, poles, fences and tool handles; sticks for supporting cultivated plants;
- uses of *sawn wood*, including for house construction, furniture and boats;
- uses of *carved wood*, including for mortars, pestles, boats, musical instruments, carvings and household utensils;
- *other uses of wood*, including for chewing sticks, toothpicks, matches, veneer and chipboard;
- uses of *woody weaving materials*, including for baskets, granaries, fishing equipment and rattan furniture;
- uses of *fibrous materials or fibres extracted from plants*, including for making cloth, bark cloth, paper, rope and twine;
- *fuels*, including fuelwood and charcoal;
- *exudates*, including resin, rubber and gum;
- *oils*, including vegetable oils and fats;
- *food*, including fruits, leafy greens and mushrooms; spices and flavourings;
- *beverages*, including herbal teas, coffee, cocoa and fermented drinks;
- *botanicals*, including fragrances, cleansing and colouring agents, and soaps;
- *herbal and pharmaceutical medicines*;
- *charms*;
- *poisons*, including for use in hunting and fishing;
- *hallucinogens*;
- *animal foods*, including herbaceous and tree fodder;
- *veterinary medicines*; and
- *honey and wax* (indirectly from plants via bees).

---

for medicine alone – see Table 7.3) and there are likely to be many more uses awaiting discovery (see Chapter 7). Box 1.3 lists some of the major uses of plants for material purposes and types of plant-based products.

Scale is an important consideration in the conservation of plant resources. Plant resources can become scarce or extinct in several ways:

- globally extinct – the resource is no longer available anywhere;
- locally extinct – the resource is unavailable locally;
- commercially extinct – resource supplies are insufficient for commercial harvest;
- culturally extinct – a plant or plant part that was once valued is no longer regarded as a resource.

There are major concerns about the availability of some plant resources. For example, it is estimated that fuelwood shortage affects 2400 million people worldwide (Dounias et al, 2000). This blights people's lives (especially women's) because they must spend many hours and much energy collecting resources that, in some cases, are increasingly hard to find.

Because people use so many plant resources and these are found in so many types of habitats, questions relating to the supply of plant resources are often intimately linked to those relating to conservation of plant diversity. In some cases, the conservation of plant diversity depends upon the continuation of certain practices of plant use and management – for example, practices associated with traditional agriculture in Europe. In developing countries, rural people rely upon many types of wild plant resources. It is often the financially poorest members of communities who have greatest dependency upon these resources and therefore stand the most to gain if they are regularly available. Wild plants provide a green social security for many rural people in developing countries in the form of low-cost housing, fuel, subsistence income, food supplements and herbal medicines (Cunningham, 1997a). Focusing on wild plant resources that are of local concern is a key element of community-based approaches to plant conservation (see Chapters 11 and 12).

Plant resources should be used in ways that can be sustained. Sustainable use is defined in the Convention on Biological Diversity (CBD) as 'the use of components of biological diversity in a way and at a rate that does not lead to the long-term decline of biological diversity, thereby maintaining its potential to meet the needs and aspirations of present and future generations' (CBD, 1992). Sustainability is an essential ideal in searches for ways for people to live in greater harmony with the environment. Sustainable use is most simply conceptualized for plant resources as the balance between rates of harvest and regrowth in a demarcated management area (see Chapter 11). In this case, a simpler definition of sustainable use can be adopted: 'the use of plant resources at levels of harvesting and in such ways that the plants are able to continue to supply the products required indefinitely' (Wong et al, 2001). For practical purposes, sustainable use refers not only to the biological and ecological aspects of plant use and management, but also to economic, cultural and social viability. In order for systems of resource use and management to persist, they must be economically viable, of sufficient interest to engage people's attention, practical in terms of people's knowledge and skills, and socially acceptable in the ways that the benefits and costs resulting from the operations are distributed through society.

There are questions about how systems should be defined in relation to sustainability, including matters of scale. For example, no one would expect a city to be ecologically sustainable within its own legal boundaries; but how wide should the geographical area considered as relevant be drawn? Sustainability for a forest department could mean that its network of forest resources is designed to produce continuing supplies of forest produce overall, but with supplies from any particular reserve being periodic. These and other complications make challenges in designing practical schemes for greater sustainability; however, they do not detract from the value of sustainability as an ideal towards which we should strive.

Striving for sustainability will require persistence and determination. There can be unexpected changes in systems that require approaches and procedures to be dramatically rethought. For instance, great efforts might be put into developing a management system for a forest in order to produce sustainable supplies of forest products that may be used by a community; but it is possible that members of the community will later adopt quite new ways of supporting their livelihoods – for example, by migrating to cities for seasonal employment and abandoning the collection of forest plants. A forest manager may dedicate many years to making the extractive use of a forest sustainable; but then a decision may be made at higher level – perhaps for good reason in the wider picture – to convert the forest to agricultural land. Managing land for sustainable agriculture will require a quite different set of knowledge and skills from managing the same land for sustainable forestry.

It is not only the available quantity of plant resources that merits attention in conservation projects. The *quality* of plant resources is normally also a concern – for example, with reference to the dietary quality of food plants, the effectiveness of herbal medicines or the strength of timber. In rural communities in several parts of the world, there has been a marked decrease in the diversity of wild plants eaten as food over recent years, resulting in a deterioration in people's diet (Alcorn, 1995; Diáz-Betancourt et al, 1999; Maundu et al, 2002). The quality of plant resources is a consideration when communities gather wild plant resources for sale, especially when there is competition between collectors (locally or perhaps even between different countries) or where certain regulations or standards must be met (see Chapter 14).

CONSERVATION CONCERNS RELATING TO ECOSYSTEM SERVICES

One reason to conserve plants is to maintain the services that they provide. These services derive from the physical, chemical and biological roles of plants in ecosystems. Examples of ecosystem services include the regulation of water flow in rivers, the treatment of waste, the regulation of climate and the provision of insects to pollinate crops. A fuller list is provided in Table 1.2, which includes estimates of the monetary values of ecosystem services for three of the major types of global ecosystem (Constanza et al, 1997). It is startling just how high the total values of these services are – for example, one hectare of tropical forest delivers ecosystem services worth an estimated US$2007 per year. Unfortunately, conventional economists rarely consider the true values of biodiversity and environmental services in calculations made for economic development (Tuxill and Nabhan, 2001).

Plants contribute to ecosystem services on many different scales, as illustrated by the case of climate. At the very *local level*, individual specimens of trees can be much appreciated for the shade, cool air and protection from the

Table 1.2 *Average global values of ecosystem services
for some major biomes*

Ecosystem service	Value (US$ per hectare per year in 1993)			
	Tropical forest	Temperate and boreal forest	Grassland and rangeland	Wetlands
Gas regulation			7	133
Climate regulation	223	88		
Disturbance regulation	5			4539
Water regulation	6		3	15
Water supply	8			3800
Erosion control	245		29	
Soil formation	10	10	1	
Nutrient cycling	922			
Waste treatment	87	87	87	87
Pollination			25	
Biological control		4	23	
Habitat/refugia				304
Raw materials	315	25		106
Genetic resources	41			
Recreation	112	36	2	574
Cultural	2	2		881
Total	2007	302	232	14,750

Note: These figures are based on such considerations as willingness to pay or the costs of providing the same services by other means. However, as the authors point out, in many cases the same
services cannot be provided by other means – they just won't be available if the biomes are lost.
Blank entries indicate a lack of adequate data.
Source: adapted from Constanza et al, 1997

rain that they provide. At a somewhat *larger scale*, managers of tea estates, for
example, have long realized the value of retaining patches of forest between
fields of tea, helping to maintain a cool moist 'tea climate'. *Regionally*, the
vegetation type present can strongly influence the climate, perhaps especially
in areas with convectional rainfall – for example much of the tropics and subtropics (Hulme, 1996; Bonell, 1998; Laurance et al, 2002). If a tropical forest
is replaced by agricultural land, then both the radiation and water budgets at
the Earth's surface can be considerably modified. More solar radiation will be
reflected back to the atmosphere and less water returned through evaporation
and transpiration (water loss through the leaves). The overall result can be a
warmer and drier climate. Very extensive areas can potentially be affected by
this type of climatic change, related to the movement of water-laden airstreams
inland from the oceans and the way in which water can be recycled back to
the atmosphere after falling as rain. If rain happens to fall onto land covered
by tropical forest (rather than grassland), then more water will be recycled
back into the atmosphere and so continue its journey downwind. In this way,
much of the water falling far inland in the forested parts of South America and

Africa actually consists of water that has been recycled several times through the forests. If the forest cover is lost or reduced, then less water is returned to the atmosphere and therefore the destruction of forest near the coast can cause desiccation of the climate thousands of kilometres inland. The conservation lesson is to retain as much forest and as many trees as possible, especially in areas of convective rainfall.

On the *global scale*, the world's plant cover plays a significant role in the global carbon cycle; therefore, its fate has an influence on greenhouse climatic warming. The amount of carbon associated with different vegetation types varies greatly. The world's tropical forests are especially significant as a carbon store, altogether holding about 20 per cent of the total global pool of organic carbon (Körner, 1998; Mahli and Grace, 2000). This 20 per cent is comprised of about 42 per cent of all carbon contained within living things (biomass carbon) and about 11 per cent of organic carbon contained within all soils. Another habitat type associated with a sizeable carbon store is peatland. The destruction of peatland over recent years has contributed significantly to greenhouse climate warming (Page et al, 2002).

The type of vegetation present in a catchment area has a major bearing on the quantity and quality of water supplied by the catchment. Under most conditions, the most practical way to ensure that supplies of water in rivers and wells are reliable and of good quality is to maintain or restore vegetation in catchments to as close to its natural state as possible. The clearance of forest from a catchment that is naturally forested will likely lead to several detrimental developments from the water supply point of view: increased flash flooding, reduced dry season flows and increased quantities of sediment in rivers (Hamilton, 1984; Bonell, 1998). Forest clearance is often gradual so that these unfortunate trends may not be apparent to many people. Disasters can bring greater awareness and spur ameliorating action. A wake-up call for China was the disastrous flooding along the Yangtze River in 1998 that affected 240 million people. This event led directly to a banning on logging in the catchment area of the Yangtze and initiation of a programme of forest restoration. Water is considered everywhere such a key resource and extensive flooding so serious that there should be many opportunities for plant conservationists to use the 'water argument' to lobby for retention or restoration of natural vegetation in catchment areas, a win–win scenario for conservation and livelihoods.

APPROACHES TO PLANT CONSERVATION

The Convention on Biological Diversity (CBD) is the single most important international conservation agreement (see Chapter 3). It has been ratified by most countries, which means that they have agreed to conserve biological diversity found within their territories and to strive for sustainable use of biological resources. One of the stipulations of the CBD is that an ecosystem approach should be used as the primary framework for its implementation

(CBD, 1992). The precise meaning of an ecosystem approach (in the sense of the CBD) is a matter of continuing debate; but broadly it encourages people to think in terms of systems and to keep in mind that humans are major elements of ecosystems in nearly all cases. Taking an ecosystem approach means that all dimensions of ecosystems should be considered in analysing conservation issues – biological, ecological, social, economic, cultural and political. It also means working on different geographical scales, as relevant to the problems at hand. Another consideration is that an adaptive approach to managing natural resources is recommended in view of the many uncertainties about how ecosystems function and about the actual effects of interventions designed to control them (see Chapter 11). A set of principles has been developed to help people better understand the CBD ecosystem approach (see Box 1.4).

One of the great advantages of the ecosystem approach for plant conservation is that analyses of the relationships between people and plants can be approached in a balanced way. Traditionally, conservationists have tended to see people as the problem and have concentrated on threats when analysing people–plant relationships. It certainly is necessarily to identify threats (see Chapter 2); but analyses of these negative influences need to be accompanied by analyses of positive influences if practical conservation is to be achieved. People should be seen as part of solutions, as well as causes of problems. If this is not done, the impression can be conveyed that the Earth would be much better off had people never existed (Schama, 1995), which is psychologically unhelpful for gaining public support for conservation. People respond best to positive messages.

Budowski (1988) has emphasized that conservationists need to become engaged with issues of development. He has pointed out that, on the one hand, development is perceived by the governments and people of developing countries as essential, while, on the other hand, too many botanists do little more than complain about the fate of disappearing forests and species, moan about the continuing use of shifting agriculture and complain when exotic species are used in plantations. In Budowski's opinion, botanists would be much better advised to forge partnerships with politicians, foresters, agriculturalists and others to find ways of achieving conservation in the context of meeting people's aspirations for development. Budowski holds the view that, while the aspiration to develop is a fact of life, there is considerable latitude in how development is defined. He suggests that botanists can play roles through introducing terms such as appropriate land use, diversity, sustainability and the maintenance of options.

A problem with a holistic approach to conservation – as encompassed in the ecosystem approach – is that matters can seem very complex and it may not be at all clear what is best to do. How should work be prioritized? Two suggestions are made here: place the *local* ecosystem at the centre of attention and focus on the *links* between people and plants. Anywhere on Earth can be local in this context. The local is key because this is where plants live, where people interact directly with them and where practical arrangements can be made that balance conservation and use. In the final

---

BOX 1.4 THE 12 PRINCIPLES OF THE ECOSYSTEM APPROACH, AS DEFINED BY THE CONVENTION ON BIOLOGICAL DIVERSITY

1   The objectives of managing land, water and living resources are a matter of societal choice.
2   Management should be decentralized to the lowest appropriate level.
3   Ecosystem managers should consider the effects (actual or potential) of their activities on adjacent and other ecosystems.
4   There is usually a need to understand and manage the ecosystem in an economic context. Management should reduce market distortions that adversely affect biological diversity, align incentives to promote biodiversity conservation and sustainable use, and internalize as feasible costs and benefits in the given ecosystem.
5   Conservation of ecosystem structure and functioning, in order to maintain ecosystem services, should be a priority target of the ecosystem approach.
6   Ecosystems must be managed within the limits of their functioning.
7   The ecosystem approach should be undertaken at the appropriate spatial and temporal scales.
8   Objectives for ecosystem management should be set for the longer term.
9   Management must recognize that change is inevitable.
10 The ecosystem approach should seek the appropriate balance between, and integration of, conservation and use of biological diversity.
11 The ecosystem approach should consider all forms of relevant information, including scientific and indigenous and local knowledge, innovations and practices.
12 The ecosystems approach should involve all relevant sectors of society and scientific disciplines.

Source: CBD, 2004

---

analysis, all work aimed at plant conservation needs to produce positive results on plants as they actually grow – which is always local. This means that eventually all actions taken in favour of plant conservation remotely from this core must feed back to the local level in order to be of conservation use (see Figure 1.2).

The ecosystem approach advocates precaution in transforming nature for human purposes. Reasons for precaution include the limited state of human knowledge about how ecosystems work and the inevitability of side effects to human interventions, some of which will not have been predicted. The most fundamental conclusion of the precautionary principle, as applied to nature, is that, as we develop, we should try to transform natural systems as little as possible. This contrasts with the traditional approach to development which sees no problems with transforming nature liberally for immediate human benefit, leaving little bits of nature here and there in odd corners that are of no interest to any one. In other words, *see development as something that must*

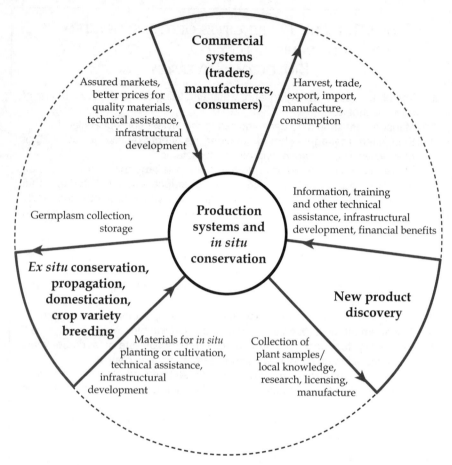

Figure 1.2 *The local as the hub of plant conservation*

Note: all actions taken remotely from this core must feed back to the local level to be useful for conservation in the final analysis.

happen within nature, not nature as something remaining after development has had its will.

Empirically, there are numerous cases that demonstrate the hazards of failing to take a precautionary approach to development. The burning of fossil fuels is leading to global changes in climate that were unforeseen, creating massive problems for people (and plants) in the coming decades. Presumably, food scientists saw nothing wrong with developing intensive systems of farming that involved feeding cattle with protein from miscellaneous sources, including animal carcasses; but a consequence has been the spread of Creutzfeldt-Jakob disease (CJD), a disease that has wreaked havoc with farming. In Africa, the drilling of boreholes to supply water for livestock in arid lands must have seemed a good idea at the time; but today each borehole is surrounded by a widening circle of desertification through overgrazing.

The precautionary principle is a reason for advocating caution with respect to the introduction of genetically modified organisms (GMOs). GMOs are organisms containing genes that have been specifically inserted by people from other organisms, not necessarily ones that are closely related. Historically, GM technology is yet one further step along a path of human ingenuity applied to the control of nature and – as with all such previous technical developments – carries promises of benefits. However, there will certainly be downsides, at present poorly understood. On the positive side, one of the potential environmental advantages of GMOs is the possibility of breeding resistance against specific pests into crops. This could reduce the use of pesticides, currently an environmental problem with non-GM intensive agriculture. On the negative side, a major worry about GMOs is that genes from GMOs or the GMOs themselves will move into natural ecosystems, becoming invasive. This would magnify the already worrying problem of invasive plants (see Chapter 2). The movement of introduced genes out of GMOs into wild relatives has already been documented for the potato in the Andes. In this case, scientists have recommended the use of male-sterile cultivars of potatoes to prevent further out-crossing with wild species (Celis et al, 2004). Another problem with GMOs is that a process commonly used to develop them (involving production of 'genetic constructs') can lead to the incorporation of antibiotic-resistant genes within their DNA. This may spread antibiotic genes into the general environment with, as yet, unknown impacts.

2

Threats to Plants

This chapter examines some of the threats facing plants. Over geologic time, there have been several spectacular mass extinctions on Earth associated with meteoritic impact. On a more modest scale, natural events, such as volcanic eruptions, continue to be threats to some plants. However, these days the great majority of threats that plants face are due to people. Some human threats are direct (for example, forest clearance), while others arise indirectly from the high and growing human population, and associated economies and cultures. Human influence has caused fundamental change to the state of the environment over substantial regions. Widespread human fragmentation of habitats will make it hard for many plants to respond successfully (through migration) to greenhouse or Ice Age climatic change.

THREATS POSED BY BIOLOGICAL AND ECOLOGICAL PROCESSES

Certain threats to plants stem from the intrinsic features of biological and ecological systems. These features are natural; but the fact that they can be threats owes much to the influence of people.

Populations of species having few individuals are vulnerable to extinction. This is because their numbers are more likely to fluctuate to zero as they undergo the natural variations in numbers to which all biological populations are subject. They are also more likely to experience genetic drift and inbreeding. Genetic drift and inbreeding reduce the likelihood that populations will be able to cope with new pressures, such as attacks by pests and diseases (Fischer and Matthies, 1998; Keller and Waller, 2002). Assessment of the sizes of interbreeding plant populations need to take into account whether populations that are *physically* separated are also *biologically* isolated. Spatially separate populations may be genetically connected through cross-pollination or the ways in which seeds are dispersed. Wind-pollinated species are more likely to interbreed between physically isolated populations than species pollinated by insects (Bruna, 1999).

The concept of a minimum viable population has been suggested as useful for conservation, this being the size of a population with an acceptably low probability of extinction. Estimates of its size for various species of plants range from 1000 to 1 million individuals (Menges, 1991). However, some populations – such as many in regions of Mediterranean climate – appear to be genetically adapted to a small population size since they have persisted in this condition for tens to hundreds of thousands of years (Cowling et al, 1996). More light on these matters is expected in the near future from studies of the genetics of small plant populations using molecular techniques.

If the extent of a habitat becomes small, then some species that are associated with the habitat are expected to face problems because of the disruption of certain ecological processes. The prairie grasslands of North America provide an example. At one time, fires swept regularly through the wide open spaces of the prairies, creating conditions suitable for the survival of diverse types of plants. However, today the prairie has been widely destroyed, surviving only in scattered patches. Burning has become less frequent and this has contributed to causing a reduction in the number of plant species found in the surviving patches by an estimated 8 to 60 per cent over a 32- to 52-year period (Cully et al, 2003).

Another problem with a small size of habitat is that the influence of edge effects becomes magnified. The ratio of edge to area increases. Edge effects are caused by the marginal zones of habitats being ecologically distinct. Edge effects in tropical forest can include drier soils, increased seed predation, decreased seed germination and an increased frequency of gaps, all unfavourable to many species of more primary forest stages (Bruna, 1999; Porembski and Barthlott, 2000).

Left to its own devices, vegetation at any one place will tend to change in its structure and floristic composition towards certain more stable (climax) forms (see Chapter 5). Today, very large areas of vegetation are intrinsically unstable due to anthropogenic disturbance. Once impeding factors blocking succession (such as agricultural practices, burning or the grazing of domestic stock) are removed, then the vegetation will start to change towards a more stable form and many plant species currently present will become lost. This process of succession is often welcome from the viewpoint of plant conservation because, very generally, more climax types of vegetation are considered to be of greater conservation worth than more secondary forms. However, there are cases when secondary vegetation is prized and succession presents a threat. This is widely so in Europe – for example, in the case of some types of (semi-natural) grassland. Activities carried out for plant conservation can then consist largely of trying to impede the natural course of vegetation events – for instance, through destroying colonizing trees and shrubs.

The presence or absence of certain animals can greatly influence plants. Plant species that depend upon single pollinators are at a higher risk of extinction than those that can be pollinated by many (generalists). Specialized pollination systems are commoner in some floras and parts of the world than others – for example, in the tropics and the fynbos (South Africa), compared with much of Europe and North America (Johnson and Steiner, 2000). People have caused the loss of large seed-dispersing mammals in many areas of tropical forest. There are varying reports about the effects of this on plants. In the case of stable, long-established fragments of tropical forest in Latin America, there is evidence that the loss of these large mammals has eventually resulted in the loss of those species of trees with seeds adapted to dispersal by them (Brokaw, 1998, based on Kellman et al, 1998). On the other hand, a recent drastic reduction in the number of large herbivores in African forests (caused by the bushmeat trade) does not seem to have yet had a drastic effect on plant populations (Hawthorne and Parren, 2000), although in this case it is still early days.

Large numbers of big wild herbivores can cause problems for plants, seriously damaging vegetation or impeding regeneration. In modern times, such high concentrations often have a human cause – for example, because of restrictions imposed on natural routes of migration or the eradication of the large carnivores that once would have limited their numbers. Locally high populations of elephant are a problem in parts of Africa (Timberlake and Müller, 1994). They have destroyed most of the best mopane woodland (dominated by *Colophospermum mopane*) in the Luangwa and Zambezi valleys of Zambia (Bingham, 1998). Large populations of deer in some nature reserves in Britain prevent the regeneration of trees. However, managers can be loath to reduce their numbers, fearing to alienate an 'animal loving' public (Idle, 1996).

Invasive animals – introduced into new places by people and then becoming self-sustaining and spreading – are a major threat to plants in some parts of the world, such as Australia (for example, rabbits), New Zealand (red deer) and many oceanic islands (goats, pigs and rats). The problem is not confined to larger animals. Nearly half of more than 300 species of Proteaceae in the fynbos of South Africa are threatened by the spread of the Argentinean ant *Iridomyrmex humilis*. The seeds of these Proteaceae are naturally dispersed by native species of ants, but *I. humilis* does not disperse the seeds. There is very poor recruitment of seedlings in invaded areas (Bond, 1994).

THE HUMAN FACTOR: ROOT CAUSES OF PLANT LOSS

In a conservation context, a root cause is a basic human condition or trend that underlies the more immediate causes of biodiversity loss (Stedman-Edwards, 1998).

A root cause of plant loss is the huge and growing size of the human population, creating pressure to destroy natural habitats, expand and intensify agriculture, and collect more resources from wild plants. Intensive farms contain little botanical diversity compared to their more traditional counterparts. Not only is the population growing, but the average person is consuming more – in fact, 460 per cent more today, on average, than a century ago (Laurance, 2001). It is predicted that the human population will rise from 6.1 billion (2005) to 8 billion (2025).

Many cities in developing countries are expanding rapidly, fuelled partly by in-migration from the countryside. This can create huge demands for charcoal, herbal medicines and other goods derived from wild plants growing in surrounding areas. Despite the sometimes high cost of these items compared with people's incomes, they can still be relatively affordable for poor people, compared, for example, to electricity or pharmaceutical drugs.

In some areas, population increase has contributed to mass movements of people for resettlement, which can cause major loss of natural habitat. Some such movements are planned, as in Indonesia, where since 1947 the government has pursued a policy of moving people from densely populated areas, such as Java, to outlying islands, such as Borneo. The result has been large-scale conversion of rainforest to fields. Likewise, the Brazilian government has supported systematic settlement in parts of the Amazonian rainforest, such as in the state of Rondônia, with intensive settlement starting during the 1970s. There have been some mass movements that have not been centrally planned. Many poor people in India and Central America have been displaced by agricultural intensification or the creation of plantations. Some have moved to land barely suitable for agriculture, others into cities. A rapid growth in soybean farms in Brazil, catering to genetically modified (GM)-averse markets in Europe, has displaced ranchers and slash-and-burn farmers, who have then moved deeper into the forest (Laurance et al, 2004).

There have been great expansions in monetary economies and trade systems in recent years. One effect has been to distance people increasingly from the plant resources upon which they depend. This distancing is mental, as well as physical. Unlike people with subsistence economies, consumers can no longer easily see the direct consequences of their patterns of consumption on the natural world. It can be hard for them to know the environmental impacts of their lifestyles. In richer countries, supermarkets have come to dominate trading systems in food, exerting considerable control over patterns of purchase by consumers, as well as over farming. Modern industrial economies rely upon large-scale use of energy, mostly from fossil fuels, with consequent contributions to greenhouse climatic warming (see the section on 'Climatic change').

The fate of the environment in poor countries can be much influenced by the richer word. With agriculture, for example, farmers in poor countries can find it difficult to compete against agricultural produce from rich countries, which is estimated to receive subsidies of more than US$300 billion annually (Mutume, 2004). The forcefulness of Western interests in plant-related matters can extend to the illegal economy. The demand in rich countries for narcotic substances (such as cocaine and heroin) can make the growing of illegal plants, such as the coca shrub *Erythroxylum coca* and opium poppies *Papaver somniferum*, difficult to prevent. The criminalization of major aspects of farming and plant trade in countries such as Colombia and Afghanistan compounds the problems of attaining orderly management of natural resources generally.

A major cultural trend over recent years with a bearing on plant conservation has been a global decline in local knowledge about plants (see, for example, Wade, 1987; Diáz-Betancourt et al, 1999; Luoga et al, 2000; Martin et al, 2002). The growth of scientific knowledge of plants over the same period has typically proved little compensation because, in most parts of the world, most of the limited scientific knowledge that is available about particular local plants is of little relevance for practical purposes. Apart from loss of traditional knowledge directly useful for managing plants, the decline of local knowledge about plants is a conservation concern more generally, because it reduces the foundation of interest in local plant diversity upon which new conservation initiatives can be built.

A related social trend is loss or weakening in traditional authority. Governments today assign much authority over natural resources to organizations such as forestry departments; but sometimes these have limited effectiveness because of shortages of money and trained staff. With governments lacking effectiveness and customary institutions ignored or weakened, the result can be political anarchy and, therefore, devastation of natural resources (Clarke et al, 1996).

Customary conservation can sometimes be easily undermined by modern forces. When the bark of the African cherry *Prunus africana* started to be collected commercially on Mount Oku, Cameroon, in 1972, this was done carefully by the only company then operating. Plantecam Medicam took steps to promote sustainability in harvesting by prescribing that the bark was to be removed from opposing quarters of the trunks – in order to avoid killing the trees through girdling – followed by periods of rest of four to five years before taking a second harvest (Cunningham et al, 1997; Schippmann, 2001). Then, in 1985, the government of Cameroon issued 50 new permits for bark collection to other businesses and the system of controlled harvesting broke down. Complete girdling now became the norm or else trees were simply felled so that they could be easily stripped of their bark:

> *For generations, the secret society, known as the* kwifon, *oversaw the uses of forest plants and animals. The* kwifon *had sophisticated rules for managing watersheds, catchment areas and fragile forest ecosystems. Certain sections (sacred forests) were reserved for the gods. On certain days of the week, called 'country Sundays', no one was permitted to enter the forest for any purpose. Violators were threatened with serious illness or death unless an expensive cleansing ritual was performed. As long as the taboos were in place, the gods protected the forest and no one species was overexploited... With issuing of the new permits in 1985 ... the traditional authority had no control over these outsiders. They violated the local norms with impunity, entering the forest even on country Sundays, and suffered no adverse consequences. The fear of sanctions by the forest gods dissolved. Thus, harvest of* Prunus africana *bark contributed to the erosion of the resource preservation ethic that continues to this day.* (Stewart, 2003)

HUMAN ACTIVITIES DIRECTLY CAUSING PLANT LOSS

Table 2.1 shows the most frequently recorded direct threats to plants mentioned at 233 sites worldwide identified as global centres of plant diversity. It can be seen that agriculture is the most frequent threat, followed in order by logging, grazing by domestic stock, burning, tourism, road construction, mining and collection of non-timber forest products (NTFPs). This picture is generally supported by other, more local, analyses (see, for example, IUCN and WCMC, 1983; Bingham, 1998; Oldfield et al, 1998; Cowling and Hilton-Taylor, 1994).

With respect to tropical forest, it has been reported that the biggest causes of its loss (in terms of area) are clearance for crops (55 per cent), clearance for cattle pasture (20 per cent), expansion of logging (12 per cent) and expansion of shifting agriculture (12 per cent) (Mahli and Grace, 2000) (see Figure 2.1). In actuality, several of the major factors responsible for loss of tropical forest are interlinked. Roads built for logging, oil exploration or other purposes can lead to the spread of agricultural settlement, the destructive harvesting of NTFPs, rampant hunting and increased burning. Logged forests are more susceptible to fires than un-logged forests (Woods, 1989; Siegert et al, 2001). It has been predicted that current plans in Brazil to dramatically expand highways in Amazonia will prove disastrous for forest conservation (Laurance et al, 2004).

Figure 2.1 *Forest clearance for agriculture,*
East Usambara Mountains, Tanzania

Source: Alan Hamilton

Table 2.1 *The most frequently recorded direct threats to plants at sites selected as global centres of plant diversity*

Threat	Notes on threat	Sites (%)	Notes on sites
Agriculture	Both extension and intensification	60	The most frequently mentioned threat in most tropical regions.
Logging	Includes side-effects	51	Clear-felling is relatively infrequent; but logging practices in the tropics are poor almost everywhere. Other types of threats are introduced along logging roads.
Grazing by domestic animals	Overgrazing and under-grazing	33	Overgrazing is generally the problem, especially in relatively dry areas, where it can be the most frequent threat. Under-stocking is a more local threat – for example, in parts of Europe.
Burning		30	Generally, the problem is too much burning.
Tourism	Includes recreation	29	This is the most frequently mentioned threat in Europe and North America.
Roads	Results of construction	27	The main threat globally is the opening up of previously remote forestlands to settlement and agriculture.
Mining	Excludes peat mining	25	Can be a major threat to species confined to metalliferous soils or limestone.
NTFP collection	Excludes fuelwood	20	Serious problems of over-harvesting medicinal plants are recorded from Central, South and East Asia.
Fuelwood collection		20	
Hunting		20	
Invasive plants		19	A major threat in oceanic islands, Australia, New Zealand and South Africa, and locally elsewhere.
Plant collecting	Ornamental and rare plants	18	Different types of plants are targeted in different places – for example, orchids in Southeast Asia and Central America, and bulbs in Turkey.
Building	Urban spread and other construction	12	
Dams		12	
Invasive animals		9	Places threatened are often the same as those threatened by invasive plants.

Sources: WWF and IUCN, 1994–1997; Hamilton, 1997

Based on projected trends, it has been calculated that 350 million hectares of natural ecosystems will be converted to cropland over the next 50 years (Laurance, 2001). Although it is possible that more benign forms of intensive agriculture will be developed and become widely adopted (Macilwain, 2004), according to present trends it looks as if intensive agriculture (based on high doses of chemical fertilizers and pesticides) will expand.

Although traditional agriculture is typically much superior in biodiversity value to intensive agriculture, it is likely that many areas of traditional agriculture will soon lose some of their wealth in plant diversity. One of the forms of agriculture likely to often suffer is swidden, which has a fallow phase (period of re-growth) incorporated within the agricultural cycle (see Chapter 6). It is likely that a trend seen over recent years will continue – involving a progressive shortening of the fallow stage of the cycle as a response to higher human populations (Lebbie and Guries, 1995; Frost, 1996; Naughton-Treves and Weber, 2001). The consequences are likely to be reduced soil fertility and the spread of pernicious weeds. Two potential end results are either replacement by permanent cultivation or abandonment of the land if the soils become too impoverished. Recolonization of abandoned agricultural land by forest can be permanently impeded under these conditions because of degraded soils, frequent burning, the destructive activities of domestic stock and lack of sources of seeds.

It has been estimated that the area of pasture used for domestic stock will increase by 540 million hectares by 2050 (Laurance, 2001). If current trends continue, overgrazing will occur on a massive scale, accompanied by severe soil erosion. Although overstocking is the major problem with livestock globally, there are cases where grazing pressure is too light for optimal conservation purposes. For example, it has been recommended that livestock should be reintroduced into certain areas of heathland in Britain to maintain species of rare plants that depend upon disturbance (Byfield, 1995a).

Inappropriate regimes of burning can be a major threat to plants. Although some fires in vegetation are natural, the great majority today are set by people. Fires are a mounting conservation problem with tropical forest. Very extensive fires occurred in tropical forests around the world during 1997 and 1998, associated with a spell of dry climate related to the El Niño oscillation. For months, a large part of Southeast Asia was enveloped in a dense cloud of smog caused by immense forest fires in Indonesia. Twenty-five per cent of Kalimantan was burned, many of the fires being started deliberately and often illegally by companies and individuals clearing land for plantations and farms (WWF, 1997; Jepson, 2001). In some rich countries, the setting of fires by vandals is a major problem. Many of the destructive fires that periodically rage in parts of Australia, France and the USA are started in this way.

The effects of fires on ecosystems depend upon their frequency and intensity. Both over-burning and under-burning can produce conservation problems. Over-burning increases the abundance of more fire-tolerant plants, such as grasses compared with trees, as well as fire-resistant species of trees, compared with those that are more susceptible. Over-burning thus tends to reduce overall

Figure 2.2 *Mechanical logging in tropical forest,*
East Usambara Mountains, Tanzania

Source: Alan Hamilton

plant diversity and can assist in the spread of invasive plants (Goldammer and Price, 1998). Examples of invasive species encouraged by burning include some species of pines *Pinus* in tropical mountains, *Melaleuca quinquenervia* in Florida and *Hakea sericea* in South Africa. Burning in the Caribbean islands is reported to be a particularly severe problem because of the scarcity of fire-adapted life forms (Woods, 1989). A major reason for the destructiveness of fires in parts of Australia, California and France is the poor management of fire regimes, which should prescribe more frequent, controlled, burns, rather than the unplanned, more occasional and fierce fires that actually occur.

Logging operations are a major cause of forest loss and degradation (see Figure 2.2). Some types of timber fetch high prices on international markets – for instance, that of the American mahoganies *Swietenia*. Illicit operations can be rampant; a study by the Brazilian government in 1997 concluded that 80 per cent of Amazonian logging was illegal (Laurance, 1998). The large profits to be made have led to some companies moving from country to country, devastating forests one by one and often taking few or no steps to encourage forest regeneration.

Tropical forests are frequently carelessly logged, leaving them riddled with tracks and with many fallen trees lying on the forest floor. The few tall trees left standing are exposed and often doomed to die. A study in Kibale Forest, Uganda, showed that hardly any canopy trees were regenerating in forest that had been heavily logged 20 years earlier. Rather, a once tall forest carrying excellent timber had been reduced to a smothering tangle of herbs, shrubs and climbers. Poor regeneration of canopy trees was blamed on a shortage in the supply of seeds, seed predation by rodents and damage by elephants (Kasenene, 2001).

The harvesting of non-timber products is a major threat to some plants, especially when the products are traded. Sometimes the destruction is obvious, as with the collection of fuelwood in the Saharan and Sahelian regions of Africa, where this is the principal cause of tree loss and desertification (Dounias et al, 2000). However, the adverse impacts of harvesting non-timber products are often rather hidden, which is a reason why governments have tended to underestimate the conservation problems connected with this sector.

The collection of plants for the horticultural trade or by specialist collectors is a serious threat to species in some plant groups, such as bromeliads, carnivorous plants, cycads, orchids, palms and succulents (including cacti) (IUCN/SSC, 1996; Johnson et al, 1996; Oldfield, 1997). Historically, there have been fashions in ornamental plants, such as a craze for collecting ferns in Victorian Britain, which drove some species to local extinction. Botanical experts can themselves pose very direct threats to certain rare plants if they combine an interest in collection with expert knowledge of where they can be found. Three species of palms in Madagascar have been reported as threatened through specialist collecting. In two cases (*Marojejya darianii* and *Voanioala gerardii*), all seed produced by the trees at the single sites at which these species were then known was removed, making regeneration impossible. In the other case (*Beccariophoenix madagascariensis*), one out of a total of only ten known specimens was felled for its seed after the species had been mentioned in a palm journal (Johnson et al, 1996). An unusual case of the perils of scientific collecting is that of *Takhtajania perrieri*, a tree only known from Madagascar and regarded as a 'living fossil'. Until its rediscovery in 1994, the only known specimen of this species was one that had been destructively collected in 1909 to obtain preserved specimens for science. Science, it was feared, may have driven the species to extinction!

INVASIVE ALIEN SPECIES AND GENES

Invasive alien species are organisms that have been introduced by people across major geographical barriers, have become established and then have spread of their own accord. Globally, about 10 per cent of invasive plant species cause substantial harm to native species or transform the characteristics of natural ecosystems over substantial areas (Richardson et al, 2000). It is predicted that the menace of invasives will grow as people cause more movements of plants and continue to destabilize natural vegetation.

There are several ways in which invasive species can adversely influence native species or ecosystems. Some species, such as the strawberry guava *Psidium cattleanum* on Mauritius, form dense thickets under which native species cannot regenerate. *P. cattleanum* is also believed to produce allopathic chemicals that suppress the growth of other plants. Some alien species impede the succession of vegetation that normally follows cessation of cultivation, logging or other disturbances. The shrubs *Chromolaena odorata* and *Lantana camara* and the small tree *Leucena lecuocephala* – all widely introduced – can cause serious delays to regeneration of tropical forest. Subtle adverse influences on native species are known. The invasion of Turkey oak *Quercus cerris* in Britain has resulted in the spread of several introduced species of gall wasps *Andricus*, the presence of which has resulted in a great decline in the reproductive fertility of the native oak *Quercus robur*. The wasps have a sexual generation in spring on the catkins of *Quercus cerris* and an asexual generation in autumn on the fruits (acorns) of *Q. robur*, turning them to galls (Schrönrogge et al, 2000).

The African grasses that have so widely invaded the savannas of South America have increased the frequency of fires. Through doing so, they have reduced the abundance of seedlings of native plants, modified the cycling of nutrients and reduced water yields (Porembski and Barthlott, 2000). The invasive tree *Miconia calvescens* is regarded as a major threat to more than 40 of the 107 plant species that are endemic to the island of Tahiti. *Miconia* suppresses native ground cover and is suspected of increasing soil erosion and the frequency of landslides (Meyers, 1998).

There are several serious invaders of aquatic and other wetland habitats. The water hyacinth *Eichhornia crassipes* has invaded many tropical dams and lakes, including Lake Victoria, where it has become a serious pest. The green alga *Caulerpa racemosa* is one of a number of organisms that migrated from the Red Sea into the Mediterranean following the building of the Suez Canal. It contains a toxin that has been found in the guts of rabbit fish in the Mediterranean and which may be affecting people who eat it (Lundberg et al, 1999). A spectacular example of a wetland invader is the cordgrass *Spartina* x *townsendii* on coastal mudflats. This is a hybrid between a British species, *Spartina anglica*, and an introduced American cousin, *S. alterniflora*. The hybrid was first detected in Britain in sterile form during the 1870s; but some time before 1892, plants with double the number of chromosomes arose, proved fertile, produced seed and rapidly spread around the British Isles. This new hybrid was intentionally introduced into San Francisco Bay in order to stabilize mudflats, but then spread of its own accord along the western coast of the US, where it has crossed with native American species, which are now difficult to find in their pure forms.

There are other examples of invasive genes diluting the purity of native species. The black poplar *Populus nigra* is in danger of disappearing from Europe through hybridization with hybrids (for example, *Populus* x *canadensis*) between itself and American species. The American plane *Platanus occidentalis*, introduced to Greece, is hybridizing with the oriental plane *Platanus orientalis*. The columnar form (var *pyramidalis*) of the cypress *Cupressus sempervirens* interbreeds in Greece with the natural broad-crowned variety (Papageorgiou, 1997). In the case of the much planted perennial

ryegrass *Lolium perenne*, research has shown that genes from modern cultivars have readily invaded wild populations in Britain, which accordingly have
completely lost their genetic distance from them (Hawkes et al, 2001). The
alacrity with which genes can move between varieties of a species, as in this
case, is one of the reasons for concern about the introduction of genetically
modified organisms (GMOs) (see Chapter 1).

Fungi can be invasive too. The fungus *Ophiostoma ulmi*, responsible for
Dutch elm disease, has virtually eliminated larger specimens of the small-leafed
elm *Ulmus minor* from the English countryside. *Cryphonectria parasitica*,
causing chestnut blight, has effectively eliminated the American sweet chestnut
Castanea dentata from woods in the eastern US. The Oomycete (fungus-like
organism) *Phytophthora* is a terrible pest in Australia, threatening many native
species.

Coping with the threat of alien invasive species through preventative
measures and programmes of eradication would be easier if more was known
of their ecology. Oceanic islands tend to be particularly badly invaded, which
suggests that somehow their isolation has resulted in their native species being
uncompetitive. This 'isolation hypothesis' receives some support from the
seriousness of invasion in some other parts of the world that have been
relatively isolated over evolutionary time, such as Australia and New Zealand.
Likewise, further backing is provided by the fact that the forests of the East
Usambara Mountains in Tanzania have a greater number of species of woody
plant invaders than recorded for any other area of primary tropical forest
(Rejmánek, 2000). These forests, like others of the Eastern Arc Mountains of
Tanzania, have been largely isolated biologically for millions of years from the
main block of tropical forest in Africa in Congo and West Africa. However,
the high number of invasives on the East Usambaras also owes something to
the presence of a substantial arboretum on the mountain, established in 1902.
In any case, the number of serious plant invasions on the main continental
land masses of Africa, Eurasia and the Americas is increasing. It was reported
in 1999 that, since 1960, there has been a tenfold increase in the numbers and
influence of invasive species, such as *Acacia dealbata*, *A. mearnsii* and *Pinus
patula*, in Nyanga National Park, Zimbabwe (Childes, 1999).

In principle, the threat of invasive species can be combated at different
stages – the prevention of new introductions, the early detection and elimination of nascent populations, and continuing efforts to control the worst
invaders (Rejmánek, 2000). Cowling and Hilton-Taylor (1994) are of the
opinion that biological control is the only effective long-term approach for
controlling invasive plants in the fynbos of South Africa. 'Invasive species' is a
domain of conservation about which there is little public awareness in most
countries and many scientific uncertainties. Public sentiment can even be in
favour of some plant invaders. The strawberry guava *Psidium cattleianum* is a
serious invader on Mauritius; but many people enjoy eating its fruits (*goyave
de chine*). There should be restrictions on the introduction of certain species
into designated types of environment based on the empirical evidence of their
behaviour under similar conditions elsewhere. Thus, it would be foolish to

introduce the serious invader *Syzygium jambos* into a moist tropical island if, by some stroke of fortune, it had not already arrived. The ability to predict which species will become invasive on theoretical grounds is at a nascent stage. However, there does seem to be a tendency for fast-growing, nitrogen-fixing, trees, such as *Mimosa nigra* and mesquite *Prosopis* (including *P. glandulosa*), to be invasive (see, for example, Khan, 1994b; Brigham et al, 1996). Disturbance of natural habitats by people or their livestock appears to make invasions more likely. In general, *undisturbed* tropical habitats seem to have some resistance to invasions, even in the case of islands (Rejmánek, 2000).

CLIMATIC CHANGE

Climatic change can be caused by many factors, operating on various scales. Regionally in the tropics, forest and tree destruction can disrupt the cycling of water between vegetation and the atmosphere, causing desiccation downwind from deforested areas. This is believed to have contributed to the warmer and less humid climates that have developed in several montane areas in the tropics since about 1970 (Lawton et al, 2001). Ecological consequences suspected to be related to this type of climatic change are recorded for the East Usambaras in Tanzania. They include an upward movement in several agricultural crops, an increased frequency of tree falls, reduced luxuriance of epiphytes and an increase in malaria (Hamilton, 1999).

Greenhouse climatic change is predicted to become a major threat to many plants over coming decades. This change is due to an increase in heat-trapping gases in the atmosphere, the most important of these being carbon dioxide (CO_2). The concentration of CO_2 has risen from 280 parts per million (ppm) in 1860 to 360 ppm today, mostly due to the burning of fossil fuels. The destruction of certain vegetation types (especially tropical forest) and associated organic matter in the soil has also made a significant contribution, as has the destruction of peat (Mahli and Grace, 2000; Page et al, 2002). The concentration of greenhouse gases is predicted to continue to rise over the coming century, causing major changes in climate. Temperatures generally will be much higher than they would have been otherwise and rainfall patterns will shift. Extreme climatic events will become more frequent. It is predicted that many species of plants will be unable to migrate fast enough to prevent their extinction – especially given the discontinuity of many habitats caused by human influences (Bawa and Dayanandan, 1998). Certain guilds of plants will be particularly vulnerable – for example, large climbers and tall trees in the case of tropical forest (Phillips and Gentry, 1994; Coley, 1998; Körner, 1998). Climatic change will cause massive disruption of human societies, further compounding problems for plants.

While modelling indicates that most parts of the world will become warmer with greenhouse climatic change, north-west Europe may be an exception. Greenhouse climatic change is implicated as a cause of the shrinking in extent and thickness of the permanent ice cap that covers much of the north polar ocean, as has been measured over recent years. It is predicted that the permanent ice cap may disappear completely in 50 to 100 years. If this happens, then the pattern of circulation of currents in the North Atlantic Ocean could change dramatically, including disruption of the Gulf Stream, which is currently responsible for the unusually mild climate of north-west Europe. North-west Europe could then enter a big chill. A similar anomalous cooling in climate in Europe occurred 11,000 years ago, in that case due to flooding of the North Atlantic by a massive influx of cold water from melting ice masses accumulated during a preceding Ice Age.

At some time during the next few thousand years, the Earth will enter another Ice Age with catastrophic consequences for many populations of plants. No warmth-loving plants will be able to survive at high latitudes, and large parts of the tropics which currently support forests will become too dry to do so. Conservationists need to think ahead to this event, which is not too distant in terms of the grand sweep of human history. The main implication for plant conservation is the desirability of putting extra efforts into conserving natural habitats and their species at the sites of *former* Ice Age plant refugia – which are likely to be refugia during the *next* Ice Age as well. These places are often priorities for the attention of plant conservationists because of their concentrations of species and infra-specific genetic diversity (see Chapter 8).

POLLUTANTS

Apart from carbon dioxide, the burning of fossil fuels produces other air-borne pollutants that influence plants. Oxides of sulphur and nitrogen from vehicles and power stations react in the atmosphere to form acids, the resulting acid rain being seriously damaging to many types of plants and fungi (Eggleston and Irwin, 1995; Hallingbäck and Hodgetts, 2000). Plants growing in lakes or on soils that are naturally acidic are the most likely to suffer since in these cases there are few chemicals to buffer the acids. Serious damage to plants by acid rain is reported from Europe, the north-east US and Canada, and China (Szechwan and Kweichow provinces). Another pollutant which owes its origin to emissions from vehicles and power stations is ozone, which is damaging to plants when present in high quantities in the lower atmosphere (Matyssek and Samdermann, 2003).

Change in ozone levels in the *upper* atmosphere is also an environmental problem, in this case because of reduction, rather than augmentation, in ozone concentration. Ozone in the upper atmosphere serves as a shield against ultraviolet radiation from the sun, which is damaging to plants (Caldwell et al, 1989). Ozone reduction is linked to chemical reactions induced by certain pollutants, such as chlorofluorocarbon (CFC) gases, used as coolants in refrigerators. Seasonal 'ozone holes' have been appearing regularly in the upper atmosphere in both terrestrial hemispheres during recent years.

Humans are responsible for enriching many ecosystems with nutrients (eutrophication). Man-made sources of nutrients include agricultural fertilizers, detergents (phosphate) and acid rain (nitrogen compounds). Since 1970, the worldwide rate of application of nitrogen fertilizers has increased seven-fold so that today contributions from humans to the global nitrogen cycle are equivalent to the sum total of all inputs from natural processes. One half to two-thirds of nitrogen fertilizers are unused by crops and enter non-agricultural ecosystems. It has been predicted that, by 2050, the global use of nitrogen fertilizers will rise by 270 per cent and phosphorus by 240 per cent (Tilman, 1998; Tilman et al, 2001). The fertilizing effect of acid rain on plants is believed to be responsible for various major vegetation changes recorded in Europe, including the killing of hummock-forming mosses on raised bogs, the rank growth of certain grasses in calcareous habitats (overpowering species-rich turf), a decline in the floristic diversity of acid grasslands and the spread of purple moor grass *Molinia caerulea* at the expense of the heather *Calluna vulgaris* in some heathlands (Bobbink et al, 1998). Nitrogen fertilization is believed to be the cause of the luxuriant growth of some rampant herbs, such as stinging nettle *Urtica dioica*, leaving less room for shorter species.

3

Actors and Stages

The basic truth is that conservationists do not conserve nor make the basic fundamental decisions. First, these decisions are made by people at the grassroots level: the person with an axe or a hoe. The question is: 'Should I or should I not cut this plant?' At a deeper level the decisions are made by the local and national political and economic leadership, who by their policies affect the decision to cut or not to cut. (Rodgers, 1998)

ROLES FOR PLANT CONSERVATIONISTS

Plant conservation needs plant conservationists – that is, people dedicated to the conservation of plants. Plant conservationists may come from any walk of life, have all sorts of educational backgrounds and may be amateur or professional. There are many roles in plant conservation, with room for many types of people. What matters for the movement as a whole are the commitment of individuals to the cause and a measure of commonality in the direction of efforts so that they are mutually reinforcing. The desirability of synergy is one of the reasons why attention needs to be given to trying to identify priorities in plant conservation pertaining to particular circumstances (see Chapter 9).

Plant conservationists can affect the fates of plants through their own behaviour and their influence on the behaviour of others. Human behaviour is strongly influenced by the norms, customs and rules of social groups. Therefore, if plant conservationists can influence the ways in which social groups function, then they can potentially influence the ways in which many people behave in relation to plants, not just the relatively few whom they can reach individually. What is needed is the institutionalization of social change in favour of plant conservation. There are many types of social group that can be relevant to initiatives in plant conservation – for instance, ones based around family, faith, community, school, occupation and nationality.

One way to influence social groups can be through influencing their policies. Policies are the directions that the members of social groups have deliberately chosen for their joint affairs. As far as public institutions are concerned, it is useful if plant conservationists can anticipate the issues that may become debated and prepare their own cases in advance. Opportunities for successful intervention in public policy debate can arise unpredictably. There are a number of attributes useful for plant conservationists if they wish to participate effectively in public policy debate. They include skills in policy analysis, and knowledge of environmental laws and how regulations are created and updated, as well as being known in policy circles (Meffe et al, 1997).

Taking an ecosystem-based approach to conservation (see Chapter 1), it follows that local people must form a key social element to engage in many initiatives in plant conservation. Plant conservationists who are involved in field projects need to find out which particular local residents are particularly relevant to their work (see Chapter 12). Key groups of local people will likely include the owners and managers of land and plants, and the users of plant resources – especially those people who are interested in a diversity of local plants or who harvest plant resources from the wild. Social groups who have long been settled in rural areas tend to have close ties to the land and local botanical diversity. There can, therefore, be a significant overlap in the agendas of plant conservationists and people who are more indigenous, especially when these people are faced with intrusions of outside interests (such as logging operations), threatening cultures and habitats alike. On the other hand, plant conservationists must generally work with *all* types of local residents, not just with those who are more indigenous. Even in the Amazon, only 5 per cent of the entire population consists of Amerindians or rubber tappers (people living by tapping wild rubber trees), out of a total population of 23 million (Zerner, 1999).

For certain purposes, the *public* (a concept contrasted with that of the *community*) will be a social element of interest to plant conservationists. The terms imply that the sections of the population of greatest significance for conservation purposes are identified more with reference to their interests in relation to plants than to where they physically live. Sections of the public, such as natural history enthusiasts and lovers of the countryside, have played pivotal roles in the development of plant conservation in the UK and are likely to continue to do so. Plant conservationists who wish to foster 'greener' trade will probably be interested in 'targeting' those sections of the public most likely to be responsive to their campaigns.

ROLES FOR GOVERNMENTS

States are the most significant modern political entities. How they are organized and run have major influences on the fates of plants. Governments can promote plant conservation through:

- passing favourable laws and regulations and enforcing them;
- beneficial management of state-owned assets, such as land, plant resources and industries using plant resources;

- setting favourable financial incentives and disincentives;
- providing advisory services (for example, for farmers) to promote practices favourable to conservation;
- support for helpful scientific research;
- accreditation of useful educational courses and curricula;
- raising public awareness; and
- setting a favourable moral tone.

States vary in their degree of centralization of government authority and in how the tasks of government are organized in ministries and departments – all of which can significantly influence how well plants are conserved. Two global trends in governance need watching – the creation of supranational political entities, on the one hand, and decentralization, on the other. Groupings of states of relevance to plant conservation include the African Union (AU), the Comunidad Andina (CAN), the Association of Southeast Asian Nations (ASEAN), the Commonwealth, the Council of Europe, the European Union (EU), the North American Free Trade Agreement (NAFTA), the South Asian Association for Regional Cooperation (SAARC) and the Southern African Development Community (SADC). Some of these are largely talking shops; but others, such as the EU, have serious powers. Member states of the EU are *required* (not just requested) to pass EU directives into national law.

Decentralization of certain government powers to lower levels in the state can be advantageous for conservation, bringing conservation closer to the people. However, in some cases it can lead to problems and therefore should only be carried out after serious analysis. Decentralization of authority over forestry resources during recent years has led to serious loss of forests in Indonesia, Tanzania and Uganda (Hamilton, 1984; Hamilton and Bensted Smith, 1989; Jepson et al, 2001). To take the example of Tanzania, authority over virtually all forest reserves was transferred from the central government to regional governments in 1972, part of a general move towards political decentralization. A rapid loss of forest followed, as regional authorities sought to maximize immediate revenues through the liberal selling of licences to cut trees. It soon became apparent that forest destruction was beginning to endanger national water supplies; just four years later in 1976, authority over key areas of forest was transferred back to central government. The major catchment areas transferred were designated as 'catchment forest reserves' (Ahlback, 1986; Hamilton and Bensted Smith, 1989).

States that are signatories of the Convention on Biological Diversity (CBD) – that is, nearly all countries – have formally acknowledged that they have a duty to conserve biological diversity within their jurisdictions. However, the extent to which states are really committed to conservation, as shown by their actions, is uneven. Some governments (of both rich and poor countries) appear to see conservation – and the environment, generally – as something of a luxury, of little relevance to mainstream national life. A more enlightened view is to see the environment as a core part and responsibility of the state, and to seek the integration of environmental concerns within every aspect of public policy (Raeymaekers and Synge, 1995). In order to create political distance from relentless calls for economic development or expansion – as found everywhere – it has been suggested that the statutory body responsible for nature conservation should stand outside the normal ministerial structure, as recommended by Arthur Tansley when he drew up recommendations for the world's first statutory conservation agency – the Nature Conservancy in the UK, founded 1949 (Tansley, 1946). Some countries take the environment very seriously. Bhutan is following a unique pathway of development, based on Buddhist philosophy. A high priority is set on biological and cultural heritage, with a target of always having more than 60 per cent of the country under forest (the current forest cover is 72 per cent).

The ideologies and economic policies of governments have significant impacts upon conservation. Nowadays, it has become widely accepted that both planning and private enterprise have roles to play in national development, and the practical question with respect to plant conservation is how best to strike the balance between them. Conservationists have traditionally set a high score on statutory controls, which seems reasonable given the broad geographical and temporal horizons needed in conservation. Recent global trends towards liberalization of national economies and reduction in trade tariffs have greatly increased pressure on some natural resources. Economic liberalization in India since 1991 has led to a very rapid growth of the woodcarving industry (expanding at 40 per cent per year), with trees being targeted at increasing distances from processing centres (Anthony Cunningham, personal communication, 2002). The collapse of communism in Eastern Europe in 1989 resulted in the dismantling of tight state controls over trade in medicinal plants. There has been a consequent mushrooming of weakly regulated private enterprises trading in medicinal plants, leading to worries about over-harvesting (Lange, 1998a, b).

Government agencies whose responsibilities relate to land or plants have key roles to play in plant conservation. Among them are the various agencies responsible for protected areas (see Chapter 10). It would often repay plant conservationists if they gave more attention to agencies charged with water supplies because of the considerable overlap in their interests. Generally, the most practical way to protect water supplies is to maintain or restore natural vegetation in catchment areas.

Most states try to protect their forest resources through systems of reserves and measures to control tree felling (see Chapter 10). Forest departments can be responsible for very extensive estates – for example, 22 per cent of the land surface of India is controlled by forest departments (Ray, 1994). A common practice is for logging to be undertaken by private contractors licensed by the state. However, logging practices are often poor (see Chapter 2) and therefore licensing and supervisory systems should often be revised. One approach introduced during the 1990s in Sabah, Malaysia (and which has been recommended elsewhere), is to reduce the number of concessionaires and to increase the lengths of their licences, the theory being that this will increase the motivation for careful logging since concessionaires can potentially benefit from more than one crop. Whether this alone will make companies more responsible is debatable. It seems to rest upon doubtful assumptions – that companies have confidence in the long-term stability of government policy towards forests and that they are prepared to tie up their money for long periods in forest capital, rather than investing it elsewhere.

During recent years, many forest departments have given more attention to social forestry and biodiversity conservation, in addition to their traditional interests in timber and forest protection (Hamilton et al, 2003). From the plant conservation perspective, it would be helpful if forest departments could devote more attention to the management of non-timber forest products (NTFPs). It has been suggested that there should be government advisory services for NTFPs similar to those often currently provided for agriculture (Shackleton et al, 2002).

Two of the ways in which governments occasionally influence agriculture are through subsidies paid to farmers for certain activities and the provision of advisory services. Subsidies are a feature of richer countries, though their effects can be widely felt, thanks to the competitive price advantage that they confer. The EU's Common Agricultural Policy (CAP) results in the payments of very high subsidies to some farmers, encouraging intensive agriculture and contributing to associated environmental problems. More welcome for conservation are EU subsidies available for organic and other, more environmentally benign, forms of agriculture.

Many governments have traditionally provided free agricultural advisory services for farmers; but during recent years farmers have often had to pay for such advice. Agricultural advice has traditionally been 'top down', with agricultural scientists drawing up recommendations for agricultural extension workers, who have then passed them on to farmers. However, a more participatory approach to agricultural advice is now being tried in some countries, including China (Song Yiching and Jiggins, 2003).

There are many connections between plants and human health (see Chapter 11). The use of plants as medicines is the most extensive use of plants in terms of numbers of species (see Table 7.3). It is estimated that 70 to 80 per cent of people worldwide rely chiefly upon traditional, largely plant-based, medicine to meet their primary healthcare needs, with the absolute number of such people increasing (Farnsworth and Soejarto, 1991; Srivastava, 2000; Pei Shengji, 2001). Only 15 per cent of pharmaceutical drugs globally are consumed in developing countries (Toledo, 1995) and it is estimated that many drugs sold in the developing world are useless fakes (for example, half of the malarial drugs sold in Africa). The high dependency of people in some developing countries upon traditional medicine is illustrated by the fact that there are over ten times as many traditional medical practitioners (TMPs) than doctors in parts of Africa (see Table 3.1).

At present, many countries provide all, or nearly all, state support for healthcare to the Western medical sector despite the fact that, in many cases, more people resort to traditional than Western medicine. Therefore, there is a strong case for governments to give more attention to traditional medicine,

Table 3.1 *Ratios of doctors (practising Western medicine) and traditional medical practitioners (practising largely plant-based medicine) to patients in Eastern and Southern Africa*

Country	Doctor-to-patient ratio	Traditional medical practitioner-to-patient ratio
Ethiopia	1:33,000	–
Kenya	1:7142 (overall)	–
	1:833 (urban – Mathare)	1:987 (urban – Mathare)
Malawi	1:50,000	1:138
Mozambique	1:50,000	1:200
South Africa	1:1639 (overall)	–
	1:17,400 (homeland areas)	1:700–1200 (Venda)
Swaziland	1:10,000	1:100
Tanzania	1:33,000	1:350–450 (Dar es Salaam)
Uganda	1:25,000	1:708

Note: Ratios of traditional medical practitioners (TMPs) to patients can be higher in urban than rural areas, related to the stresses of urban life (Cunningham, 1996a).
– = data not available
Source: Marshall, 1998

including conservation of medicinal plants. Crises can sometimes force countries to take action to develop traditional healthcare systems, as happened in Cuba during the 'special period' (*periodo especial*, 1990–1996) when the country was faced with loss of financial and trade support from the Soviet Union and difficulties associated with a continuing economic blockade imposed by the US (Hernández Cano and Volpato, 2004).

Less dramatically, governments can make steady efforts to develop national healthcare systems, drawing upon all available resources of medical knowledge and personnel. Aspects of traditional medicine which can benefit from government attention include the authentication of treatments using culturally sensitive methods, the standardization of medicines to control dosages, and the registration and training of TMPs – as well as conservation of medicinal plants. Official recognition of traditional medicine has the potential to raise the social status of TMPs within communities, which has advantages for plant conservation. TMPs are often the most knowledgeable people about local plant diversity within rural communities and traditionally have enjoyed high prestige. Enhancement of their status under modern conditions has the potential to increase the chances that their knowledge of plants will be used to improve the ways in which the plants are managed.

THE ROLES OF INTERNATIONAL AND REGIONAL CONSERVATION AGREEMENTS

Countries (parties) that have ratified international conservation conventions are obliged to pass their provisions into national law. They do so with varying degrees of enthusiasm. Some major plant-related conventions are described briefly below. There are several others of relevance, among them the Convention on Long-Range Transboundary Air Pollution (1979), the Vienna Convention for the Protection of the Ozone Layer (1985), the Framework Convention on Climatic Change (1994) and the Convention to Combat Desertification (1996). There has been a call for a new United Nations body to coordinate work on combating invasive species, including for the creation and maintenance of a global database (Rejmánek, 2000).

The 1992 Convention on Biological Diversity (CBD) is the single most important international convention for plant conservation. It has been ratified by nearly all countries, a notable exception being the USA. Under the CBD, parties are recognized as having sovereign rights over their biological resources, although recognizing that conservation of biological diversity is a legitimate concern of all humankind. By signing the CBD, governments have agreed to its three objectives:

1 the conservation of biological diversity;
2 the sustainable use of the components of biodiversity; and
3 the fair and equitable sharing of the benefits arising out of the utilization of genetic resources (see Chapter 4).

Parties to the CBD are required to prepare national biodiversity strategies and action plans. Representatives of parties to the CBD meet biennially at Conferences of the Parties (COPs) to assess progress and see how best the convention can be taken forward. They are assisted in their deliberations by advice from a Subsidiary Body on Scientific, Technical and Technological Advice (SBSTTA). The CBD has seven thematic work programmes (for example, forest biodiversity and agricultural biodiversity) and pays attention to selected 'cross-cutting' issues, including access to genetic resources and benefit-sharing, alien species and the Global Strategy for Plant Conservation (GSPC).

The GSPC was agreed by the parties to the CBD in 2002. It is unique for the CBD in dealing with a single taxonomic group (plants) and in having outcome-orientated targets. Its 16 targets cover information, conservation, sustainable use, education and capacity-building (see Box 3.1). Parties are obliged to incorporate the requirements of the GSPC within their national biodiversity strategies and action plans. The GSPC represents a big advance in international plant conservation and should be promoted by all plant conservationists. Its targets constitute a common agenda for uniting plant conservationists worldwide. The GSPC has already been taken forward at the regional level in Europe, with agreement on a European Strategy for Plant Conservation, worded to conform to the GSPC and developed under the auspices of the Council of Europe and Planta Europa. Botanic Gardens Conservation International (BGCI) has been instrumental in encouraging botanical gardens worldwide to make support for the GSPC a fundamental part of their missions.

Box 3.1 Targets of the Global Strategy for Plant Conservation (to be met by 2010)

Understanding and documenting plant diversity

1   A widely accessible working list of known plant species as a step towards a complete world flora.
2   A preliminary assessment of the conservation status of all known plant species at national, regional and international levels.
3   Development of models with protocols for plant conservation and sustainable use based upon research and practical experience.

Conserving plant diversity

4   At least 10 per cent of each of the world's ecological regions effectively conserved.
5   Protection of 50 per cent of the most important areas for plant diversity ensured.
6   At least 30 per cent of production lands managed consistent with the conservation of plant diversity.
7   Sixty per cent of the world's threatened species conserved *in situ*.
8   Sixty per cent of threatened plant species in accessible *ex situ* collections, preferably in the country of origin, and 10 per cent of them included in recovery and restoration programmes.
9   Seventy per cent of the genetic diversity of crops and other major socio-economically valuable plant species conserved, and associated indigenous and local knowledge maintained.
10 Management plans in place for at least 100 major alien species that threaten plants, plant communities and associated habitats and ecosystems.

Using plant diversity sustainably

11 No species of wild flora endangered by international trade.
12 Thirty per cent of plant-based products derived from sources that are sustainably managed.
13 The decline of plant resources, and associated indigenous and local knowledge, innovations and practices that support sustainable livelihoods, local food security and healthcare, halted.

Promoting education and awareness about plant diversity

14 The importance of plant diversity and the need for its conservation incorporated within communication, educational and public-awareness programmes.

Building capacity for the conservation of plant diversity

15 The number of trained people working with appropriate facilities in plant conservation increased, according to national needs, in order to achieve the targets of this strategy.
16 Networks for plant conservation activities established or strengthened at national, regional and international levels.

Source: CBD, 2002

The 2000 Millennium Resolution of the United Nations encourages UN members to adopt a new ethic of conservation and stewardship. Eight Millennium Development Goals (MDGs) (to be met by 2015) have been agreed as part of a roadmap to implement the resolution. Goal 7 is concerned with environmental matters and reads: 'Ensure environmental sustainability', with one of its targets as 'Integrate the principles of sustainable development into country policies and programmes and reverse loss of environment resources'. Two of the indicators of this target are 'Proportion of the land area covered by forest' and 'Ratio of area protected to maintain biological diversity to surface area'. The MDGs are significant with respect to the channelling of international development assistance because they are used by many donor agencies to target their aid. At their sixth COP in 2002, parties to the CBD adopted the target: 'Achieve by 2010 a significant reduction of the current rate of biodiversity loss at the global, regional and national level as a contribution to poverty alleviation and to the benefit of all life on Earth.' This target was subsequently endorsed by the World Summit on Sustainable Development (WSSD, the Earth Summit) held in Johannesburg in 2002. All in all, it is reasonable to conclude from these various international resolutions, goals and targets that biodiversity conservation has been moving higher up the international development agenda.

Several international conventions and programmes have scope for the designation of sites of special conservation concern. The United Nations Educational, Social and Cultural Organization (UNESCO) started its Man and the Biosphere (MAB) programme in 1970. By 2001, over 400 areas in nearly 100 countries had been designated as biosphere reserves. Each biosphere reserve is intended to fulfil three functions: protect biological diversity; foster sustainable economic and human development (in ways that are compatible with the first function); and facilitate education, training, research and monitoring (in support of the first two functions) (see Figure 6.3). The 1972 Convention for the Protection of the World's Natural and Cultural Heritage, or World Heritage Convention, also associated with UNESCO, is a binding legal instrument concerned with the protection of cultural and natural heritage. Sites designated as World Heritage Sites are regarded as so exceptional that their protection is deemed a matter of international, not just national, concern. As of mid 2001, 690 sites had been inscribed on the World Heritage list: 529 cultural sites, 138 natural sites and 23 mixed sites, altogether situated in the territories of 122 parties to the convention. There is some overlap with biosphere reserves, some 60 of which have been fully or partially designated as World Heritage Sites (Hadley, 2002).

Parties to the Convention on Wetlands of International Importance especially as Waterfowl Habitat (the 1971 Ramsar Convention) are required to consider wetland conservation within their national land-use policies and to establish wetland nature reserves. They can propose sites for inclusion in the Ramsar List of Wetlands of International Importance, which now includes 1037 sites worldwide. Parties must manage Ramsar sites so as to avoid changing their ecological character.

BOX 3.2 APPENDICES OF THE CONVENTION ON THE INTERNATIONAL TRADE IN ENDANGERED SPECIES OF WILD FAUNA AND FLORA (CITES)

Appendix I

Includes species threatened with extinction through international trade. No trade for commercial purposes is allowed for specimens taken from the wild. This covers parts of plants (for example, pollen and seeds) and products from plants (including scientific materials such as herbarium specimens). Trade in artificially propagated specimens is allowed, subject to permit. Two hundred and ninety-eight plant species as well as four subspecies are listed, including a number of orchids, such as the Asian slipper orchids *Paphiopedilum*.

Appendix II

Includes species not necessarily threatened now, but may become so unless trade is regulated. This Appendix also contains species similar in appearance to some in Appendix I – for example, cacti and orchids – which have been included to help with enforcement. Trade in both wild and artificially propagated specimens is allowed, subject to permit. Permits should only be issued if the specimens have been obtained legally and once the scientific authority has indicated that the trade is unlikely to be detrimental to the survival of wild populations. Trade in parts or derivatives also requires a permit, except for certain stipulated items that can be traded freely. A total of 28,074 plant species, as well as three subspecies and six geographically separate populations of species, are listed, including all orchids and cacti not in Appendix I.

Appendix III

Includes species that are threatened in at least one country, which has asked other CITES parties for assistance in controlling the trade. Trade is allowed provided that specimens have been obtained legally, but CITES documentation is required (certificates of origin, except in the case of those parties that have listed the species, which require export permits). Forty-five plant species, as well as one subspecies and two geographically separate populations of species, are listed.

Source: Alan Hamilton and Patrick Hamilton with the assistance of Dr Alison Rosser, formerly Wildlife Trade Programme Officer of the World Conservation Union (IUCN)

The aim of the 1976 Convention on International Trade in Endangered Species of Wild Fauna and Flora (CITES) is to ensure that international trade in specimens of wild animals and plants does not threaten their survival. In 2004 there were 167 parties to CITES. Requirements relating to trade vary according to the appendices upon which species are listed (see Box 3.2). Operation of the CITES system depends upon intending exporters and importers of plants, plant parts or plant products being aware of the species on CITES appendices, and upon procedures being in place for issuing and checking permits (see www.cites.org). Each party is required to appoint one or more Management Authorities to administer the licensing system and one or more scientific authorities to provide scientific advice to the management authorities. Permits are required for both wild and artificially propagated specimens of listed species; in principle, these are granted easily and without fuss for the latter.

There are various ways in which CITES needs to be developed (Schippmann, 1995; Oldfield, 1997; Sheldon et al, 1997). There are often deficiencies in data about the status of threatened species, which creates problems for listing and issuing permits. It is suspected that some exporting countries are too ready to issue permits for economic reasons. Trade in artificially propagated specimens should be made easier. Most experts believe that trade in artificially propagated specimens of CITES species is beneficial for conservation, based on the theory that this will take the pressure off wild populations. CITES also requires improvement regarding the effectiveness of border controls. Customs officers have many duties, with botanical matters not necessarily high on their list of priorities (except for certain illicit narcotic plant products!). Guidelines have been produced for customs officers in some countries to help them recognize CITES-listed plants and to distinguish between wild-collected and artificially propagated specimens.

Various declarations, regional agreements and programmes are significant for plant conservation. The African Union has declared 2001–2010 the Decade for African Traditional Medicine. The Council of Europe, with 46 member countries, has two significant accords in favour of plant conservation: the Convention on the Conservation of European Wildlife and Natural Habitats (the 1982 Bern Convention) and the European Strategy for Plant Conservation. The Bern Convention provides for the conservation of wildlife and wildlife habitats in Europe. It establishes lists of plant species and sites (the Emerald Network) that must be afforded special protection. The Council of Europe is establishing a Pan-European Ecological Network (PEEN) based on the principle that the European landscape should be managed to conserve the full range of ecosystems, landscapes, habitats, species and genetic diversity of species considered to be important on a Europe-wide scale.

The European Union (EU) exerts considerable authority over conservation in Europe, both with respect to its 25 member states and among countries aspiring for membership. The EU implements CITES as a single entity, but has additional rules so that, overall, the rules governing cross-border movements of plants are stricter for the EU than under CITES alone. However, the customs border for CITES is that of the EU as a whole, which creates inefficiencies because intending malefactors may seek to route their goods through weak sections of the barrier. Once within the EU, there is little to stop the free movement of goods. The EU Directive on the Conservation of Natural Habitats and of Wild Fauna and Flora (the 1992 Habitats Directive) obliges member states to conserve certain threatened habitats and species. They must establish Special Areas for Conservation (SACs) to maintain and, if necessary, restore certain habitat types and species listed in its annexes. Those species that are listed must be maintained at a 'favourable conservation status'; this is deemed to occur if data on species population dynamics indicate that the species is maintaining itself, its natural range is not being reduced and its habitat is judged sufficiently extensive. The network composed of the SACs is known as Natura 2000 (Natura 2000 sites will also be Emerald Network sites in the case of EU members). Natura 2000 will become the dominant network for protecting species and habitats in the EU. Over 10,000 sites have been proposed as SACs, covering 13 per cent of EU land.

ROLES FOR BOTANICAL INSTITUTES, NETWORKS, SOCIETIES AND NGOS

The scientific underpinning of plant conservation consists of expert botanists, together with their associated networks, institutes and departments – including herbaria, botanic gardens, seedbanks, research organizations and university departments (see Figure 3.1). Today, many institutes concerned with botany have plant conservation among their objectives. Colombia set a wonderful example in founding the Instituto Alexander von Humboldt in 1993, following passage of the CBD. Devoted to biology and conservation, the work of the Humboldt Institute covers biological inventory, conservation biology, the use and valuation of biological resources, policy, legislation, communication, cooperation and education.

Herbaria are key parts of the plant conservation infrastructure. Their collections of preserved plants are used to check the identities of plants encountered in the field and are the core materials used by taxonomists in their studies of plant types and their relationships. The efficient operation of herbaria depends, in part, upon their networking skills. National herbaria are central facilities, serving as points of reference for herbaria 'lower' in the system, such as at field stations, and linked internationally through a few major herbaria that contain specimens from extensive regions, allowing international comparisons. The Royal Botanic Gardens, Kew, in the UK, now progressively requires its taxonomists to include conservation assessments in all new taxonomic descriptions. If plant specimens are being collected for herbaria, then it is good practice to distribute duplicates to other herbaria. Herbaria can be damaged by neglect or misfortune. A priceless collection of orchids at the Jardin des Plantes in Paris was destroyed by German artillery fire in 1871. The herbarium at the Botanical Museum in Berlin was burned down during an Allied bombing raid in 1943, causing the loss of one of the world's fundamental scientific collections of plants.

Botanic gardens and arboreta are distinguished from leisure gardens and municipal parks in being founded upon scientific principles. Their specimens of plants should be correctly named scientifically, with their origins recorded (Wyse Jackson and Sutherland, 2000). There are over 2500 botanic gardens globally, some engaged in activities (nationally or internationally) well beyond their gates. Botanic gardens typically have large collections of plant species

Figure 3.1 *Department of Botany, University of Cluj-Napoca, Romania*

Source: Alan Hamilton

(altogether about one third of all vascular plant species are grown in botanic gardens) and are of obvious value for *ex situ* conservation. Botanic gardens are particularly well suited to educating the public about plant conservation. They can be very popular – for example, with over 400,000 visitors annually in the case of Xishuangbanna Tropical Botanical Garden in China. Both the International Association of Botanical Gardens, founded in 1954, and Botanic Gardens Conservation International (BGCI), founded 1987, are devoted to plant conservation, with BGCI also having a strong focus on environmental education and sustainable development. BGCI links together 550 member botanic gardens in 116 countries into a cooperating network for plant conservation. Its members have agreed to a common International Agenda for Botanic Gardens in Conservation and BGCI engages in various supporting activities, such as the production of publications on plant conservation, sustainable development and environmental education, provision of technical guidance and maintenance of a database on plants found in botanic gardens, and supporting conservation and education projects in botanical gardens worldwide.

Several countries have networks devoted to plant conservation. The Australian Network for Plant Conservation (founded in 1991) runs courses on techniques in plant conservation for government agencies, land managers, community groups and others. In France, the Conservatoires Botaniques Nationaux (founded in 1988) strives to conserve rare plants, with programmes of monitoring, research and *ex situ* cultivation, with reintroduction or reinforcement of field populations in the case of the most threatened species (Akeroyd, 1999). One member, the Conservatoire Botanique National de Brest, is especially concerned with the floras of oceanic islands, both those that are parts of France (for example, Martinique and La Réunion) and others that are not (such as Madeira and St Helena).

Among regional networks, the Red Latinoamericana de Botanica (founded in 1988) links institutions in six Latin American countries to promote training, joint projects and meetings (Maldonado et al, 2003). The Red de Herbarias de Mesoamérica y El Caribe (founded in 1995) plays an instrumental role in plant conservation in its region. The Grupo Etnobotánico Latinoamericano (GELA, founded in 1986) promotes scientific exchanges between ethnobotanists active in Latin America, with national coordinators in many countries. The Southern Africa Botanical Diversity Network (SABONET, founded in 1996) is a project, now ending, aimed at enhancing scientific and technical capacity in Southern Africa, with the ultimate goal of achieving conservation and sustainable use of plant resources (Huntley, 1999). Planta Europa (founded in 1996) was established to promote plant conservation across Europe. Its members include government agencies, botanical societies and non-governmental organizations (NGOs), with a coordinating secretariat at Plantlife International.

Plantlife International (founded in 1989) is an international NGO dedicated specifically to plant conservation. It is especially dedicated to promoting targets 5 and 13 of the Global Strategy for Plant Conservation (GSPC) (see Box 3.1) through its programmes on Important Plant Areas and Plant Conservation and Livelihoods, initially focusing on a Medicinal Plants Conservation Initiative. People and Plants International (founded in 2004), also concerned with target 13, espouses the adoption of more effective practices in community-based plant conservation. Some of the large international conservation NGOs have run campaigns, programmes or projects in plant conservation. The Global Trees Campaign (started in 1993) is a partnership of Fauna and Flora International (FFI, founded in 1903) and the United Nations Environment Programme (UNEP) World Conservation Monitoring Centre. The People and Plants Initiative (1992–2005) was a partnership of the World Wide Fund for Nature (WWF, founded in 1961), UNESCO and the Royal Botanic Gardens, Kew. Its aim was to build international capacity in applied ethnobotany (see the section on 'The People and Plants Initiative', page xiv). See Chapter 13 for a description of the network associated with the International Plant Genetic Resources Institute (IPGRI).

Apart from specifically plant-related groups, there are various voluntary groups and initiatives that can be great allies for plant conservationists. The award of the Nobel Peace Prize in 2004 to Professor Wangari Maathai has drawn international attention to the close links that often exist between women and plants, and the great potential roles of women's groups in conservation. Professor Maathai fought for the environment in Kenya, working closely with women's groups, and founded the Green Belt Movement, which has planted 25 million trees in Africa to combat forest loss and desertification.

The Alliance of Religions and Conservation (ARC, founded in 1995) is a secular body that helps the major religions of the world to develop their own environmental programmes, based upon their core teachings, beliefs and practices. Religious organizations have the potential to be close allies for plant conservationists. Beliefs associated with religion underlie many traditional conservation practices, while there are theologians in all the major world religions who teach that people have a sacred duty to care for the environment. Churches, temples and mosques can be powerful influences on village life in developing countries.

Botanical and natural history societies play important roles in conservation in some countries. Examples include the Bombay Natural History Society, the Botanical Society of the British Isles, the Botanical Society of South Africa, the California Native Plant Society and the Malaysia Nature Society. Ethnobotanical societies, such as the Ethnobotanical Society of Nepal (ESON) and the Tanzanian Society on Ethnoscience (TSE), concentrating on links between people and plants, can have great value for engaging the interest of botanists and others concerned with rural development in developing countries. Some subjects covered by ethnobotany are of great economic or cultural concern to communities.

Networks that link government conservation bodies or botanical institutes with the public have considerable potential for plant conservation. Flora Guardians, a programme managed by the Swedish Threatened Species Unit with financial support from WWF-Sweden, brings together 250 people and 16 botanical societies to monitor populations of threatened plants. One hundred and eighty-five Endangered and Vulnerable vascular plants are covered through mostly annual visits to each of over 5000 sites.

Botanists at the Western Australian Herbarium in Perth, part of the Department of Conservation and Land Management (CALM), have developed a network that links their herbarium with community organizations and individual members of the public throughout Western Australia. Western Australia has a large flora, with many plants that are difficult to name and with few up-to-date field guides. The Western Australian Herbarium contains 500,000 specimens, all data based and bar coded with their names, classificatory positions, geographical locations and habitats. Through advice and training workshops, the herbarium encourages members of the network to collect duplicate specimens of unknown species, together with notes on their habitats. One specimen is retained in a local herbarium and the other sent to the Western Australian Herbarium, where it is identified and data based and a copy of its unique bar code is sent to the local herbarium. In this way, each specimen in local herbaria is linked to a voucher specimen in the Western Australian Herbarium, which has the advantage that any changes in the scientific names of plants can easily be captured and updated even in remote areas. This system is allowing many more members of the public to become actively engaged in plant conservation. It has led to the discovery of several new species and numerous new populations of rare plants, enabled the survey of hundreds of previously unrecorded areas and added thousands of specimens to the

Western Australian and local herbaria. It has the potential to develop into a comprehensive weed-watch programme, involving local people in the front line of the battle to combat invasive plant species – a big problem in this area (Marchant, 2000).

The World Conservation Union (IUCN – formerly the International Union for the Conservation of Nature and founded in 1948) has a unique position on the global conservation stage. Its members include states, international bodies and NGOs. Its major services include the establishment of certain global conservation standards, such as the definitions of categories of protected areas and degrees of endangerment of species, and the running of five commissions concerned with major aspects of conservation, including the Species Survival Commission (SSC) and the World Commission on Protected Areas (WCPA). The SSC is a network of some 7000 volunteer members from around the world, mostly deployed in more than 120 specialist groups and task forces. Specialist groups dealing with plants are organized variously on lines of taxonomy (for example, orchids and palms), geography (for example, Madagascar and the Mediterranean Islands) or themes (such as invasive species and medicinal plants). They are encouraged to prepare action plans. IUCN has a Plants Programme with objectives and activities set by a plant conservation committee and guided by an agreed strategy.

4

Information, Knowledge, Learning and Research

Information and knowledge form essential foundations for plant conservation. Information in published form is much more available in richer countries than in the developing world; but rural people in developing countries can be especially knowledgeable about plants. Whatever the bases of information and knowledge available, there will always be uncertainties of how best to tackle conservation problems. Research and learning should be integral parts of the process of conservation.

INFORMATION

Information useful for plant conservation is available in written form (for example, books and scientific papers), as audio-visual materials, on the internet and in people's heads. As far as written information is concerned, conservationists working on projects should not confine their searches for information to formally published works. Much useful information can often be gleaned from unpublished documents (grey literature). Many types of information can be relevant for plant conservation, as suggested by the broad contents of this book.

There are great gaps in information relating to plant conservation. In the fields of taxonomy and ecology, for example, a review of the Acanthaceae (one of the major families of flowering plants in Africa) revealed that the habitats of most African species are unknown and, for 370 out of 1393 species, it is not even clear whether they are forest or non-forest plants. The taxonomy of most groups of African Acanthaceae needs revision, not having been seriously studied since 1901 (Balkwill and McCallum, 1999).

Oral information is immensely valuable in plant conservation. It can have the advantages of being up to date, site specific and transmitted in forms tailor made to the issues at hand. Knowledgeable botanists can be mines of information about the plants, places, conservation concerns and development issues at sites where conservation projects are being contemplated. The inclusion of local people who are especially knowledgeable about local plants (key knowledge holders) in project teams is strongly recommended in community-based plant conservation initiatives (see Chapter 12).

Published information relevant to plant conservation is distributed highly unevenly, with very little available in many developing countries (Huntley and Matos, 1994; Ruddle, 2000; White, 2001). Books useful for plant conservationists are not only scarce, but also prohibitively expensive in terms of financial incomes in poorer countries. The internet has helped greatly to make information widely available in some parts of the world (for example, in parts of Asia) and has great potential for development. However, it is still largely inaccessible elsewhere (such as in much of Africa) and many people appreciate a hard copy. There can be an element of farce in the information that actually is available in developing countries. When working on the East Usambaras in Tanzania in 1986, Alan Hamilton noticed that the *only* publication in one of the field offices of the Tanzania Forest Department was a brochure for the same logging machinery that, ironically, was just then devastating a nearby forest for which the office was responsible.

More written and archived information about the plants of some countries exists *outside* the countries than within them. This can be partly due to colonial history. Today, some efforts are being made to redress this imbalance. For instance, the Royal Botanic Gardens, Kew, in the UK, is repatriating data on Brazilian plants back to Brazil (Prance, 2003). Foreign researchers active today in other people's countries have a responsibility to ensure that their findings are available within the countries that have hosted their studies. This should be partly through supplying copies of technical papers and reports to collaborating national scientists and institutions. If they have worked with communities, then they should make their findings known to the communities in formats that will be appreciated.

Information is ineffective for practical purposes unless it is available in an appropriate form to those who can make use of it. Reports with plenty of technical detail are good for scientists; but policy-makers are likely to respond best to brief, attractively presented documents, with the main arguments and conclusions succinctly drawn. Oral presentations, interactive sessions at workshops, drama, booklets and posters can all work well with communities, depending upon the subject matter, cultures and levels of literacy. Technical publications are much needed in many aspects of plant conservation, varying from more theoretical treatises to practical guides for those who just want to get on with the job. The Australian Network for Plant Conservation publishes 'how to' guides for practitioners (see www.anbg.gov.au/anpc). Field guides for the identification of plants, including specimens lacking flowers or fruits, are invaluable for field projects (Rejmánek and Brewer, 2001). The People and Plants Initiative is writing a manual for producing user-friendly guides (Lawrence and Hawthorne, in preparation).

Today, computer databases are widely used in plant conservation for storing, ordering, classifying and analysing information. They include shared information systems, with formal requirements about how new data can be entered. An example is the Species Information Service of the World Conservation Union's (IUCN's) Species Survival Commission. The fields used include: species; authority; taxonomic position; references; population size;

geographical range; habitat; Red List assessment (see Chapter 9); conservation actions taken; threats; and whether the species is traded. Standardization in data entry is promoted by incorporating pull-down lists of allowable entries for some fields, which also reduces input errors. Efforts are made to manage the data as close to their sources of origin as possible (preferably by the specialist groups of the IUCN) to ensure good quality data and encourage feelings of ownership (IUCN, 2004).

Conservationists will often wish to create their own databases in order to meet their particular requirements. An advantage of designing the structure of one's own database is that there is then the potential for more control. A manual is available from the People and Plants Initiative on how to design small databases for plant conservation (Berjak and Grimsdell, 1999; see also www.peopleandplants.org). There are some points of principle to follow in database design in order to increase flexibility in data analysis and to make it easier to share data between databases. First, data fields should be broken down as far as reasonably possible – for example, with separate fields for the various parts of the scientific names of species (that is, genus, species and authority). Furthermore, standard fields and standard lists of allowable entries for fields should be used where feasible.

Two People and Plants publications contain recommendations about the construction of databases associated with ethnobotanical surveys (Martin, 1995; Tuxill and Nabhan, 2001). Tuxill and Nabhan's *People, Plants and Protected Areas*, referring specifically to crops, suggests several extra fields that may be useful, in addition to those normally used when collecting information about plants. They concern the utilitarian and cultural aspects of plants, and include palatability, aroma, cooking time, resistance to pests (in field and storage), time to reach maturation, storage longevity and ceremonial value.

KNOWLEDGE

Knowledge is a personal matter, composed of beliefs, cognitions, models, theories, concepts and other products of the mind (Röling, 1988). Knowledge is not just a matter of cold data; it is also about associated feelings and impulses. Knowledge is acquired throughout the course of life, unconsciously and through conscious effort. Although everyone's knowledge is unique, people do share certain aspects of their knowledge with others, being members of social groups with common cultures. A useful concept is that of knowledge systems, associated with particular social groups and cultures. Box 4.1 provides some examples of knowledge systems relating to medicine, incorporating the significant involvement of plants. Since people can be members of different social groups for different aspects of their lives, they may also be associated with several knowledge systems. Knowledge systems are dynamic, since they are associated with living people, and are subject to continuing processes of creation and adaptation, even if they seem to be apparently unchanging over many years.

BOX 4.1 EXAMPLES OF KNOWLEDGE SYSTEMS INVOLVING
PLANTS: SOME SCHOLARLY MEDICAL TRADITIONS

Each knowledge system listed includes personnel (medical practitioners and
other experts), bodies of written information (including pharmacopoeias) and
facilities for training and networking:

- Traditional Chinese Medicine (TCM); one of its earliest texts is *Sheng-Nong's
  Herbal*, which records plants used possibly as far back as 3000 BC, though
  written later;
- Kampo, traditional Japanese medicine;
- Tibetan medicine, associated with Bon and Buddhism;
- Mongolian medicine, associated with Buddhism and with similarities to
  Tibetan medicine;
- Dai medicine, associated with Theravada Buddhism and practised in south-
  ern Yunnan (China);
- Ayurveda, normally associated with Hinduism, but with Buddhism in Sri
  Lanka (the earliest mention of the medicinal use of plants in Hindu culture is
  in the *Rigveda*, written before 1600 BC);
- Siddha, associated with Tamil-speaking areas of India;
- Unani, an Islamic system of medicine common in the Indian subcontinent;
  Unani, like Western herbal medicine, is rooted partly in the medicine of
  Ancient Greece (Unani is an Arabic word derived from Ionia, part of Ancient
  Greece);
- Uigur medicine, practised in north-west China and associated with Islam;
- Western herbal medicine, practised in Europe, North America and elsewhere
  (it is poorly standardized compared to some other medical traditions, with
  much variation between countries and individual practitioners);
- Homeopathy, developed by Samuel Hahnemann (1755–1833); this differs
  from the above systems in that it uses minute quantities of botanical and
  other ingredients in its medicines – substances that are often poisonous
  when taken in larger quantities;
- Western medicine, the principal officially recognized system in most
  countries; Western medicine is sometimes referred to as orthodox, conventional or allopathic medicine, in contrast to terms used for the other
  traditions, such as traditional, herbal, complementary or alternative.

Source: Hamilton et al, 2003

Knowledge networks can be a useful tool in plant conservation to allow the
exchange and development of ideas. They can be formed at various social
levels and can make use of a variety of methods of communication. The
'remote workshop' is a modern development, consisting of a group of people
exchanging and developing views on a given subject by email over a limited
period of time (for example, one to four weeks). A central coordinator periodically sums up the state of the 'conversation' and guides the workshop towards
useful goals.

The World Wide Fund for Nature (WWF) Programme Office for Nepal has been encouraging the development of a knowledge network among doctors (*amchis*) of Tibetan medicine (*sowarigpa*). A workshop in January 2004 was the first time that *amchis* from various Himalayan countries had been able to meet to discuss concerns about their profession, including the conservation of medicinal plants. This initiative has considerable potential for the development of conservation in the Himalayas because some of the *amchis* live in scattered villages and are almost the sole deliverers of healthcare services across substantial regions. They are the most knowledgeable local people about plants in the Himalayas and have an intrinsic interest in their conservation.

Tracing the origin of knowledge in knowledge systems can tell us much about the evolution of social groups and cultures. There are many examples of the movement or mixing of knowledge traditions in herbal medicine, which helps to explain the considerable similarities between Western herbal medicine and Unani, and between Unani, Ayurveda and Tibetan medicine. Sometimes the mixing of traditions is deliberate. During the eighth century AD, Trisong Detsen, King of Tibet, called a conference of medical experts – from China, Dolpa, India, Nepal, Persia, Tibet and other parts – to discuss how to improve medical treatments by drawing upon all traditions. This conference was instrumental in the evolution of Tibetan medicine (Lama et al, 2001). Folk medicinal traditions can also draw upon several sources. The medicine of eastern Cuba is a blend of knowledge from African, Amerindian, Antillean, French-Haitian and Spanish sources (Hernández Cano and Volpato, 2004). European herbs are conspicuous in the herbal medicine of *afrodescendente* communities in Choco, Colombia, presumably originally derived from Spain.

People use and manage plants according to knowledge derived from their personal experiences and thoughts, as well as information from outside sources. People's knowledge is variously influenced by understandings associated with modern science. Folk knowledge – that is, knowledge strongly based upon local, orally transmitted traditions – forms the traditional basis of plant use and management all over the world (Nakashima, 2000).

Much folk knowledge is down to earth and practical – for example, concerning which crops must be rotated to retain soil fertility. However, such concrete knowledge blends imperceptibly into the metaphysical domain – for instance, 'this type of seed is worth planting because the experts say so' or 'the continuing flow of water in the irrigation system depends upon due respect being accorded to the forest deity on the hill'. Much folk knowledge has been shown to have real value for practical purposes. Many studies, for example, have demonstrated that folk knowledge of the medicinal uses of plants is a good guide to their physiological activity (see Holmstedt and Bruhn, 1995; Lewis, 2003). The Kenyah of the Apo Kayan, a tribe living in a remote forested plateau in Borneo, use 17 malarial remedies derived from natural sources. A statistical test has shown that the remedies regarded as most important by the Kenyah have the greatest ability to inhibit the growth of malarial parasites *in vitro* (Leaman et al, 1995). On the other hand, there are some medicinal uses of plants that seem to be purely psychological, such as the use of plants in charms or divination.

Many studies have demonstrated the wealth of knowledge about local plants found in rural societies. Members of longer-established groups tend to be the most knowledgeable (see, for example, Van den Eynden et al, 2003). Brent Berlin has estimated that, among two Amerindian groups which he has studied, biological knowledge (much about plants) is greater than the sum of all other domains of knowledge combined (Berlin, 1978, cited in Alcorn, 1995).

There are many aspects of folk knowledge of plants and the environment that can be relevant to initiatives in plant conservation (see Figure 12.1). Local people have their own ways of classifying types of land use, vegetation and soils, the phases of growth in vegetation successions and the seasons of the year. They will know ways of influencing the development of vegetation for different purposes. Finding out about their understandings of the geography of their neighbourhoods will throw much light on how people have moulded its features and the potential for different kinds of development. Techniques used to study traditional ecological knowledge have been described in *Ethnobotany: A Methods Manual*, a People and Plants book (Martin, 1995, especially pp138–154).

Knowledge about plants is unevenly distributed through communities according to such factors as personal interest, gender, age, social roles and membership of particular families (Berlin, 1978; Ladio and Lozada, 2000). Studies of gender differences in knowledge are useful in conservation projects because they reveal much about the different roles of men and women in economic production, their associations with different resources and the contrasting social constraints that they face (Clarke et al, 1996). As a broad generalization, it seems that, globally, women are more involved with plants than men, although there are reported to be significant regional variations – for example, women are said to be responsible for 80 per cent of agricultural production in sub-Saharan Africa, but only 30 per cent in Latin America (Ruddle, 2000). The collection of fuelwood and wild greens is generally the preserve of women. In their interactions with plants, it has been argued that women tend to be more concerned with aspects that touch on domestic matters – family health, nutrition, healing and security – while men deal more with aspects relating to economic provision for the household and external relation-ships (Richards and Ruivenkamp, 1997; Kothari, 2003; Song Yiching and Jiggins, 2003). It is the women of the Wola highlanders of Papua New Guinea who produce the basic foods that form the backbone of the diet, while men cultivate speciality crops, which are exchanged to forge social alliances (Sillitoe, 2003).

Special knowledge can convey special privileges, which can take various forms – such as high social status, financial wealth or power. Expert female basket-makers in indigenous Californian societies are highly regarded and their baskets prized in ceremonies (Bissonnette, 2003). Men in Anatolia, Turkey, take pains to conceal from each other the places where they collect the fungi that they sell – in this case, knowledge is wealth (Ertug, 2003). Gabrielle Hatfield (Hatfield, 1999) has pointed out that, in the British Isles, everyday healthcare over the centuries has largely been the provenance of women, administering herbal remedies in practical ways devoid of superstition; their medicines have tended to be freshly made from locally available plants. She believes that it is when specialist healers have become involved that remedies have become clothed in ritual and obscurity. She sees this as a two-way process – sufferers started to cast themselves in the roles of patients, expecting more from the healers than they would if they had just used home remedies, and the healers responded, developing systems of healing which erected psychological barriers between themselves and their patients. This reinforced the belief among patients that healers had superior knowledge.

An element of mystification of knowledge is also found in modern professions. Science, rather than the spirits, has now become the mythological source of special knowledge and power. The forestry profession is traditionally divided into higher and lower ranks, the former deemed to understand the theory lying behind forest management and the latter expected to carry out forest working plans with little or no deviation year on year (Tsouvalis, 2000). Similarly, the conventional model of agricultural development places the agricultural scientist in a privileged position with respect to agricultural extension workers and farmers (Röling and Wagemakers, 1998).

LEARNING

Learning through observation, experiment and teaching is a normal part of managing natural resources. Foresters and farmers take note of what works or doesn't work and adjust how they manage their trees or farms accordingly (Alcorn, 1995). Common features of how knowledge about managing natural resources is transmitted between generations in traditional societies include the following (Ruddle, 2000; Bissonnette, 2003):

- Training is provided by particular kinsfolk, usually one or the other of the learner's parents.
- Training is arranged by age and gender, with teaching by members of the appropriate sex.
- Tasks are taught in a systematic manner, with complex tasks being broken down in sequences, starting with simple activities.
- Tasks are taught at the locations where they are normally performed.
- Fixed periods are set aside specifically for teaching.
- Rewards or punishments are given for success or failure at certain tasks.

Most learning about resource management is informal. A study of rice cultivation in Sierra Leone found that individual farmers were growing two to six varieties of rice *Oryza sativa* at any one time, with 20 to 40 varieties in a village. Farmers were found to constantly change their varieties – in one survey, 53 out of 59 households were found to have changed some of their varieties over a four-year period, the new varieties being obtained through non-market exchanges with family and friends (55 per cent), purchase from rice merchants (19 per cent), purchase from local sources (19 per cent) and provided by projects (6 per cent). Farmers actively collected and screened unfamiliar types of rice found growing in their fields, being well aware that such varieties tend to occur on field margins, where spontaneous out-crossing with adjacent varieties is most common (Richards and Ruivenkamp, 1997). Likewise, research in Ethiopia has revealed that farmers growing landraces of sorghum *Sorghum bicolor* know that they can cross with wild relatives, sometimes producing progeny with useful features. They therefore tolerate the presence of wild relatives of sorghum in or around their fields (Teshome et al, 1999).

People vary in their tendency to adopt new practices (Röling, 1988). This is why conservation projects sometimes seek to identify more enterprising members of communities prepared to try out new techniques. The idea is that other people will 'watch over the fence' and, if they see success, might follow suit. Times of personal or social crisis can provide special opportunities for change. These are when taken-for-granted views of the world and related routines can become uncertain, strange or unfamiliar. Such 'problematization' can be critical for the adoption of new mental attitudes and new social structures better adapted to the new realities (Tsouvalis, 2000).

The most deep-seated perceptions of reality are acquired during the earliest years of life, when babies and young children give form to, and learn to interact with, the world around them. It would seem self-evident that children who have substantial, direct, physical contact with nature are most likely to develop the feelings deep within them that nature really is an important part of reality. Historically, many people who have developed a strong interest in field botany became interested in nature when they were young. It is hard to see how children wrapped up constantly in the virtual worlds of television and computers can ever become fully 'environmentalized' (the environmental equivalent of socialized).

Schools play a pivotal role in learning. It is unfortunate that the teaching of biology is often boring and puts off many students (see, for example, Lock, 1996; Smith, 2002). Gabriell Paye, a teacher in Boston, US, has developed a course in botany which, she says, has better captured the attention of her students (Paye, 2000). Activities include ethnobotanical interviews (learning about plants from family, friends and neighbours), learning how to collect and identify plants, and studies of plants as sources of foods and medicines. The work includes making simple tests on products purchased from stores. She writes: 'by entering the world of science through a personal experience of ethnobotany, students gain a real sense of how important plants are in their lives. I find that when they are fuelled by this enthusiasm, they are more willing to then delve into other aspects of botany'. At Ayubia, Pakistan, a People and Plants Initiative project has been fostering nature clubs in local primary schools. Children have been encouraged to ask their elders about the uses of plants, bring specimens of plants into schools to serve as the bases for discussion, plant school gardens and embark on nature trips with their teachers to local forests. In Sri Lanka, another WWF project has encouraged children to raise the highly priced and endangered medicinal plant *Munronia pumila* in their homes. Apart from raising awareness about the values and conservation of plants, this has provided opportunities for households to make money from selling the plants.

The key aspects of the academic plant sciences relevant to plant conservation are those concerned with types of plants, how they live, and how people manage and use them. This covers fields such as plant taxonomy, plant systematics, ecology, ethnobotany, economic botany, forestry, agriculture and horticulture. There are, of course, many other aspects of botany (as well as non-botanical subjects) that are also useful (see Figure 12.1).

In some countries, it is difficult to find pathways through tertiary education that are really suitable for aspiring plant conservationists. National trends in education can be unhelpful, such as (in some cases) trends towards broad, generalized courses lacking much detail (liberal arts-type courses) or (in other cases) botanical or biological courses that concentrate on microbiology and make little reference to 'whole plant botany'. 'The production of botanists who can't tell a composite from an orchid or leaf anatomy from stem anatomy is unconscionable. And common... Regardless of speciality, botanists should have a broad background in all major phases of botany' (Botanical Society of America, 1995). There are serious shortages of people worldwide with some of the key skills needed for plant conservation, such as in taxonomy (see, for example, Oteng-Yeboah, 1998; Godfray, 2002).

Universities can be low on the list of budgetary priorities for governments, with botany perhaps seeming inconsequential to some politicians. Actually, botany *should* have an important role to play in national development, especially in those countries where the everyday lives of many people revolve around plants. There is a need to identify how scientific botany can best contribute to development. In the words of Patrick Mucunguzi of the Botany Department of Makerere University in Uganda, 'Botany needs to be taken to the community' (Hamilton et al, 2003). Applied ethnobotany is a particularly useful subject for plant conservation, dealing with the relationships between people and plants, and related questions of conservation and development (see Chapter 12). Recommendations for the development of courses in applied ethnobotany are available in a publication of the People and Plants series (www.peopleandplants.org) (Hamilton et al, 2003) (see Figure 4.1).

RESEARCH

Research is fundamental to success in conservation. For example, research is needed for the initial evaluation of plant resources before management plans are constructed and later (in the form of monitoring) as a routine part of resource management. In principle, research for conservation is a never-ending process – there will always be new questions to answer. Therefore, research for conservation can be seen as a cycle of continuing enquiry (see Figure 4.2). Research aimed at achieving more immediate conservation results tends to be

Figure 4.1 *Training in market survey techniques near Mount Kinabalu, Sabah, Malaysia – part of a certificate course of the People and Plants Initiative*

Source: Alan Hamilton

problem based and participatory. 'Problem based' means that research is aimed fairly directly at finding answers to practical problems, such as how to manage plants in the field. For the sake of efficiency and in order to retain people's interest, it is important that priority issues are selected for attention (see Chapter 9). The choice of topics for research and ways to pursue them will benefit from thinking across the boundaries of traditional scientific disciplines and taking full account of both scientific and local perspectives. Members of project teams should be creative in their approaches to research, making efforts to focus on real problems and not just apply 'standard' methods because they are familiar.

There are many types of people who can potentially be involved in participatory research (see Figure 4.3; also Table 12.2 in the case of communities). Various forms and degrees of participation are appropriate according to circumstances. An advantage of more participatory approaches is that more people are involved in learning and there is a better chance that the recommendations from the research will be taken seriously, with appropriate practical actions to follow. In the most participatory mode, stakeholders (such as local people) are intimately involved in every stage of the research – setting the research questions, devising research methods, gathering data, analysing results and presenting the results to those who might use them (Pretty, 1998). The relative merits of two types of participatory research – participatory

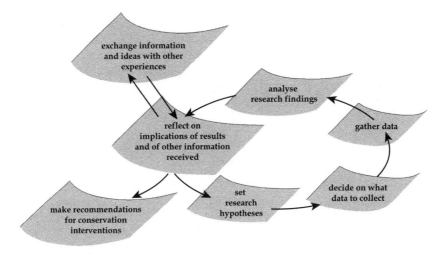

Figure 4.2 *Stages in applied conservation research*

Note: Notice similarities to stages in adaptive management (see Figure 11.3) and the project management cycle (see Figure 12.3). The three cycles should be closely linked in projects (see Figure 12.4).
Source: Alan Hamilton and Patrick Hamilton

appraisal and participatory action research – are discussed in Chapter 12. Participatory appraisal involves a larger number of people and is useful for revealing general contexts, while participatory action research can provide more precise information. Participatory action research has been recommended in numerous conservation and development contexts, for example improved management of natural habitats, the development of new cultivation practices, the development of improved healthcare systems, pest control in agriculture, and the selection or breeding of new varieties of crops (Berlin and Berlin, 2000; Cunningham, 2001a; Jones and Garforth, 2002; Song Yiching and Jiggins, 2003; Uniyal, 2000).

There is a need to identify more effective approaches and methods in plant conservation (see target 3 of the Global Strategy for Plant Conservation in Box 3.1). Since plant conservationists often work in isolation, a useful first step is the sharing of experiences between those facing similar circumstances. Drawing upon a parallel with medicine in which there is said to have been an 'effectiveness revolution' during recent years, Pullin and Knight (2001) have recommended the development of 'evidence-based conservation', by which they mean drawing up recommendations for conservation practices based upon systematic reviews of evidence relating to the success or failure of different approaches and methodologies, and setting and testing formal hypotheses to find out which methods are most effective.

Figure 4.3 *Field workshop, Gunassa, Dolpo, in Nepal*

Note: Project team and local resource users discussing the management of medicinal plants in Himalayan pastures. See also the case study on pp212–214.
Source: Alan Hamilton

HOW CAN BIODIVERSITY RESEARCH BENEFIT CONSERVATION? A DISCOURSE AND SOME SUGGESTIONS

By the mid 20th century, many governments had come to accept that it was their duty to support basic research into the natural environment and were engaged in national surveys of plants, vegetation, forests, soils and land-use potential, while maintaining facilities such as herbaria, botanic gardens, and agricultural and forestry research stations. The modern scientific foundation for plant conservation owes much to this traditional positioning of governments and an associated openness of scientists and research institutions globally to the exchange of knowledge and information. Our understanding of the types of plants on Earth, their distribution and their conservation status could not have been achieved without a global network of plant collectors, taxonomists and herbaria cooperating willingly in free exchanges of plant specimens. Even today, the basic science underlying plant conservation depends upon the willingness of scientists internationally to cooperate. For example, the quality of information associated with national collections of plants cannot be maintained unless there is easy contact with taxonomic specialists internationally, ready to check the identity of specimens.

The same open attitude to enquiry and exchanges of knowledge once pervaded crop breeding. Until about 1980, most crop breeding was undertaken in government stations, exchanging germplasm freely between them. It never occurred to most crop breeders that their work was not a public good, nor did they question the presumption that the crop-breeding system as a whole served the best interests of humanity (Hawkes et al, 2001).

The sharing of knowledge is encouraged by some philosophies and religions. Writing in the context of a People and Plants project on medicinal plant conservation and healthcare (see 'Case study: *Amchi* medicine and conservation of medicinal plants in Nepal', Chapter 11), Yeshi Choden Lama and her colleagues have explained (2001):

> *Transfer of this knowledge [about the medical uses of plants] to the global community does not pose any ethical problem to the* amchis *[traditional doctors following the Tibetan medical tradition], except regarding specific compounds that have not been fully tested and therefore cannot be used by non-specialists. It is also to be noted that in the context of Buddhism and Bon, the* amchis *see this knowledge as an asset to be used for the good of all sentient beings – that is, human welfare.*

Especially since 1992, the 'ethical and legal envelope' within which biodiversity research takes place has become considerably more complex and, moreover, is forever changing. There are now a number of international agreements, laws, regulations and ethical admonitions to which researchers working on biodiversity must be sensitive (Laird, 2002). Box 4.2 contains some suggestions that may be of assistance to plant conservationists as they navigate these tricky waters. It would be helpful if more plant conservationists were to take time to review the biodiversity policies that affect their work and make suggestions for improvements based on field-level realities.

And now for the complexities!

For many years, the attitude of industry to new discoveries and the sharing of knowledge has been very different (especially in some countries) from the openness traditionally associated with science and education. Intellectual knowledge has come to be seen as a type of property, and a system of patenting has been developed, providing inventors with monopolies over the use of their inventions for specified periods of time, often 20 years. A common justification given for granting such intellectual property rights (IPRs) is the need to provide incentives for research.

The application of patenting to living things is relatively new, the USA consistently being at the forefront of developments (Kevles, 2001). The first American patent act dealing with life was the 1930 Plant Patent Act, which provided protection to plant breeders, but was narrowly drawn. It only allowed patents to be granted for human-created varieties of plants that were commercially multiplied by asexual means. New varieties that were reproduced commercially by sexual means were excluded because it was known that such varieties did not necessarily breed true and therefore could not be said to be inventions in a *fixed* sense. It also excluded plants that were multiplied by

---

BOX 4.2 SUGGESTIONS TO GUIDE PLANT CONSERVATIONISTS IN THEIR APPROACHES TO RESEARCH

- The key level at which benefits from research on plants should accrue to actually achieve plant conservation is the *local* level – that is, where plants actually grow.
- The national level is a secondary level at which benefits for conservation from research can usefully accrue, partly because of the existence of key national biodiversity institutions (for example, herbaria, training colleges and forest research institutes).
- Collaborative research is to be encouraged and will often be necessary in order to achieve practical conservation results. There are several potential partners in collaboration – for example, foreign and national scientists working together, or researchers and communities.
- The ways in which research collaborations start are important for setting their tone. Relationships may evolve with time. Written agreements will sometimes be appropriate or required. They are useful to clarify mutual expectations and obligations.
- No written agreements are perfect and adjustments will always be necessary. The most important products of a successful negotiation are levels of trust and commitment on both sides, which will facilitate later adjustments, as well as agreement on other topics (Barsh and Bastien, 1997, cited by Laird, 2002).
- Researchers should be transparent about their motives. Research aimed at directly finding solutions to conservation problems is to be encouraged. Special efforts will be needed to achieve real conservation benefits from research carried out basically with other motives – for example, to identify new commercial opportunities.

---

tubers due to unease about allowing the patenting of such economically important food crops as the potato *Solanum tuberosum*.

Over the years, the types of plants for which such plant breeding rights have been granted in the USA have been extended and some other countries have adopted similar legislation. In 1961, the International Union for the Protection of New Varieties of Plants (UPOV) was formed, creating a system which continues to the present day, now with 51 member countries. The UPOV system provides legal recognition and protection for named varieties of plants that have been developed by plant breeders for a minimum of 15 years (a minimum of 25 years for trees and climbers) (Hawkes et al, 2001). Among the numerous plants covered by UPOV are commercial varieties and breeders' lines of African violets *Saintpaulia*, *Begonia*, maize *Zea mays*, Rose *Rosa* and wheat *Triticum* (lists of registered plants are published in issues of the *Plant Varieties Journal*). Meanwhile, patenting has been extended in some countries to cover other aspects of life, with the US Patent Office being more prepared than the European Patent Office to grant patents over new areas, such as genetically modified organisms (GMOs) and genes (considered to be 'discovered' chemical sequences). There are many areas of ethical dispute – for instance, about various types of inequities, claims about the destructive consequences on developing countries of granting patents for agricultural crops, whether humans should be 'playing around with nature' through creating GMOs and – fundamentally – whether forms of life should be privately owned anyway.

The Convention on Biological Diversity (CBD, 1992) has proved instrumental in causing re-examination of the purposes of biodiversity research and how benefits from research should be distributed. The third objective of the CBD promotes the fair and equitable sharing of benefits arising out of utilizing genetic resources. In essence, the third CBD goal is an acknowledgement that much biological diversity lies in the South; but much research and industrial power to exploit biodiversity – in terms of developing new commercial products – lies in the North. What is called for is that benefits are provided to developing countries when research is undertaken on developing new products (biodiversity prospecting). The reasoning is that this will provide an incentive to developing countries to conserve biodiversity and is the right thing to do anyway on the grounds of social equity. Article 8(j) of the CBD states that:

> *Subject to its national legislation, respect, preserve and maintain knowledge, innovations and practices of indigenous and local communities embodying traditional lifestyles relevant for the conservation and sustainable use of biological diversity and promote their wider application with the approval and involvement of the holders of such knowledge, innovations and practices and encourage the equitable sharing of the benefits arising from the utilization of such knowledge, innovations and practices.*
> (CBD, 1992)

Other articles of the CBD are also relevant – for instance, Article 15 (entitled 'Access to genetic resources'). This calls for parties to facilitate access to genetic resources for environmentally sound uses by other contracting parties on mutually agreed terms and subject to prior informed consent. The sixth meeting of the Conference of the Parties (COP) to the CBD adopted guidelines (the 2002 Bonn guidelines) which provide some clarification about the meaning of the terms 'mutually agreed terms' and 'prior informed consent', the types of benefits that might be appropriate and the role of intellectual property.

There are unresolved conflicts relating to intellectual property rights between the CBD and a 1994 agreement known as Trade Related Aspects of Intellectual Property Rights (TRIPS), part of the trade negotiations leading to the establishment of the World Trade Organization (WTO) (Masood, 1998). Under TRIPS, the 148 members of the WTO have agreed to provide a level of protection to the intellectual property (IP) of fellow WTO members. Developed countries, which produce most of the world's IPRs, have been advocates of strong international IP protection. They have been concerned about the unlicensed production of pharmaceutical drugs, videos and computer programmes in some developing countries. Contrary to the provisions of the CBD, there is no requirement under TRIPS for applicants for patents to involve or consult with local communities or governments from source countries before patenting compounds based on natural products. Nor do they have to share any benefits accruing from the sale of these compounds, or recognize the contributions of local or indigenous peoples to the innovations. What this means is that a company in a developed country can freely take a traditional crop variety from a developing country, make some minor genetic modifications, patent the novel 'discovery', copyright its traditional name, export the product in competition with the natural variety, and prevent the natural variety from being sold in importers' markets using its traditional name. This may seem far fetched; but it is more or less what happened in the cases of Basmati rice from India and Jasmine rice from Thailand, variants of which have been patented and copyrighted by firms in the US (Kerr et al, 1999). The reason why traditional crop varieties can be claimed to be 'natural varieties' is because no inventors can be identified and, in any case, no IPR protection has been provided by the countries concerned. Similar problems arise in relation to the exploitation of local knowledge of the medicinal uses of plants to develop and patent new pharmaceutical drugs.

It is not yet clear how a compromise will be reached between the commitments to accessibility and equity enshrined in the CBD and the pressures for private ownership and profit-based systems of reward represented by TRIPS. One view is that the granting of IPRs over life forms is incompatible with the conservation of biological and cultural diversity and should not be allowed, while others believe that the best solution will be to reform TRIPs in the light of the CBD. The dispute has revealed radical differences in attitudes to knowledge and the purposes of scientific research.

Some state governments, institutes and non-governmental organizations (NGOs) in India have started to document local knowledge of biodiversity in community biodiversity registers (CBRs). Apart from its value for promoting conservation of local cultures and biodiversity, the compilation and publication of these registers is intended to demonstrate 'prevalent use' or 'prior art' as a defence against the granting of patents to companies exploiting traditional knowledge. Additionally, India has taken legal steps to protect local biodiversity and related intellectual knowledge and to try and ensure that traditional inventors receive shares of benefits when new commercial products are developed. The 2001 Indian Plant Variety Protection and Farmers Rights Act refers not only to breeders' rights (as required by TRIPS), but also farmers' rights (not required by TRIPS). This law requires farmers to benefit when their knowledge is used to breed new crop varieties. It establishes a national gene fund into which breeders must pay for using crop varieties from farmers. The legislation requires a full disclosure of source and origin of varieties, and complete passport data from breeders (Sahai, 2004).

The third objective of the CBD relates to research aimed at the development of new commercial products. This objective – and other policy developments over recent years (especially concerning human rights) – have had major impacts on official approaches to research on biodiversity. These impacts can extend into fields of research which are quite unconnected with biodiversity prospecting. Many people have felt lost in the complexities that have developed. Two publications of the People and Plants Initiative have been produced to provide some help (Cunningham, 1993; Laird, 2002). Laird's *Biodiversity and Traditional Knowledge* gives case studies, analyses and suggestions to assist various groups in formulating their policies and operating practices regarding:

- terms of agreements between researchers and local communities or indigenous peoples' organizations;
- codes of practice for commercial companies and trade associations on standards required when contracts are made relating to biodiversity prospecting;
- ethics and research guidelines for members of scientific and professional societies;
- the formulation of national regulatory frameworks for research by governments, including setting minimum terms and standards relating to biodiversity prospecting;
- conditions of research permission in protected areas;
- policies to guide the work of scientific and related institutions (for example, regarding the use of *ex situ* collections of botanic gardens); and
- requirements placed on grantees relating to their research by funding agencies.

Even before passage of the CBD, questions had begun to arise in the international community about access to traditional varieties of crops for research and how benefits deriving from the use of crop varieties should be distributed. There has been a big increase in private crop breeding since 1980, which has fuelled these concerns. Policy towards access and benefit-sharing in relation to those crop genetic resources held under the Consultative Group on International Agricultural Research (CGIAR) system was covered by the 1983 International Undertaking on Plant Genetic Resources for Food and Agriculture (with subsequent resolutions), an instrument of the United Nations Food and Agriculture Organization. The concept of farmers' rights was endorsed by the FAO in 1989 to recognize the rights of 'informal' plant breeders (contrasted with 'formal' plant breeders). Farmers' rights refer to the:

> ... *rights arising from the past, present and future contributions of farmers in conserving, improving and making available plant genetic resources, particularly those in the centres of origin/diversity. These rights are vested in the international community, as trustee for present and future generations of farmers ... [and should] allow farmers, their communities and countries in all regions to participate fully in the benefits derived, at present and in the future, through the improved use of plant genetic resources, through plant breeding and other scientific methods.* (FAO, 2004)

In other words, farmers who develop or maintain crop landraces are seen just as much as inventors of plant varieties as commercial plant breeders.

A re-look at the International Undertaking on Plant Genetic Resources for Food and Agriculture was needed after passage of the CBD, including examining whether anything practical (rather than rhetorical) could be achieved in relation to farmers' rights. The ownership of *existing* collections held by CGIAR centres was clarified in 1994, when agreements were reached between the centres and FAO stipulating that they were being held in trust by the centres on behalf of the international community. This germplasm was to be made available without restriction to researchers around the world on the understanding that these researchers would not apply for intellectual property protection to it. For new accessions provided to the centres from 1994, a requirement was made that these must be accompanied by stipulations from their source countries regarding the terms on which they could be used. Various matters remain unclear in relation to access and benefit-sharing for crops – for example, how the requirement not to take out intellectual property rights on material supplied by CGIAR centres until 1994 can be enforced and how support through access agreements can be provided for the *in situ* conservation of crop landraces and wild crop relatives. In practice, a sharp downturn in the exchange of germplasm between crop research centres is reported, contributing to worries that this will adversely affect plant breeding (Rajaram and Braun, 2001).

At the back of some people's minds regarding biodiversity prospecting is the idea that plants contain wonderful new drugs, the discovery of which will be of great benefit to human health and generate large sums of money, some of which can be secured to conserve biodiversity. The discovery of new drugs is welcome; on the other hand, perspective is needed in judging how likely major new drugs will be and the scale and type of benefits that can be expected during the process of drug development and production. Only three major pharmaceutical drugs have been developed from plants over about the last decade: camptothecine (a cancer drug developed from the tree *Camptotheca acuminata*), galantamine (used to treat Alzheimer's disease, from the daffodil *Narcissus pseudonarcissus* and the Caucasian snowdrop *Galanthus wornorii*) and taxol (a cancer drug from yew *Taxus*).

There have been several well-publicized attempts to strike access and benefit-sharing agreements linking drug discovery to the conservation of tropical forest. One, in 1991, involved the pharmaceutical company Merck and InBio, a non-profit organization in Costa Rica. Another, in 1992, involved the establishment of five international cooperative biodiversity groups by the National Institute for Health, National Science Foundation and the US Agency for International Development (USAID) (Hawkes et al, 2001; Macilwain, 1998). One of the most dedicated efforts has been that of Shaman Pharmaceuticals, a company established specifically for the discovery and development of novel pharmaceutical products for human diseases through isolating active compounds from tropical plants used as local medicines. The Healing Forest Conservancy (HFC) was established with a donation from

Shaman for the purposes of conservation of tropical forest and the welfare of tropical forest peoples. Because drug discovery in any particular case was deemed uncertain, and because many source countries and communities have pressing problems, the approach developed by HFC was to share benefits equally between all participating parties, regardless of which particular ventures happened to be commercially successful and produce revenues. As well as a share of any royalties (a long-term possibility), short- and medium-term benefits were provided, including provision of legal assistance to communities to demarcate tribal lands and finance for conducting in-country research on traditional medicines. Shaman placed much hope on being able to develop a drug based on research into *sangre de drago*, *Croton lechleri*, a fast-growing South American tree. As a herbal, *sangre de drago* is used to treat a variety of diseases and is exported to the US for the treatment of diarrhoea, cholera and gastric ulcers. In the event, the company found that the marketing of an active principle isolated from the tree (CP303 – Provir) proved to be too expensive because of the high costs of testing to meet the requirements of drug registration in the USA. Therefore, instead, they aimed to market a standard herbal extract. Even this proved impossible and, in 2001, Shaman Pharmaceuticals declared bankruptcy.

It is difficult to know what is really happening in the field of biodiversity prospecting, including the fairness and equitability of particular ventures. The atmosphere has sometimes become polarized, sour and legalistic. On the one hand, there have been complaints that Western scientists or multinational corporations are 'stealing' biological materials or intellectual property from developing countries. On the other hand, biodiversity prospectors have sometimes been put off by a lack of clarity about terms of access to genetic resources and aggressive demands for upfront cash. There can be a lack of business awareness on the part of regulators (Holmstedt and Bruhn, 1995; Balick and Cox, 1996; Jayaraman, 1998; Macilwain, 1998; Nakashima, 2000). It is difficult to find examples where benefits of *any* kind have accrued at the *local* level from biodiversity prospecting; and even when this has happened, the benefits have tended to be social, not biological (Laird, 2002).

Related to objective three of the CBD, some countries and territories have introduced very tight conditions over research, making it difficult to obtain permission to undertake botanical research of *any* type. While terms are often especially stringent for foreigners, nationals can also be affected. This tightening up seems to have occurred because of concerns about biopiracy, unrealistic assumptions about the amounts of money to be made from bioprospecting, and politicization of the process of granting permission for research. Under a politically charged atmosphere, there can sometimes be no one in government agencies prepared to make decisions on applications for research. Research on plants, potentially beneficial to conservation and completely unconnected to biodiversity prospecting, is being seriously impeded in some countries.

5

Plant Life

The management of plants for conservation requires knowledge of how they live, individually and collectively (as vegetation). This chapter discusses a few features of plants and vegetation of significance for conservation, beginning with accounts of how people name and classify plants.

TYPES AND NAMES OF PLANTS

People have categories for things that they encounter in the world around them and give them names, assisting them to handle information mentally and to communicate with others. The placing of items in the same category – for example, types of plants – means that they are perceived to have something in common. Studies of the ways in which people classify plants can give insights into how people see the world, the things that are important to them and, indeed, how societies are organized (Nakashima, 1998; Tsouvalis, 2000). The importance of animals in the life of the Moguls is reflected in their frequent use of animal names or features in the naming of plants. Species of the dwarf shrub *zhegergen* (*Ephedra*) of different heights are called *morin* (horse) *zhegergen* (*E. equisetina*, height 100cm), *honin* (sheep) *zhegergen* (*E. intermedia*, height 20–50cm) and *imaan* (goat) *zhegergen* (*E. monosperma*, height 3–10cm) (Hasbagan and Chen Shan, 1996; Hasbagan and Pei Shengji, 2001). Every culture has its own ways of looking at plants and plant products. Foods in Vietnam are classified according to a hot–cold system, with most vegetables considered cooling. Optimal health is deemed to depend upon maintaining the right balance between hot and cold in dietary intake (Ogle et al, 2003). The Wola highlanders of Papua New Guinea classify crops into male and female, with female crops, such as *hurinj* (the sedge *Eleocharis* cf *dubia*) and *hokay* (the sweet potato *Ipomoea batatas*), being cultivated only by women and male crops, such as *diyr* (bananas, *Musa*) and *wol* (sugarcane *Saccharum officinarum*), only by men (Sillitoe, 2003). The gender distinction in crops reflects the different purposes to which they are put in Wola society. Female crops provide the main foods, while male crops provide luxuries used by men in social transactions.

Classification schemes used for plants by conservationists and resource managers also have their implications. Once a species becomes labelled as 'endangered', this can set in motion a flurry of conservation activities. Until the 1970s, the beech *Fagus sylvatica* and some other hardwoods were classified by German foresters as 'forest weeds' and were continually removed when occurring in mixtures with spruce *Picea abies* or pine *Pinus sylvestris* (Kenk, 1992). Their labels had condemned them.

The species is the basic plant type used in conservation. Being able to assign plants to species is the key to unlocking scientific information about them – for example, regarding their distributions, degrees of endangerment, uses and methods of management. The species is generally the key taxonomic entity from the legal viewpoint, and is the reference category for most laws and regulations. If a species has been officially classified as endangered, then there may be a legal duty to protect it, even in the case of inconspicuous plants that can be difficult or impossible to distinguish from one another in the field, such as the eight species of water starwort *Callitriche* in the UK (Lemars and Gornall, 2003).

Biologically, individual plants assigned to the same species are deemed to be capable of interbreeding with one another in nature, producing offspring able to survive in the wild. However, not all plants reproduce freely by sexual means and very few species have actually been tested for the abilities of plants in their scattered populations to interbreed successfully. In most cases, judgements about how species are recognized are made by taxonomists familiar with particular groups of plants, based upon morphological features apparent on preserved specimens in herbaria. Good taxonomists are the powers behind the scenes in plant conservation.

Plant taxonomy and systematics are fundamental botanical disciplines in scientifically based efforts to conserve plants or to develop their uses for people. In principle, plants placed in the same higher taxonomic categories, such as genera and families, should be related to one another in an evolutionary sense (they are the only living relatives descended from a common ancestor) and may be expected to share some common properties. There are many instances that demonstrate the importance of getting the taxonomy right. For example, in 1969, the US National Cancer Institute reported an interesting compound, maytansine, found in samples collected from an African species of *Maytenus*, a genus of trees and shrubs. The substance was active *in vitro* against leukaemia. Chinese scientists decided to pursue this lead, even though, at that time in 1972, no species of *Maytenus* were known from China. Support was found for a taxonomic study of *Gymnosporia*, a closely related genus. This revealed that all of the Chinese species previously placed in *Gymnosporia* should actually be in *Maytenus*. It was found that at least 12 species of *Maytenus* occur in China, including four species that were new to science (as of 2004, 20 species of *Maytenus* have been found in China). The next step was to look at the chemistry of seven of these species; of these, *Maytenus hookeri* proved to have the highest concentration of maytansine (see Figure 5.1). Extracts were tested, leading finally to approval of a standardized crude drug (*Maytenus* no 1 tablet). This medicine became well known during the 1970s. It was taken by Premier Chou En-lai, whose life

Figure 5.1 Maytenus hookeri *planted in Xishuangbanna Tropical Botanical Garden, China, with Professor Pei Shengji, ex-director of the garden*

Source: Alan Hamilton

was thereby extended for several months (Pei Shengji, personal communication, 2001).

Knowledge of patterns of variation *within* species is helpful in drawing up conservation plans, based on the principle of conserving as much genetic diversity as possible. Various taxonomic categories below the level of the species are recognized (for example, subspecies, varieties and forms); but of these, only subspecies are usually recorded in local Floras (Chater, 2003). Especially in better-resourced countries, infra-specific variation is being increasingly studied using DNA or isoenzymes. Very little attention has yet been accorded to tropical plants, even in the case of such an economically significant group as the rattans (Braun and Dlamini, 1994; Rao, 1997; Dawson and Powell, 1999). The importance of recognizing infra-specific variation is well known in forestry, where the selection of the right varieties of trees (known as provenances in forestry) can be critical in forest plantings. The selection of varieties conferring resistance to the potentially devastating caterpillars of the pine beauty moth *Panolis flammea* has been critical to the success of the lodgepole pine *Pinus contorta* planted in Scotland (Trewhella et al, 2000).

In conservation work with communities, it is important to know the local (folk or vernacular) names of plants, as well as their scientific names. Care needs to be taken in recording local names during fieldwork in order to make sure that the names given really do refer to the plants in question and don't mean something like 'I don't know' or 'a flower'. The government of Ethiopia

has been encouraging the planting of *Juniperus excelsa*, a highly prized indigenous tree; unfortunately for the campaign, many people have been planting
the exotic cypress *Cupressus lusitanica* instead. Confusingly, both species are
known locally as cedar and *Cupressus* is more available in nurseries (Pierre
Binggeli, personal communication, 2002).

Local names can refer to plant categories that are *exactly equivalent* to
scientific species (known as one-to-one correspondence) or local people can
recognize *more* types of plants than botanists (known as over-differentiation)
or else *fewer* (under-differentiation) (Hedberg, 1993; Luoga et al, 2000;
Cunningham, 2001a). The level of one-to-one correspondence is as high as 88
per cent in the case of the Dai of Xishuangbanna, China (Wang Jinxiu et al,
2004). A high level is also found among the Masaai of East Africa, although
there are exceptions. For instance, *empalakae* refers to 22 species of grasses
and sedges, and *olosida* to over 20 related species, mainly in four genera of the
Acanthaceae (Maundu et al, 2001). *Enchani-pos* refers to species in at least 11
genera and 9 families, some quite unrelated to one another.

Plants can be grouped in various ways in local taxonomies – for example,
according to their economic values (for instance, timber trees), cultural values
(for instance, beautiful flowers), nuisance values (for instance, weeds) or their
origins (for instance, native plants). People who specialize in particular types
of plant resources or plant-based products have their own ways of grouping
and naming plants. Conservationists working with farmers need to know what
they mean by 'corn', pharmacists by 'lignum sal' and timber merchants by
'mahogany'. Brent Berlin (1992) claims that societies worldwide follow the
same general principles in the ways that they group plants in their *most basic*
ways of naming. He believes that people everywhere recognize between four
and six major ranks of plant types, with members of lower ranks usually being
grouped into higher types in a nested fashion (see Table 5.1). The most fundamental rank is said to be the folk generic, often more or less equivalent to the
scientific category of the genus.

HUMAN INFLUENCES ON PLANT TYPES

Over large parts of the world, many plants have been influenced in how they
look or grow by people. Doubtless, human influence often extends down to
the level of the genes (Jennings et al, 2000). Oliver Rackham (1992) reports
that human activities have probably been responsible for breaking down breeding barriers between seven or more species of oak *Quercus* that grow in woods
on the Lower Pindos Mountains in north-west Greece. Individual oak trees in
these woods are often difficult to identify. The woods cover thousands of
square kilometres and have a long history of management for timber, fodder-
cutting and wood pasture.

The extent of human influence on plant types is apparent in a well-studied
area – the British Isles – where at least 6 of the 13 native genera of larger trees are
known to have been greatly influenced by people. To mention but two examples:

Table 5.1 *Ranks used for categorizing plants in folk taxonomies*

Ethnobotanical rank	Description	Examples in English, with an indication of hierarchical relationships
Folk kingdom	Contains only one member, which, in most languages, is unnamed	Plant
Folk life form	A major structural class of plants of which not more than 10 to 15 are recognized in any language	Tree Climber Herb
Intermediate (minor category)	A grouping of a small number of similar-looking folk generics, mainly falling within the same or closely related plant families as recognized by scientists	
Folk generic	The basic 'plant type', of which generally less than about 500 are named in any language. They are usually recognized by their overall morphology, rather than by their usefulness or symbolic significance	Willow Oak Beech
Folk specific	Subdivision of a folk generic. Commonly in agricultural societies about 20 per cent of folk generics are so divided. Often given a binomial name, part of which is the relevant folk generic	Goat Pussy Weeping willow willow willow
Folk varietal	Subdivision of a folk specific, mostly plants of major cultural significance	

Note: People often recognize more types of plants than they actually name; such hidden categories are known as covert taxa. Furthermore, while most plant types in lower ranks belong also to higher categories, there are exceptions (unaffiliated taxa) that stand alone; these are often odd-looking plants or plants of high cultural importance and are predominantly cultivated or protected.
Source: Alan Hamilton and Patrick Hamilton, based on ideas in Berlin, 1992

- Many willows *Salix* are difficult to identify, with numerous hybrids recorded. This genus has long been used for basketry, with much selection and movement of desirable strains (Meikle, 1984).
- Most specimens of lime *Tilia* in Britain are hybrids *Tilia* x *vulgaris* between the two species that are native to Britain (*T. cordata* and *T. platyphyllos*), although the stock from which these hybrids arose did not actually originate in Britain, but rather from continental Europe, being introduced in the mid 17th century (Mabey, 1997).

One of the most drastic ways in which people can influence plants is through domestication. Domesticated plants have been defined as plants that are useful to people, differ morphologically and genetically from related wild species, and are unlikely to survive for long away from human protection. Maize *Zea mays* is an extreme case of domestication; it is so highly modified by selection that its seeds cannot be dispersed without direct human help. Domestication is not something that just happened in the past. Plants at an incipient stage of domestication are common today in traditional agricultural settings. They are the subjects of such practices as the transplantation of specimens of wild plants back to home gardens, the retention of useful wild plants when clearing bush to make fields (including, occasionally, selecting specimens of higher quality, such as those with better tasting fruits) and leaving wild plants that provide useful commodities when weeding crops.

There is a continuum between plants that are fully wild and fully domesticated. Many useful weedy species, which are deliberately retained when crops are weeded, are genetically influenced by people and difficult to classify along the wild to domesticated spectrum (Asfaw and Tadesse, 2001). Numerous crops, including barley, maize, oats, rice, sorghum and wheat, are known to exchange genes with related wild species (Lenné and Wood, 1991; Tuxill and Nabhan, 2001; Arnold, 2004). Individual plants can therefore be difficult to identify precisely. For taxonomic purposes, it can be useful to recognize the concept of the crop–weed species complex. Some crop–weed complexes have provided genetic traits to modern crop varieties that have resulted in major economic benefits, such as resistance to the grassy stunt virus of rice provided by *Oryza nivara*, a member of the crop–weed complex of rice O. *sativa* (Lenné and Wood, 1991).

Some crops that have become economically important originated as weeds. Oats *Avena sativa* and rye *Secale cereale* were once weeds in fields of barley *Hordeum vulgare* and wheat *Triticum* in the Middle East. They evolved with these crops through mimicry as people selected certain morphological and physiological traits (Hawkes et al, 2001). It was only later that people realized that oats and rye were useful crops in their own right and started to cultivate them deliberately.

Traditional crop varieties (landraces) are the products of observation and selection by generations of farmers. They typically have more restricted geographical distributions and higher phenotypic and genetic variability than modern crop varieties (Clement, 1999). 'Modern' (that is, scientifically based) crop breeding can be said to have originated with the rediscovery of Mendel's laws of genetic inheritance at the beginning of the 20th century; but modern crop varieties only became widely grown after 1950, even in Europe and North America. Breeders of modern crop varieties work by identifying a character in landraces, wild crop relatives or modern varieties that they think might be useful if introduced into a commercial variety. They then try to introduce it genetically by crossing the commercial variety with the plant type having the useful character and then repeatedly back-crossing the progeny with the commercial variety to restore the other characters of interest. This process typically takes 10 to 20 years. Wild crop relatives can contribute greatly to the production of new commercial varieties, including for conferring resistance to pests and diseases. Eleven wild species of *Rubus* contributed to the production of modern crop varieties in the breeding programme for raspberries (based on *Rubus idaeus*) at the East Malling research station in the UK between 1948 and 1979 (Knight, 2004). The types of characteristics in which crop breeders tend to be interested are high yield, resistance to pests and diseases, and marketability – the definition of the latter being increasingly dictated by supermarkets. There has been a tendency in crop breeding programmes towards a rapid elimination of phenotypic and genetic variability (Lenné and Wood, 1991; Clement, 1999). Modern crop varieties are bred especially for intensive agricultural systems and are typically grown in extensive monocultures with high inputs of chemical fertilizers and pesticides.

Plant breeding in parts of the world is increasingly based upon genetic engineering, which involves the movement of genetic material by scientists between organisms that may not be closely related. Some of the environmental pros and cons of introducing genetically modified organisms (GMOs) are mentioned in Chapter 1. There are also ethical concerns. Some people argue that humans are going too far in transferring genes between totally different species. It is also asserted that GMO technology will often work to the advantage of multinational agricultural companies and against farmers in developing countries. The worry here is that farmers will lose some of the independence that is so vital for their economic security, becoming trapped into the purchase of new seed for planting with every sowing because the companies have introduced genes imparting sterility to the seeds (terminator technology).

Crop breeding does not have to be a solitary matter for agricultural scientists isolated in research stations. Plant breeding and the related process of plant variety selection can also be undertaken through participatory methods, involving plant breeders working closely with farmers or the users of plant resources at village level. The advantages include close interaction with local knowledge and concerns, and a greater possibility of contributing towards maintaining plant genetic diversity across the landscape (see Chapter 13 for more details) (Leakey, 2003).

THE LIVING PLANT

Plant species can be considered to consist of individual plants, grouped into populations. *Within* each population there is free interbreeding, while *between* populations there is reduced, though variable, amounts of genetic exchange (see Figure 5.2). Populations of plants are subject to constant selection pressures, including in response to interactions with other organisms, both mutualists and enemies (co-evolutionary systems). It is known that species can show genetically based adaptations to environmental variations over very short *distances* and also that populations of plants can be subject to strong selection pressures over short periods of *time*.

Individual species of plants possess certain basic adaptations to the environment and have evolved their own particular ways of tackling the businesses of living and reproducing (life history strategies). Plants display a very large number of reproductive strategies (Barrett and Kohn, 1991). If species require particular pollinators or seed dispersers, then obviously they will be in trouble should these animals disappear. Some species of plants put all their seeds in one basket because they only flower once in their lives and then die (monocarps). Such species are vulnerable if people collect them before they have distributed

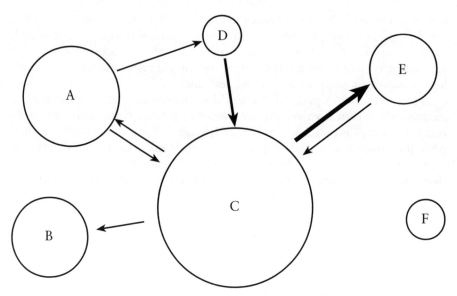

Figure 5.2 *Example of a population model of a plant species, showing six populations of the species (A–F), with differing numbers of individuals (sizes of circles)*

Note: A population of a species is a group of interbreeding individuals of a species living in the same area. Here, the populations belong to two meta-populations (A–E and F). The sizes of the populations are indicated by the sizes of the circles. A meta-population is a set of populations with occasional connections between them. Selection pressures can vary in type and strength between populations, and so can the amount of genetic flow between populations (indicated by the arrows). Periodically, populations may become extinct or new populations may become established.
Source: Alan Hamilton and Patrick Hamilton

their seeds, as can happen in the case of the high-altitude Himalayan medicinal plant *Saussurea lasiocarpa*. Unfortunately, there is no published information on the ecology and life history strategies of most species of plants so that those who wish to manage them for conservation purposes must rely upon information that happens to be available on species presumed to have similar properties and any knowledge that local people have on their ecologies, and use adaptive approaches for their management (see Chapter 11).

Individual plants pass through a series of life history stages as they progress from their origins, perhaps as seeds, through to their deaths (see Figure 5.3). The lengths of time that individual plants are resident in life history stages and the probabilities of moving from one life history stage to others are important considerations in managing plants for conservation and sustainable use (see Chapter 11). A significant feature of plants, from the conservation viewpoint, can be their reactions to harvest. It makes a lot of difference for management whether stumps of trees are capable (or not) of re-sprouting after trunks have been cut or whether perennial herbs can re-grow from root fragments remaining in the soil after the plants have been harvested by uprooting.

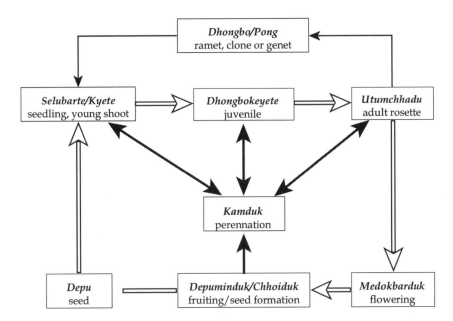

Figure 5.3 *The life cycle of a perennial medicinal plant from the understanding of some traditional medical doctors (*amchis) *at Dolpo, Nepal*

Note: A vegetative propagation cycle is seen as progressing from a seedling stage through a juvenile stage to an adult rosette stage, which gives daughter ramets, thus forming a clone (pong). A sexual propagation cycle progresses from a seed through to an adult rosette, which flowers, fruits and produces seeds. In winter, perennial herbs go into dormancy or perennation, a stage that is identified by the *amchi* as *kamduk*.
Source: Ghimire and Aumeeruddy-Thomas, 2004

There are some categories of life history stages that are usefully recognized for conservation purposes, but whose significance is either seen differently or ignored in conventional systems of managing natural resources. For example, deadwood has traditionally been regarded as a source of infection or as waste by foresters, who have often encouraged the 'sanitation cutting' of dead standing trees to discourage pests and diseases, or allowed the free collection of dead branches on the forest floor by villagers for use as fuel. However, from the biodiversity standpoint, dead standing trees and fallen deadwood represent vital habitats for a number of animals, plants and fungi – in some cases species that are endangered (Samuelsson et al, 1994; Du Plessis, 1995). The rotting of deadwood adds organic matter to the soil and helps to maintain soil fertility. Therefore, deadwood can be an important life history stage from the conservation perspective.

The example of African miombo woodland demonstrates how knowledge of the ecology and life history strategies of species can be useful in management. Many tree species in the miombo have seeds that are poorly dispersed and unable to remain dormant for long in the soil. Following germination, the seedlings of these trees experience many years of seasonal growth, gradually gaining strength in their rootstocks, but dying back above ground during the dry seasons. During this growth-and-dieback phase, the seedlings and later the saplings will die if exposed to excessive cutting or grazing, or if fires in the dry season are too fierce. Eventually, the saplings attain a height at which they are able to survive moderate grazing or fires without dying back in the dry season. On the basis of these understandings of ecology and life history strategies, it has been recommended that the best way to manage miombo for a continuing supply of produce from its trees is to burn its lower vegetation stratum (dominated mainly by grasses) early in the dry season. At this time, the fires will be relatively cool (due to a high water content in the plant fuel) and thus not overly destructive. If trunks or branches are harvested for produce, then they should be cut as high above the ground level as possible, as this raises the chances that coppice shoots will be produced and the trees recover quickly (Chidumayo and Frost, 1996; Chidumayo et al, 1996).

Species known or predicted to share certain key ecological features can be considered to belong to the same guild. The management of tropical forest for timber is commonly based on a classification of species of trees into three guilds according to how they regenerate (Peters, 1994). One guild consists of **early pioneers**, which are fast-growing, light-demanding, trees that only become established when canopy gaps form (whether through natural tree falls or logging), allowing light to flood down to the forest floor. These species have wood of low density and die young. An example in central Africa is the umbrella tree *Musanga cecropioides*. **Late secondary species** are also light demanding, but less stringently so than early pioneers. They have wood of intermediate density and can persist for long periods in the canopy. An example in central Africa is the mahogany *Entandrophragma utile*. Finally, **primary forest species** are able to germinate and survive for long periods as small plants growing very slowly in the shade while they await liberation through formation of a light gap. They have heavy wood and can live for many years. An example in central Africa is ironwood *Cynometra alexandri*. There are variants. Early pioneers are absent from *kerangas* forest, a dwarf forest type found on poor lateritic soils in Borneo (Newberry et al, 1986). Forestry operations commonly aim to create conditions suitable for late secondary species, which include many of those yielding the most valuable timber, such as the African mahoganies belonging to the genera *Entandrophragma* and *Khaya*. There have been various suggestions for forestry operations designed to maintain timber yields of these species, while ensuring that their genetic diversity is conserved. They include increasing the amount of light in the forest understorey through opening up the canopy (in order to allow the seedlings of late secondary species to become established and persist) or to restrict logging to short periods just after individual trees of these timber species have had 'fruiting events' (when there should be good stands of seedlings that can be 'liberated' through logging) (Jennings et al, 2000).

VEGETATION TYPES

As with types of plants, both scientific and local names for types of vegetation are of interest to the conservationist. Scientific systems for recognizing vegetation types are based upon structural features of the vegetation (physiognomy) or the types of species present (floristics) or combinations of these two. Very often, a greater use of physiognomic characters is made on classification schemes drawn up for larger geographical scales and floristic characters for smaller ones (White, 1983; Granger, 1996). Physiognomic characters include such things as the height of vegetation, the number of plant layers (strata), whether plants are evergreen or deciduous, leaf size and degree of thorniness.

Standardized systems of scientific classification for vegetation types have some advantages from the conservation perspective. They provide objective bases for assessing the commonness or rarity of vegetation types, which can be useful information when deciding how important it is that they are conserved (Rodwell et al, 2002). They also provide frameworks for pigeon-holing lessons about management. However, there is no universally acknowledged scientific system for recognizing and naming vegetation types on the global scale, parallel to that used for species. One of the practical problems is that many areas of vegetation are highly irregular and patchy – often due to human influence – making recognition of 'typical' areas of vegetation difficult. Plant geographers sometimes resort to trying to recognize and map 'natural' types of vegetation (see 'Vegetation dynamics') or else choose more 'ideal' patches of vegetation for their studies. On the theoretical side, there are no processes (equivalent to sexual breeding and evolution in the case of species) to provide a firm theoretical foundation for a universal system of vegetation classification. Classification schemes for vegetation based on **floristic** characters are inevitably limited in the extent of their geographical application because the ranges of species and high categories of plant taxa are limited for historical reasons. Using *combinations of species* to delimit vegetation types also has its limitations, because it is known from pollen analysis that species have reacted individually, rather than in groups, to changes in the environment (such as in climate). There is nothing particularly long-enduring about particular associations of species as seen today (Livingstone, 1967; Davis, 1976; Hamilton, 1981; Hart et al, 1996). On the **physiognomic** side, is true that there are certain similarities in the appearance of vegetation types occurring under similar environmental conditions in different parts of the world, thanks to convergent evolution; therefore, physiognomy does have some value for devising systems of vegetation classification on the global scale. Convergent evolution is the process through which plants of unrelated ancestry, living in different parts of the world, come to resemble one another morphologically as they become adapted over geological time to similar environmental conditions. The physiognomic equivalents of the cacti (Cactaceae) of the Americas are succulent species of *Euphorbia* in Africa and the Didiereaceae in Madagascar. However, convergent evolution is an inefficient process because timescales for evolution are limited and because certain aspects of the physiognomy of plants are believed to be quite firmly fixed genetically, limiting the extent to which they can be modified through evolution (Wright and Calderon, 1995).

Partly because of the limitations of scientific schemes for classifying vegetation, in practice it is common for conservationists to use vernacular terms to classify the vegetation at their field sites. Terms found in widely spoken languages such as English, French and Spanish are commonly used – for example (in the case of English), the terms forest, grassland and scrub. Attempts have sometimes been made to define such terms fairly precisely (for instance, for Uganda by Langdale-Brown et al, 1964); but most often they are employed quite loosely, which gives potential for misunderstandings. One example is the use of the term 'forest' to describe both a woody vegetation type dominated by sal trees *Shorea robusta* in India and montane tropical forest in central Africa, which actually are quite dissimilar types of vegetation ecologically. A perceived conservation danger in this case is that the commonality of the name might lead to uncritical transfer of conservation techniques used in joint forest management in India to montane forest in Africa, where they would not be ecologically appropriate (Aumeeruddy-Thomas et al, 1999).

People around the world seem to recognize variations in vegetation in similar ways (Ellen, 2002; Hamlin and Salick, 2003). This is probably because people everywhere are interested in the same sorts of things about vegetation, such as their values as sources of plant resources and their uses for predicting certain environmental factors – for example, soil fertility, liability to flooding and history of management. Conservationists need to understand how local people perceive the vegetation at their field sites, including their terms used for vegetation types, what the vegetation types indicate environmentally and why the vegetation types exist. As an example of locally recognized categories, a survey of vegetation on Mount Kinabalu in Malaysia found the following types to be recognized by the local Dusun community (with their approximate English equivalents): *timbaan* (primary forest with large trees), *puru* or *talun* (primary forest), *temulek* (secondary vegetation), *geuten* (thicket), *tume* (cultivated fields), *butur* (grassy area) and *natad* or *liwan* (home garden) (Martin, 1995). Not all types of vegetation recognized by local people are necessarily given names (Ellen, 2002).

The attributes that people mentally associate with vegetation types can make a big difference to how they are treated. Areas of natural vegetation can be given *derogatory* labels, such as the terms 'bush' or 'jungle' applied to tropical thorn forest in the Punjab, Pakistan, by foresters – a factor that has contributed to the destruction of this now highly endangered vegetation type (Khan, 1994b). The institutionalization of new labels associated with *positive* feelings, on the other hand, can be a major achievement for conservation. The influence of nomenclature on changing attitudes to vegetation is provided by the example of native woodland in Britain (Tsouvalis, 2000). Traditionally, native woodlands in Britain have been managed for a variety of purposes, serving both villagers and gentry (Rackham, 1976). However, traditional woodland management collapsed extensively between the 18th and early 20th centuries, related to the loss of rights to woodland resources by ordinary villagers, the widespread replacement of fuelwood and charcoal by coal, the rise of timber imports and the spread of 'scientific forestry'. Scientific foresters derided the knowledge and practices of traditional woodsmen as irrational and emotional. Instead, they promoted methods of selection of species and varieties, planting methods, thinning regimes and harvesting times to maximize yields of timber according to the type of site. Scientific forestry became widely adopted following national anxiety caused by a shortage of home-grown timber and difficulties of timber importation during World War I (1914–1918). The shock of the war triggered the establishment of the Forestry Commission in 1919 and a massive spread of plantations of fast-growing coniferous species of largely foreign origin, such as Sitka spruce *Picea sitchensis*. There was widespread planting of conifers on land classified as 'wasteland' or 'marginal land' (low-grade agricultural land on hills and sandy areas) or 'derelict land' or 'scrub' (old broadleaved woodland). Many areas of native woodland were lost, which is considered today to be a conservation tragedy, rather than just the elimination of 'useless vegetation', as it seemed at the time. The thinking about the value of broadleaved woodland changed completely within the Forestry Commission from around 1990. This transformation was a hard-won victory on the part of certain conservationists, who were initially labelled by foresters as emotional rather than rational, just as the villagers had been before them. What happened is that, during the 1970s, Oliver Rackham and George Peterken developed a new system of classification for woodlands, with, most significantly, invention of a new category called 'ancient woodland'. Ancient woodland eventually became accepted as a useful concept by foresters, partly because Peterken could talk their language. Its successful marketing involved the development of protocols and methods by which ordinary foresters could recognize and designate ancient woodland, based upon such features as being long enduring (woodland should have stood on the site since at least 1600 AD) and the presence of 'indicator species' (species taken to be indicators of long-standing woodland) (Peterken, 1974, 1996; Rackham, 1980).

VEGETATION DYNAMICS

It may be decided for conservation reasons that it would be best if an area of vegetation remained in its existing state or perhaps, alternatively, changed in a certain way. In order to try and achieve such ambitions, it is necessary to have some understanding of why vegetation, as it currently exists, has its particular features. An historical perspective is invaluable. The memories of long-term residents, documentary records and fossils can all reveal something of how the vegetation and the factors that influence it have changed over time and thus help in the understanding of how modern conditions have arisen. They will likely also give some clues as to how the vegetation will evolve in the future. In exceptional cases, data may be available from scientific experiments designed to assess the influence of key environmental factors on the vegetation. For instance, there may be data from exclusion experiments (which involve excluding certain factors, such as fire or grazing, from sample areas of vegetation and observing how the vegetation reacts).

Vegetation types are living entities that are always changing – very locally, as plants die and new ones become established, and very often also on grander scales. Some understanding of this dynamism can be gained from careful studies of modern vegetation, allowing predictions to be made about the state of the vegetation in the future. For example, it may be found that a species of tree that is common today in the canopy of a forest is unrepresented by smaller plants, which suggests that the species may eventually die out in the forest, perhaps being replaced by other species. Very often, a range of vegetation types can be found in an area, some of which can be interpreted as stages in series of vegetation types that, given time, will naturally replace one another at particular sites. To experienced ecologists, landscapes are open books in which the past and future can be read.

Numerous studies have demonstrated that people have profoundly influenced the vegetation of most parts of the world. This influence can be very ancient. People are known to have had the use of fire for at least 500,000 years; today the vegetation of extensive parts of the world, such as much of Africa and Australia, has been profoundly influenced by long histories of burning (Bird and Cali, 1998; Langton, 2000). Burning increases the relative representation of fire-adapted (pyrophytic) plants at the expense of fire-sensitive species. Another profound and often ancient way in which people have influenced extensive areas of vegetation is through the changes that they have wrought to faunas. Humans have caused the extinction of many animals – for example, 73 to 86 per cent of species of the large mammals that once lived in Australia and the Americas (Barnosky, 1989). The domestication of certain animals, such as cattle, goats and sheep, was followed by their widespread introduction around the world, with subsequently massive impacts upon many areas of vegetation.

The concept of **natural vegetation** is a useful one in ecology, defined as the vegetation that would have existed had there never been any people. Although its precise features can be difficult to determine, the conceptual reconstruction of the natural vegetation of a project area is a useful exercise. It assists in making estimates of the productive potential of the land and in providing a baseline against which human impact can be assessed. **Semi-natural vegetation** is another valuable concept in vegetation analysis, defined as vegetation that is an artefact of human intervention, but retains some natural features (Tansley, 1946). It is possible to conceive of some types of landscapes as progressing through a series of stages as a result of conscious manipulation by people, from pristine conditions (with no human influence) to intensive agriculture – a process that has been termed landscape domestication (Clement, 1999).

The profound influence of people on vegetation is well established for Europe, where (it is generally accepted) there are very few areas of completely natural vegetation (Birks, 1996). Human influence in Europe is ancient and continues apace. Major changes in vegetation over the last century in the UK have included large-scale loss of semi-natural grassland and heathland, major expansion in the areas of intensively managed grassland and planted conifers, a huge increase in the range of some introduced plant species, and major changes in the floristic composition of plant communities in hedgerows and woodlands and along watercourses (Preston et al, 2002). It used to be supposed by ecologists that vegetation in the tropics has been less influenced by people than that of Europe. However, it has become clear that humans have had major influences on the structures, floristic compositions and dynamics of many areas of tropical vegetation (Hamilton, 1981; Prance, 1995; Hamilton et al, 2001; Hart, 2001; White, 2001).

Vegetation types are subject to both internal and external processes, influencing the patterns that they show. The sum of all types of disturbance that affect an ecosystem is known as its **disturbance regime**. Some types of disturbance are due to external factors, such as those resulting from fires, storms and outbreaks of pathogens, while others result from internal processes, related to the establishment, growth and death of individual plants. Disturbances can also be classified as either natural or man-made. Since there are many types of disturbance influencing vegetation, operating on many different spatial scales, the end result can be that some types of vegetation consist of intricate mosaics of many vegetational sub-types, forming patterns of various size (Newberry et al, 1986).

Vegetation succession is a valuable concept for evaluating the status of vegetation and predicting how it will change. The basic concept is that certain types of vegetation, left to their own devices, tend to change naturally in certain directions. For example, abandoned agricultural land will often pass through several types of vegetation sequentially, dominated, in turn, by annual plants, perennial herbs, shrubs and finally trees. At one time, it was believed that all successional series on normal soils in a given part of the world will converge eventually to the same end-point vegetation type (known as the **climatic climax**). However, there is empirical evidence that this is not always so; in any case, there are normally forces influencing vegetation types that keep them in states of constant change (Connell and Slayter, 1977; Rejmánek, 1999; Agee et al, 2000). The concept of a **blocked (checked) succession** can be useful in explaining the dynamics of some vegetation types. For example, places in the tropics that have been cleared of forest and are either subject to frequent fires or else have experienced degradation of their soils can become covered by persistent fern-dominated vegetation (*helechal* in Spanish), dominated by genera such as *Dicropteris* and *Gleichenia* (García et al, 1994; May, 2000; Slocum et al, 2000). This can be regarded as a conservation problem because the smothering layer of ferns prevents the return of tree-dominated vegetation. On the other hand, not infrequently, conservationists wish to block the natural course of succession to maintain interesting, potentially unstable, vegetation types. It is common in Europe for conservationists to seek ways of maintaining certain types of species-rich grassland that are intrinsically unstable by encouraging grazing or other disturbances.

Areas of semi-natural vegetation in many parts of the world need to be actively and carefully managed in order to maximize their conservation benefits. Under modern circumstances, they cannot be left just to care for themselves (Tansley, 1946). It is not infrequent for conservationists to know that new management measures are needed, but for them to be less certain of exactly what they should be.

Traditional ways of managing the land and plant resources have often resulted in landscapes of relatively high biodiversity value, compared with those associated with normal modern practices, and it is quite common for conservation managers to contemplate returning to more traditions ways. Following this line of thinking, Aboriginal burning practices are being reinstated in an increasing number of protected areas in Australia in recognition of their value for maintaining biodiversity-rich landscapes in the wet–dry tropics and the arid zone (Nakashima, 1998; Langton, 2000). In southern England, attempts are sometimes made to reintroduce traditional livestock grazing to the heathlands as a way of preventing their reversion to woodland (Northwood et al, 2001). One of the problems with this approach is lack of knowledge about exactly what the traditional practices were. There are also concerns that traditional practices may be reinstated uncritically for sentimental reasons, rather than based on objective assessments of their actual conservation advantages (Pullin and Knight, 2001).

6

The Management of Plants and Land

Issues of land management often arise in conservation. This chapter introduces some aspects of the ways in which people manage plants using physical, chemical and biological means, set within the broader context of wider systems of land management and ways of living. The role of plants in providing food is so fundamental that there is a close connection between the ways in which societies manage plants for food and the ways in which societies, as a whole, are organized.

RESOURCE ACQUISITION AND SOCIAL MODES

People obtain a wide variety of material resources from plants that have grown in a variety of ways. Some plants that people use have been deliberately planted, while others are self-established. Again, some plants are carefully tended, while others are left to grow on their own without any attempt to manipulate them. Taking the world as a whole, large quantities of goods from plants are today obtained from high-yielding varieties grown intensively in monocultures. Even so, people still rely substantially upon plant resources that have originated in other ways. A farmer in England, growing hundreds of hectares of wheat *Triticum aestivum* (all for sale) may still enjoy picking blackberries *Rubus fruticosus* from plants growing in pockets of wild land, buy mahogany furniture harvested from wild trees of *Swietenia macrophylla* in Brazil and rub down his bruises with ointment made from the flowers of *Arnica montana*, picked from plants growing in traditional hay meadows in Romania. Rural people in developing countries tend to be much more dependent upon plant resources that have originated locally than their counterparts in richer parts of the world. However, these resources can come from a wide variety of sources in terms of intensity of management. A farming household may obtain rice *Oryza sativa* from intensely managed irrigated fields, mangos *Mangifera indica* from planted but otherwise untended trees, twine from the wild plant *Smilax* that is lightly managed, and fuelwood from deadwood originating from all sorts of wild and tended trees.

Knowledge of social interactions is important for identifying how people obtain their plant resources. Neighbours everywhere and members of extended families in developing countries can exchange plant resources extensively between them so that households can have easy access to a greater variety of resources than may be at first apparent. The livelihoods of 'pygmy' people, such as the Aka of central Africa, are closely tied to those of their agricultural neighbours with meat and other forest goods and services provided in exchange for iron, salt and agricultural produce (Naughton-Treves and Weber, 2001; Noss, 2001).

It is probable that, on the global scale, greater quantities of resources are collected from wild plants today than has ever been the case before. In developing countries, wild plant resources are especially significant in the lives of the financially poorest rural people, many of whom rely upon wild plants to supply them with free or relatively low-priced produce – fuels, foods, medicines, construction materials and much else – and, beyond that, may also provide them with a source of income through their sale.

There are many types of agricultural systems. For present purposes, it is useful to distinguish between two major categories: traditional and intensive. **Intensive agriculture** is typically marked by large fields, monocultures of single varieties of crops developed by scientific plant breeders, and extensive use of fossil energy, inorganic fertilizers, pesticides and sometimes water (through irrigation). As mentioned in Chapter 2, there is much concern about the effects of intensive agriculture on the environment, including loss of plant diversity. One of the greatest challenges in conservation today is to find ways of producing the massive quantity of food required to support the huge human population, while minimizing environmental damage. Research is needed to develop high-yielding systems of organic and integrated (intensive plus organic) agriculture (Reganold et al, 2001; Macilwain, 2004).

Traditional agriculture refers to organic, long-established farming systems with a high reliance upon human labour. These systems are currently responsible for meeting 15 to 20 per cent of the world's food supply. Irrigation is used in some cases, as are beasts of burden. Land managed under traditional agriculture is usually a mosaic of many different types of land, containing a wide variety of plants, both cultivated and wild (Alcorn, 1995; Gómez Pompa and Jiménes-Osornio, 1999; Jenkins and Kapos, 2000; Tuxill and Nabhan, 2001). Traditional agriculture, as practised now or formerly, has lessons to teach agricultural developers today. Take the case of agriculture in the tropical forest region of Amazonia, where many soils on well-drained sites are sandy and naturally infertile. Recent settlers, clearing forested land for farms, have difficulty in maintaining agricultural productivity. Yet, scattered across Amazonia there are patches of humus-rich dark soil, known as *terra prêta dos indios* in Portugese ('black Indian soil') or *tierra negra* in Spanish ('black earth'). This most cherished soil for farming is the product of pre-Colombian agriculture. Its existence shows that there are other ways of managing the land apart from those demonstrated by many contemporary agricultural projects, which generally leave trails of ecological disaster and erosion (Prance, 1995; Hamlin and Salick, 2003).

To an extent, the major modes of acquiring food from plants lie in an historical sequence, proceeding from wild resource collection to unsubsidized organic agriculture, then to subsidized organic agriculture (with irrigation and/or manuring) and finally to intensive agriculture. All plant resources were wild-harvested before the invention of agriculture, which dates back to about 9000 BC in the Middle East, Mexico and New Guinea. Irrigation was developed around 5000 BC in the Tigris–Euphrates valley, with other early irrigation systems known from China, the Indus valley and (later) the Andes. The use of draught animals, such as oxen, for ploughing probably started a little before 2000 BC. Intensive agriculture is essentially a post-AD 1950 phenomenon.

Once a more 'advanced' mode (in an historical sense) of resource acquisition had arrived in an area, it would have tended to expand and the societies that practised it become dominant, thanks to competitive advantage. It has been estimated that the population density increased from around 0.01 to 0.1 individuals per square kilometres with hunter-gathering to 10 to 100 times as many with unsubsidized organic agriculture and by 1000 to 10,000 times (above hunter-gatherer levels) as agriculture became more intensive (Cavallli-Sforza et al, 1994). From an individual's viewpoint, agriculture would seem to offer no great advantages over hunter-gathering: it leaves less time for social activities and its introduction is known from archaeological evidence to have led to a decline in human health as a result of repetitive hard manual labour and an impoverished diet dominated by a few carbohydrate-rich foods (Karlen, 1995). Sometimes, changes in the dominant mode of resource acquisition may have been catalysed by environmental circumstances. In Madagascar, a transition from unsubsidized agriculture on hill slopes (based on swidden – see 'Management processes, tools and systems') to irrigated agriculture in valleys may have become virtually inevitable once the easily erodible forest soils had been destroyed. An island, which was once almost completely covered by forest, has today been converted into a landscape largely of unproductive coarse grassland on hill slopes, with agriculture now concentrated in paddy fields in the valleys, ironically founded on the products of soil erosion.

Exceptionally, there are reversions to earlier dominant modes of resource acquisition. Following the arrival of Europeans in the Americas, it has been estimated that the population of indigenous people in Amazonia fell from 6 to 9 million (in 1500) to less than 200,000 (in 1992), mainly as a result of their susceptibility to deadly new diseases. There was an associated decline in the intensity of managing landscapes and species by indigenous people, with many of the (at least) 138 species of crops that had previously been cultivated becoming lost (Clement, 1999).

The introduction of major new modes of acquiring food from plants was associated with changes in many other aspects of human societies. Along the way, people have developed more extensive trading systems, more types of specialist occupation, larger settlements, more pronounced social hierarchies and (until recently in some countries) a greater marginalization of women in public affairs. The adoption of subsidized organic agriculture was associated with the appearance of pre-industrial states and cities. Elites with distinctive cultures came to monopolize privileged positions in public affairs.

The lives of people in small-scale societies – practising hunter-gathering or unsubsidized organic agriculture – are intertwined closely with one another and with the local natural world. Worldviews in such societies are relatively holistic, with close connections felt between the fates of the self, other people, the natural world and the spirits. Nothing is believed to happen by chance. Plants play major roles in medicine and religion, serving in practical treatments as herbal remedies, being used in divination to determine the causes of misfortune (as with the use of the ordeal tree *Erythrophleum* in Africa) and, through the incorporation of spiritually powerful plants in ceremonies and charms, to safeguard the individual, society and natural resources against malign influences.

The great mass of people in pre-industrial states (more than 95 per cent of the population) continued to live in small rural communities and doubtless continued to follow local medical and religious practices. However, pre-industrial states were also associated with new types of worldviews – perhaps appealing particularly to members of the ruling classes – that placed emphasis on the duties of people to live their lives correctly according to the social positions to which they had been ascribed. The reward of righteous living was salvation in the afterlife. New teachings developed within pre-industrial states about how to maintain the human system in a good condition, making considerable reference to the merits of different types of plants as foods or medicines. Scholarly systems of medicine appeared in Asia and Europe, based on the ideal of achieving balance in living. Greek medicine and Unani make reference to four psycho-physiological conditions (or humours) to be kept in balance (translated as blood, bile, phlegm and choler), while Ayurveda and Tibetan medicine refer to three (wind, bile and phlegm in the case of Tibetan medicine).

Large parts of the world have recently entered a new trajectory, especially during the last 50 years, with the development of intensive agriculture, highly monetarized economies, the industrial production of goods, a huge expansion in trading systems and strong specialization in occupations. Worldviews in these 'Western' societies can be strongly influenced by ideologies, which serve something of the roles of religions in more traditional societies, but are less all pervasive and more open to questioning. Pathways through life have become increasingly conceived as individual ventures, with a new emphasis on human rights (conceived as rights that apply equally to all people), rather than on performing one's duty according to one's social station, as in traditional societies.

Reductionist science has played a major role in the development of intensive agriculture, with the pursuit of high productivity and (especially) high financial profit being major motivations underlying the ways in which it has evolved. The financial benefits of intensive agriculture have been increased in the European Union (EU) and the US by the payment of substantial subsidies to farmers and because they do not take into account the costs of their environmental damage. Today, farmers are increasingly under the control of supermarkets, which are able to dictate their requirements and, hence, farming methods through near-monopoly positions dominating food chains.

MANAGEMENT PROCESSES, TOOLS AND SYSTEMS

The most immediate ways in which people influence plants are through direct forces – physical, chemical and biological. Box 6.1 lists some of the purposes lying behind people's efforts to influence plants and some of the processes with which they try to achieve them. Conservationists, working in their project areas, should have some knowledge of how people try to manage plants and the types of tools and machines that they use (see also Figure 6.1). More than this, they should gain some understanding of how people's direct interactions with plants fit into their wider systems of resource management and use, including what happens to the plant resources after they have been harvested. Cultures and economies influence how plants are managed. For instance, certain tools may customarily be used only by certain sections of society, people may have tools but not know how to use them, and there may be economic restraints limiting the types of tools actually available.

There are many systems of resource management and many methods used for managing plants, of which only a few are mentioned here.

Fire as a management tool

One way in which people try to manage plants is through the use of fire. Conservationists, like other people interested in plant resources, are often faced with the task of finding the optimal fire regimes for their purposes. People

Figure 6.1 *Sycamore trees* Acer pseudoplatanus *felled on Madeira to control the spread of this invasive species*

Source: Alan Hamilton

Box 6.1 Some purposes in human efforts to influence plants and some of the processes used to achieve them

Ground preparation and soil care

- *Site and soil preparation*: cutting or slashing plants; burning to clear standing vegetation; applying herbicide to kill plants; preparing the soil through digging, ploughing or harrowing.
- *Soil improvement by organic means*: fallowing; composting; manuring; mulching; liming; growing soil-improving plants (for example, nitrogen fixers).
- *Soil improvement by inorganic means*: adding artificial fertilizers.
- *Water control*: drainage; irrigation; watering.

Locating, raising and planting plants

- *Sourcing planting materials*: locating sources of plants (including finding desired varieties); being given planting materials by family or friends; purchasing from commercial sources; breeding new varieties.
- *Raising young plants*: making beds for seedlings or cuttings; breaking dormancy (for example, seed scarification); tissue culture; layering; grafting.
- *Direct planting*: planting seeds, cuttings, saplings, bulbs or other plant parts; transplanting.

Ensuring useful production

- *Encouraging the right kind of growth*: coppicing; pollarding; supporting (for example, for climbers); mowing; thinning; pruning; disbudding; deadheading.
- *Ensuring seed set*: ensuring presence of pollinators; hand pollination; planting varieties to encourage cross-breeding.
- *Discouraging competing plants* (see Figure 6.1): preventing the introduction of unwanted plants; weeding; cutting; digging out; hand pulling; burning back; use of organic or inorganic herbicides or arboricides; using domestic animals to clean fields.
- *Discouraging pests and diseases*: crop rotation; inter-planting or adjusting planting times; removing sites of pest concentration; encouraging predators; removing or burning infected material or dead plants.

Harvesting products from plants (examples)

- *Harvesting woody materials*: felling trees; cutting timber into transportable sizes.
- *Harvesting grain*: direct collection of seeds from plants or ground; cutting whole plants or parts of plants with subsequent seed removal (for example, by threshing and winnowing).

Post-harvest and pre-planting treatments (example: seeds)

- *Seed storage*: placing seed in granaries; pest control measures; use of anti-fungal agents.

burn vegetation for many reasons, among them the clearance of land as preparation for cultivation, the enhancement of soil fertility, the encouragement of fresh grass growth for livestock during dry seasons, the driving of wild animals during hunts and making it easier to collect certain non-timber forest products (for example, the fruits and flowers of mahua *Madhuca indica* in central India). Studies of traditional fire management by Aboriginal people in Australia have shown that their fires are sometimes intended to maintain certain fire-dependent plant species, while simultaneously protecting fire-intolerant plant communities (such as monsoon forest) through the creation of buffer zones (Langton, 2000).

Livestock management

Livestock can cause major changes to vegetation and, as with burning, determining optimal regimes of livestock grazing can be a major part of habitat management for conservation. Globally, the species with the greatest impact upon plants are cattle, sheep and goats, with others (such as yak and llama) influential locally. There are many systems of livestock management, varying greatly in scale. The most extensive are associated with **migrant pastoralists**, such as the *gadis* of India, moving their animals over distances of hundreds of kilometres annually following the availability of grazing. *Gadis* have established relationships with some farmers along their routes, who allow the grazing of stubble in their fields in exchange for animal manuring. Livestock owners who graze over extensive areas can have traditional rights to pastures, as do the Loita Maasai of Kenya, with specific rights of use assigned to particular villages and clans (Maundu et al, 2001). Another form of livestock management is **transhumance**, with grazing in high mountain pastures in summer and overwintering at lower altitudes. The **stall-feeding** of animals is a rapidly growing form of livestock management, with many types of natural fodder and farm produce being exploited as sources of food (see the case study at the end of this chapter). Stall-feeding is a common element of intensive agriculture, with, at the most extreme, food supplements for the livestock being manufactured from almost any form of cheap protein available, including ground-up carcasses from abattoirs and fish-meal from far-off seas.

The management of tropical forest for timber

Several methods have been devised for managing tropical forest for the production of timber. Operations can be entirely or mainly hand based, or can involve the use of heavy machinery. **Hand-logging** involves the careful selection of trees, their felling by axe or chainsaw, the sawing of the logs into planks (often using pits – hence the term pit-sawing) and, finally, the carrying of the planks manually out of the forest. This is hard work, but can have ecological benefits compared with mechanical logging, such as creating relatively little incidental damage to the forest and a greater efficiency of wood recovery. It has been found in Tanzania that pit-sawyers use 60 per cent of possible wood, compared with only 30 per cent with mechanical logging and sawmilling (Brigham et al, 1996). **Mechanical logging** involves the use of heavy machinery inside the forest, frequently with consequent major disturbance to the forest through the construction of numerous roads, tracks and loading sites. It is possible to reduce the level of damage if operations are conducted very carefully at all scales, right down to the micro-scale of exactly which trees are selected for harvest, how they are felled (for example, in which direction) and how the logs are extracted. A common arrangement is for logging to be by private companies, granted concessions to blocks of forest for fixed periods of time and required to operate under regulations set by forestry departments. Forests are frequently divided into compartments, with harvesting rotated around them, allowing intervening periods for forest recovery. Operational rules stipulate the types and sizes of tree species that are allowed (or required) for harvest and may also prescribe pre-harvest and post-harvest treatments, such as climber slashing and enrichment planting. By far the commonest form of management prescription is for logging to be highly selective in terms of species (known as 'commercial species'), with a minimum felling diameter of usually over 45cm and with about two to seven trees harvested per hectare. Prescribed rotation periods are often around 70 years (Jennings et al, 2000; Plumptre, 2001).

Forest management for the production of non-timber forest products (NTFPs)

There are many management systems for NTFPs, varying in their degree of management intensity. Although some forests may seem, on casual inspection, to be completely spontaneous and wild, in fact very often people have had a considerable hand in their creation. In Amazonia, the Kayapó have been found to subtly influence forest composition by encouraging or transplanting useful plants along trail sides (Davis, 1995). A traditional method of woodland management in Britain is as 'coppice with standards', which combines growing scattered large trees for timber known as standards (commonly oaks *Quercus*), with management of a lower stratum of trees (commonly with much hazel, *Corylus avellana*) through coppicing to supply poles and firewood. There is evidence from fossil trackways that the management of trees for coppice in Britain dates back to the Neolithic and Bronze Ages (4000–750 BC) (Rackham, 1976). At its most extreme, the management of forests for NTFPs can be so intensive that the term 'wild cultivation' seems appropriate. Hundreds of thousands of hectares of hill slopes in Anji County, Zhejiang Province in China are covered by pure stands of the bamboo *Phyllostachys heterocycla* var *pubescens*. The bamboo has not been planted; instead, it has probably achieved its present abundance through intensive forest management carried out over many years. Broadleaved trees were probably once much more abundant than they are now. Today, the management of the bamboo stands involves hand-tilling to 30cm every two to four years and the application of inorganic or organic fertilizers. Pesticides and fungicides are used when there are problems. The medicinal orchid *Gastrodia elata* is sometimes grown under the bamboo canopy to provide an additional crop. In at least one area, the author Alan Hamilton has personally seen that every standing bamboo culm is marked with an owner's sign, a level of intensity in the ownership of 'wild' plants perhaps unrivalled worldwide.

Traditional systems of agriculture with forestry

There are many variants, typically with a variety of land types serving different purposes and under various forms of tenure. The Karen in Thailand have six categories of land type (Nakashima, 1998):

1 home gardens, lying close to homes;
2 paddy fields with rice, often close to villages, owned by individual families and exploited on a continuous basis;
3 areas of swidden cultivation, much more extensive in area than the paddies and generally lying in broad bands around the villages (families have the right to continue to use the swidden areas that they have used traditionally; each family divides its holdings into plots, cleared for cultivation on a seven- to ten-year rotation);
4 community forests, usually extensive zones beyond the swidden areas, with access and use regulated on a community basis; community forests provide the villagers with wood and other forest products and serve for livestock grazing;
5 watershed forests, covering large areas encompassing the headwaters of major watersheds; these are generally maintained in states close to the natural climax and are strictly protected by the communities (watershed forests are associated with the forest spirits, to whom offerings of food and drink are provided);
6 sacred forests, which are reserved largely for ceremonial purposes, such as for rights associated with the newborn or dead; sacred forests are small in size and are off-limits to any form of harvesting.

Swidden (slash-and-burn) agriculture

This is an ancient form of agriculture, which persists in some areas of tropical forest and woodland where the human population density is relatively low (Davis, 1995; Frost, 1996) (see Figure 6.2). Occasionally, swidden agriculture involves the felling of primary forest or woodland; but more commonly the vegetation cleared to create fields is secondary, having grown up relatively recently in previously cleared areas. Rotation periods can be 5 to 40 years. There are many varieties of swidden systems. Among the Karen in Thailand, the stages are slashing trees (although leaving the larger ones), allowing the felled trees to dry for a month and then burning them, planting upland rice and some vegetables in the ashes, and eventually abandoning cultivation and letting the forest become re-established (Nakashima, 1998). *Chitemene* is the name given to the best-known swidden practice in wetter areas of the African

Figure 6.2 *Slash-and-burn agriculture at Manongarive, Madagascar*

Source: Alan Hamilton

miombo (Frost, 1996). This involves the creation of an ash garden to grow the crops. Trees are lopped and chopped over an area about ten times the size of that to be cultivated, and cut trunks and branches are brought together to burn. Cultivation is for three to four years, with an initial a crop of finger millet *Eleusine coracana*, followed by cassava *Manihot esculenta* during the second and third years. Swidden systems should *not* be thought of as consisting just of short phases of cultivation, principally to produce carbohydrate-rich crops, followed by long periods of abandonment. After the cultivated crops are harvested, the fields can continue to yield many other products, including medicinal plants, fruits, oils and firewood. The original chopping and burning stage is also not necessarily just a matter of outright destruction. Often, trees yielding valued products are left uncut, while those that are felled may be deliberately cut well above ground level in order to give them a better chance to survive the burn and re-sprout (Prance, 1995; Asfaw and Tadesse, 2001).

Home gardens

These are intensively managed lands lying close to homes. They often contain many species, grown for a diversity of vegetables, fruits, medicinal plants, culinary herbs and other produce. Their contribution to farm production can be substantial – for example, providing 40 per cent of the calories, 30 per cent of the protein and 65 per cent of the fuel used by households in parts of Indonesia (Wilson, 2003). There are many variants of home gardens in terms of their layouts and types of plants grown. The *afrodescendente* communities of Choco, Colombia, have an unusual type, involving the use of raised platforms (*zoteya*) that support plants grown in small beds or pots. In Bangladesh, men and women each have their own home gardens, with those of the women being as unobtrusive as are the women themselves – the gardens are tucked into extremely limited spaces in the corners and alleys between and behind the homesteads, although, even so, they sometimes provide a surprising diversity of vegetables (Wilson, 2003).

LANDSCAPE MANAGEMENT FOR CONSERVATION

The management of plants or vegetation can be practised on the landscape scale as well as at the level of individual units of management, such as particular farms or forest reserves. A landscape can be seen as a patchwork of different types of habitat, under diverse ownership and serving diverse purposes. Depending upon the place, the patches might consist of fields and forests, rivers, lakes, roads, villages and cities. If conservationists can agree on key features of landscape design, then a great advantage is that individuals and groups can work steadily together over many years towards the same long-term goals. Landscape planning for conservation is today becoming more of a science as experience is gathered about the procedures most likely to be successful (see Box 6.2).

While landscape design for conservation must take account of all aspects of conservation, botanical matters should almost invariably rank high. This is because the types and distributions of vegetation and plant species are fundamental parameters determining the conservation values of landscapes (McDonald and Boucher, 1999). Very often, there is rather little information available from actual records about the distributions of plant species, an uncertainty in planning that can be reduced, to a degree, through deductive modelling (see Chapter 8). The more that is known about the factors responsible for determining the patterns of distribution of vegetation and plant species, the better, since this will strengthen any predictions made about how the plants present will react to future environmental change (such as climatic change).

Landscape planning provides opportunities to consider the big picture, including in relation to:

- retaining or restoring as large areas as possible of natural vegetation (following the precautionary principle; see Chapter 1);
- retaining or restoring as large areas as possible of habitats of higher conservation value, including maintaining or creating habitat corridors between existing tracts (following the principles of conservation biology; see Chapter 2);
- maintaining or promoting ecological connectivity (see paragraph below);
- retaining vegetation types (such as tropical forest) whose loss will contribute most significantly to greenhouse climatic change (see Chapter 2);

Box 6.2 Steps used in systematic conservation planning in the Cape Floristic Region (CFR) of South Africa

Steps used in conservation planning in the Cape Floristic Region (CFR) of South Africa include the following:

- identifying and classifying habitat types;
- mapping areas of habitat types;
- assessing the conservation values of habitat types (in general and at particular places);
- assessing threats;
- assessing possibilities of success;
- devising ideal conservation networks (for example, with extensive areas of certain habitat types, corridors and micro-reserves);
- assessing management requirements for habitat types (in general) and for particular places;
- reviewing legal and other instruments and instigating new legislative and other supportive measures;
- negotiating with stakeholders;
- identifying more effective management practices, key areas for training and other capacity-building measures, and mounting capacity-building initiatives.

'Systematic conservation planning' is the name given to a new science of landscape planning being pioneered in Australia and South Africa. The aim is to develop practical methodologies that will rapidly advance the quality of the landscape from the conservation perspective. In the case of the CFR, a decision was made in 1998 to develop a systematic conservation plan in response to a number of concerns:

- The existing reserve system was seen as unrepresentative of biodiversity patterns and processes.
- There were escalating threats to biodiversity.
- Institutional capacities for conservation were diminishing.

Various lessons have been learned as the work has proceeded, the most important being the desirability of incorporating *implementation* within all stages of planning in order to avoid 'planning fatigue', when nothing practical is apparently being achieved.

Sources: Pressey, 1999; Kier and Barthlott, 2001; Cowling and Pressey, 2003

- retaining vegetation types that are important for regional to local climatic amelioration, such as forest and other vegetation types with trees (see Chapter 1);
- retaining vegetation types that protect water supplies and reduce erosion (see later in this section).

The maintenance of ecological connectivity provides benefits to both man and other species – for example:

- Research in Costa Rica has demonstrated that retaining patches of forest within 100m of coffee crops helps to ensure that the coffee is pollinated because of increased and more reliable visitation by pollinating bees (Ricketts, 2004).
- In Peninsular Malaysia, *Econycteris* bats are apparently the exclusive pollinators of durian trees *Durio zibethinus*. These bats feed preferentially on the flowers of *Sonneratia alba*, a coastal mangrove. In order to forage on the flowers, the bats fly 20 to 40km from their roosts each night, on the way pollinating any durian trees that they encounter. A problem is that mangrove forest is being rapidly reduced through coastal development (Peters, 1994).
- Similarly, long-nosed bats migrating over extensive areas in Mexico and the USA move between patches of natural vegetation and pollinate many plants useful to people on their travels. In this case, the conservation lesson is the desirability of maintaining islands of natural habitat at intervals along the migration routes (Tuxill and Nabhan, 2001).
- On the East Usambara Mountains in Tanzania, it has been found that many species of birds move seasonally between forests at lower and higher altitudes. Accordingly, a conservation objective should be to retain a forest cover over the whole altitudinal gradient in order to allow the birds to move easily (Stuart, 1981).

Having the right types of vegetation at critical places in the landscape can do much to provide ecological services important to people's lives. Among the most vital of such services are water supplies; in a well-planned world, for example, a high priority would be placed on retaining patches of surviving natural vegetation at higher altitudes on tropical mountains and hills (see Chapter 1). Some protection against landslides and mudflows, which can be serious hazards in higher rainfall regions, can be afforded by avoiding complete deforestation on vulnerable slopes or, if this has already happened, encouraging the return of a stabilizing plant cover. Vegetation in wetlands and along watercourses deserves special attention because of its major influence on flooding and erosion. 'Soft engineering' along rivers and coasts, based on retaining or restoring natural vegetation and ecological processes, is often a superior long-term option to hard engineering, based on canalizing rivers and erecting concrete seawalls. Floodplains, coastal sand dunes and salt marshes deserve special attention in landscape planning.

Landscape design for conservation should take account of likely possible future climatic change, both shorter-term greenhouse change and longer-term Ice Age events. On the **Ice Age** front, places that were biological refugia during the last Ice Age deserve special conservation attention, not just because they will generally be centres of concentration of plant species and genes today, but also because many of them will probably again be refugia in the next Ice Age

(Hamilton, 1981; Fjeldså and Lovett, 1997a, b). Regarding **greenhouse climatic change**, attention should be paid to any models available that predict how vegetation and plant species will react to projected climatic change and plans for the landscape should be adjusted accordingly. Natural habitats should be maintained along latitudinal and altitudinal gradients as far as possible to make it easier for species to migrate (Bawa and Dayanandan, 1998).

A major aspect of landscape planning for conservation is assessing the adequacy of networks of protected areas. In most countries, such networks have been developed rather haphazardly from the plant conservation perspective (for example in Africa – Rebelo, 1994; Weber and Vedder, 2001). There are reported to be exceptions, such as Indonesia, where the siting of national parks and nature reserves (but not forest reserves) is reported to have been based on the principles of conservation biology (MacKinnon and Artha, 1982; Jepson et al, 2001). A review of the current network of protected areas on the global scale (currently covering 11.5 per cent of the planet's land surface) shows that 12 per cent of species in a sample of vertebrate groups were not covered by the network (20 per cent for *threatened* species), and it is thought that representation is even worse for plants because they are more abundant and there are more narrow endemics. According to this analysis, the places where more protected areas are most needed to better cover the full diversity of global species include Yunnan Province and the mountains surrounding the Sichuan basin in China; the Western Ghats in India; Sri Lanka; the islands of Southeast Asia and Melanesia; the Pacific Islands; Madagascar; the Cameroon Highlands; Mesoamerica; the Tropical Andes; the Caribbean; and the Atlantic forest of South America (Rodrigues et al, 2004). Another general problem with the siting of protected areas is the paucity of lowland compared with highland sites because of the often greater attraction of the former for agriculture and settlement. For example, reserves in the Cape Floristic Region (CFR) of South Africa are mostly in rugged, inaccessible and infertile mountain landscapes, while there are few in the productive and more densely populated lowlands (Cowling and Hilton-Taylor, 1994; Cowling, 1999). Habitat types that are poorly represented in the European national park network include deciduous forest, Mediterranean habitat types and steppe (Synge, 2001).

There are a number of principles against which existing or potential networks of protected areas can be evaluated. These relate to the **irreplaceability** of particular protected areas (for example, in terms of their complements of species, vegetation types or other features), the extent to which the different protected areas **complement** one another, the degree of habitat **connectivity** displayed over networks as a whole and how well the range of available natural habitats is **represented** (Pressey et al, 1993). Representativeness was a major consideration in the initial selection of nature reserves in the UK, when the principal objective was to gain a sample of the range of habitat types available, with the secondary objective of biasing the selection of areas within habitat types towards those carrying rare species (Tansley, 1946).

Small protected areas (micro-reserves) are being increasingly established for plant conservation, notably where human pressure on the land is high.

There is concern that the small size of populations of some species in these reserves may make them vulnerable to extinction, although some of these populations have probably always been small and it is likely that they are genetically adapted to small size (Cowling et al, 1996). Studies of long-established fragments of tropical forest that have attained biological equilibrium have shown that they can contain a diversity of species, though not the same mix of species as found in nearby extensive tracts of forest (Turner and Crolett, 1996; Brokaw, 1998). **Micro-reserves** are becoming increasingly established around the Mediterranean, where point endemism is a marked feature, with many species having very restricted ranges, such as being confined to one or a few mountain tops, cliffs or dunes. The Regional Wildlife Service of the Valencian region in Spain has been a leader in this respect, responding to the fact that more than 97 per cent of the 350 species that are endemic to the region are only found in specialized habitats that are often small (one to two hectares in extent) and isolated from one another and do *not* grow in the widespread vegetation types of the area, such as garique and pine forest. By 2004, the service had created 230 micro-reserves, covering a total of 1440ha and including more than 70 per cent of the endemic species and 85 per cent of the priority habitats included in Annex 1 of the Habitats Directive (see Chapter 3). The micro-reserves have been established on both public and private land, and always with the agreement of the landowners. A decision to establish a micro-reserve is voluntary; but once an agreement has been made, it is irrevocable and the reserve cannot be deregistered by the landowner. A one-off payment is made to landowners to compensate them for loss of economic activities and some funding is available for certain conservation tasks.

Protected areas will be strengthened if they are ecologically and socially supported by their surrounding landscapes. This is a principle that the United Nations Educational, Social and Cultural Organization (UNESCO) has promoted through its template for **biosphere reserves** (see Chapter 3) with core, buffer and transition zones (see Figure 6.3). Although the core zones of biosphere reserves do not *have* to be statutorily declared protected areas, in practice they often are. The allocation of land for different uses is also commonly found internally *within* protected areas.

Many individuals and organizations can contribute to landscape planning for conservation in their various ways. Several large conservation non-governmental organizations (NGOs) operating internationally have adopted a landscape approach as a major part of their work. One of these is the World Wide Fund for Nature (WWF), which is promoting the concept of *ecoregions* as a basis for landscape planning. An ecoregion is an area of land or water of intermediate size (generally 100 to 10 million square kilometres) that contains a geographically distinct assemblage of natural communities (WWF, 2004). WWF has divided the terrestrial world into 825 ecoregions, about 240 of which (the Global 200) have been suggested as priorities for conservation on the basis of their exceptional biodiversity and because they capture a representative sample of the world's ecosystems (Olson and Dinerstein, 2002). There are many disputes about exactly how to recognize ecoregions or otherwise

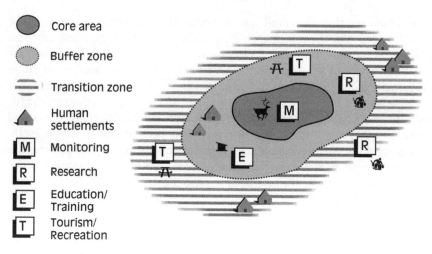

⬤	Core area
◉	Buffer zone
⬳	Transition zone
🏠	Human settlements
M	Monitoring
R	Research
E	Education/ Training
T	Tourism/ Recreation

Figure 6.3 *Zonation in a generalized biosphere reserve*

Note: The model shows several concentric zones. The purpose of the inner **core area** (or areas, since there may be more than one) is to protect the landscape, ecosystems and species that it contains. Human activities are minimal, except research and monitoring. The surrounding **buffer zone** accommodates more human activities, such as research, environmental education and training, as well as tourism and recreation. Human uses in the outer part of the buffer zone may include applied research, traditional uses, restoration, human settlements, agriculture and fisheries. The next zone is a **transition zone**, grading into the **general landscape** within which the biosphere reserve lies. The model can be implemented in many different ways to accommodate local geographic conditions and constraints, allowing creativity and flexibility.
Source: adapted from Hadley, 2002

divide up the world for the purposes of conservation landscape planning. However, this is often not really a matter of great significance, provided that those responsible for actually making conservation decisions consider as much evidence as possible before making up their minds. They should take care to consider processes operating at all geographical scales – from the very local (for example, some tenure issues), to the medium scale (for example, for water supplies) to the very wide ranging (such as some trade systems).

Landscape planning for conservation is mainly useful at the national and more local levels, although there are exceptions. Landscape planning on the larger scale is being promoted in Europe by means of a strategy known as the Pan-European Ecological Network (PEEN, 2001). Eventually, this will be a physical system in which the full range of landscapes, ecosystems, habitats, species and genetic diversity of significance on the European scale will be conserved. Its components will include core areas, corridors, buffer zones and restoration areas. A more modest multinational landscape-level initiative, again in Europe, is aimed at the conservation of natural grasslands in its centre and east. The work involves the identification and mapping of sites of semi-natural grassland, the classification of their vegetation, the identification of current management and threats, and proposals for new protected areas (Veen, 2001). Finally, border areas between countries can be a useful focus in transna-

Figure 6.4 *Experimental habitat restoration for the nationally threatened species perfoliate penny-cress* Thlaspi perfoliatum

Note: Topsoil covering in an old quarry in Gloucestershire, UK, is being removed to expose the quarry floor, suitable habitat for this species.
Source: Simon Williams; copyright © Plantlife International, 2004

tional conservation because they are sometimes places of importance for plant diversity for reasons of history or physical geography, and the establishment of transboundary parks can be welcome politically as symbols of neighbourly cooperation.

Government departments can engage in landscape-type planning for their areas of responsibility, even if these areas are widely scattered. Like many forestry departments, the National Forest Authority (Forest Department before 2004) in Uganda manages a network of forest reserves across its country. A decision was made during the late 1980s to dedicate 50 per cent of the total area of the reserves to sustainable timber production, 30 per cent to lower impact resource harvesting and 20 per cent to nature conservation. The challenge, then, was to identify which forest reserves, or parts thereof, should be placed in which category. A five-year programme was initiated to assess the values of the forests for biodiversity, timber, community uses and non-consumptive uses. The biodiversity value was calculated according to the representation of five biological groups (one of which was trees), with extra weight being given to rare species. An iterative procedure was then used to select the most suitable combination of forests for nature reserves, care being taken to maximize complementarity in the representation of species and habitats. Eventually, a preferred network for the nature reserves was identified, including 14 forest reserves of substantial size (on average, 100km^2 in extent). With 25 smaller forests added (average size of 32km^2), the total network was calculated to contain 99 per cent of species in the groups used as indicators (Howard, 1991; Howard et al, 2000).

Private companies can play their parts in landscape-level conservation. One private company in Sweden, owning 3.4 million hectares of productive forests, has tried to manage its estate by balancing environmental and productive values. One approach used has been to identify key habitats for biodiversity in landscape units of 5000 to 25,000ha and then to try to enlarge or connect these key habitats with habitat corridors, including along waterways. In addition, the company has discovered that much biodiversity can be conserved by increasing the number of small areas of un-logged land at the very local level, which requires giving attention to detail at each site before making decisions about logging. Efforts are made to mimic the dynamics of natural forest stands. All of this contrasts with the large-scale forestry practices in vogue during the 1960s to 1980s, when homogeneous operations created difficult conditions for many species (Johansson, 1995).

CASE STUDY: FODDER SYSTEMS AT AYUBIA, PAKISTAN

Ayubia National Park (3312ha in extent and 2300–3000m in altitude) in the Himalayan foothills carries one of the best remaining examples of temperate forest in Pakistan. The park attracts many summer visitors, drawn by the cool climate and a vision of pristine nature – forest, streams and wildlife. The forest is dominated by conifers, especially blue pine *Pinus wallichiana* and fir *Abies pindrow*. Broadleaved trees occur, but are rather uncommon, probably having been reduced over many years by their collection for fodder and fuelwood, and inhibited in their regeneration by the browsing of domestic stock. They include *barungi* (*Quercus dilatata*), *bankhor* (*Aesculus indica*), *kain* (*Ulmus wallichiana*) and *kala kath* (*Prunus padus*). The park is an important site for Kalij and Koklass pheasant and also carries leopard. It has a major catchment function, providing water directly to nearby villages and the town of Murree, and contributing to river flows in the Indus. Higher altitudes are snow covered in winter.

There are about 42,000 people in 6000 households living close to the park and depending upon its resources, principally for fodder and fuelwood. Many have small farms (0.25 to 0.5ha) and keep a handful of animals – buffalo, cattle, goats and horses. All of these animals are stall fed in winter, with some of the oxen and non-milking buffalos being released into the park in summer for free-range grazing. Apart from fodder and fuelwood, the people collect leafy vegetables such as *kunji* (*Dryopteris stewartii*) from the park for their own consumption, as well as morel mushrooms *Morchella esculenta* (two varieties: *kali* and *surkh guchi*) to sell.

The national park is a Reserved Forest, as are some areas of forest outside the park. Both the Forest and Wildlife Departments are involved in park administration, with some uncertainties about the relative boundaries of their

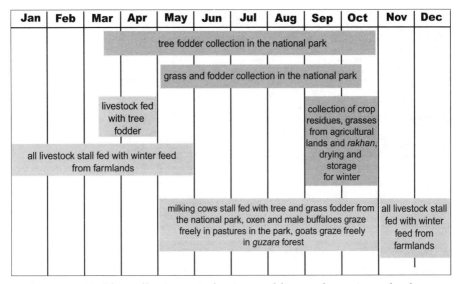

Figure 6.5 Fodder collection, production and livestock-rearing calendar at
Ayubia National Park and adjoining villages, Pakistan

Source: Aumeeruddy-Thomas et al, 2004

authority. Reserved Forest is a legal category of forest in Pakistan in which *all*
resource rights rest with the government. Villagers have no rights to use any
resources. Some areas of land outside the park are gazetted as *Guzara* Forest,
a legal category of forest reserves that are privately owned but in which permission to cut timber trees must be obtained from the Forest Department. Apart
from timber, the plant resources in *Guzara* Forests can be used freely by
villagers. There are disputes about where the boundaries of some of the
reserved and *Guzara* Forests precisely lie.

The annual calendar of fodder collection is shown in Figure 6.5. Groups
of women visit the park on a daily basis from spring to autumn to collect
fodder, which is delivered fresh to the stall-fed animals. Tree fodder collection
starts in mid March (when little grass fodder is available), with broadleaved
trees being favoured and *barmi* (*Taxus wallichiana*) also considered a useful
fodder tree. Herbaceous fodder predominates from late May to late October
and is harvested with small sickles in forest glades. The women selectively
harvest grasses (rather than herbs), looking for species such as *gha* (*Agrostis
nervosa*) and *jodra* (*Poa pratensis*). Some species are favoured for special
purposes – for example, to feed milking cows or calves. This careful selection
has probably increased the abundance of certain avoided herbs in the glades,
among them *choti rutunjot* (*Geranium wallichianum*), *masloon* (*Polygonum
amplexicaule*), *panjakha* (*Fragaria nubicola*) and *swar ohul* (*Viola canescens*).
From late autumn into spring, fodder is no longer available in the park and
fodder for stall-feeding comes from sites close to the homes, being harvested in
late September and October and stored for winter use. These sources of winter

Piles of fodder and fuelwood in a village area outside the park, the original source of these materials

Terrace fields separated by banks (*banna*) used for growing 'wild' grass fodder

Maize stems stored for winter fodder, with Asma Jabeen (fodder researcher)

Figure 6.6 *Different aspects of the fodder system at Ayubia National Park, Pakistan*

Source: Alan Hamilton

fodder include two types of managed uncultivated land: *banna*, comprising the strips between the terraced fields, and *rakhan*, which are sloping or rocky areas sometimes under an open cover of pine. *Rakhan* is managed through occasional burning to discourage herbs and shrubs. *Banna* can carry pure stands of *maniara* grass *Alopecurus*, while *dugri* (*Digitaria*) and *lundar* (*Apluda mutica*) are common in *rakhan*. Winter fodder is also available from crop residues (maize and wheat) (see Figure 6.6). Field weeds, as well as grasses and bushes in unmanaged areas near houses, provide limited free-range autumn grazing.

The collection of fodder and the grazing of domestic animals in the national park are illegal, resulting in serious conflicts between park staff and local women. From the biodiversity perspective, the collection of *tree* fodder in the park would seem to be clearly undesirable, because this inhibits the regeneration and growth of broadleaved trees, reducing biological diversity, and contributing to the declining cover of forest in the park and its surroundings. On the other hand, the collection of *herbaceous* fodder is a more ambiguous matter and a conservation case can be made for allowing it to continue, provided it is regularized. The diversity of herbaceous species in the forest glades is probably increased by the collection of herbaceous fodder. At present, the authorities cannot stop fodder collection and there are no channels for proper negotiations between the park officials and the collectors. If fodder collectors were brought into the management system, then it should be possible to make agreements that serve the dual interests of biodiversity conservation and local livelihoods. One of the obstacles is a lack of experience of joint forest management in Pakistan, with fears on the part of the Forest Department that it will lose more control over their reserves if 'concessions' are made to local people.

There are substantial areas of *Guzara* Forest and 'wasteland' outside the park that, apart from some timber production, currently serve little productive purpose. These are open-access areas with free-range goat grazing and competitive collection of non-timber plant products. However, these lands could be developed to provide fodder and fuelwood, provided that effective management can be introduced, the main obstacle to which is unsatisfactory tenure. Meanwhile, some steps to reduce pressure on the park and improve livelihoods are possible through the promotion of on-farm production of fodder and fuelwood and the encouragement of greater efficiency in the use of resources. A project of WWF-Pakistan, as part of the People and Plants programme, has assisted local residents in establishing 203 nurseries of trees for fuelwood and fodder, and provided subsidies for the purchase of fuel-efficient stoves by 480 households (figures as of 2004).

This example demonstrates the interconnectedness of 'wild', 'semi-wild' and agricultural habitats as sources of plant resources. It shows the value of gaining a holistic understanding of natural and social systems in identifying and seeking answers to conservation problems that are linked to livelihoods.

Sources for case study: Jabeen, 1999; Aumeeruddy-Thomas et al, 2004

7

Meanings, Values and Uses of Plants

This chapter explores some of the values that people place on plants and some of the main categories of use. An appreciation of the meanings of plants to people forms an essential foundation for community-based plant conservation or, indeed, for work with any social group.

SYMBOLISM OF PLANTS AND NATURE

Plants are not experienced by people merely as 'things'. Rather, they are mental concepts carrying meanings derived from personal experience, strongly influenced by culture. Plants evoke memories and feelings. All sorts of properties are ascribed to plants by people, who might see them, for instance, as belonging to certain individuals, having financial value or being beautiful.

Some understanding of how people see plants at the deepest levels of their minds can be useful for providing insights into how they see the world to be fundamentally constructed, including how people should behave. Studies of the roles of plants in rituals, ceremonies and myths can provide some comprehension of these inner meanings. Many traditional myths have a moral that describes the malign consequences of acting in socially unacceptable ways (Martin, 1995; Nakashima, 1998).

Exploring how people comprehend 'wild' in relation to plants and nature can be revealing in terms of how they fundamentally understand the nature of the world. Farming communities living in clearings in the Congo forest understand the forest as a wild and threatening place, contrasting with the secure zone of the village and its fields. In contrast, their 'pygmy' neighbours, such as the Apagibeti, regard the forest not just as a resource, but as part of their identity. They ritually protect the forest against threats emanating from the village (Almquist, 2001). Societies may view the elimination of 'wild' elements in the landscape as symbolic of progress. One of the reasons why scientists have advocated the suppression of Aboriginal burning practices in Australia was a fear that the fires were 'out of control', an advocacy today commonly considered to have been unwise (Langton, 2000). The most industrious citizens during the colonial period in North America were regarded as those who felled the most trees; they were praised for their contributions to 'improvement' as they eliminated vast tracks of forest in the Appalachians (Williams, 1997). Governments in Latin America have tended to regard forests as idle spaces requiring 'development', encouraging the felling of these 'untamed wilder- nesses' to create 'civilized' fields and plantations interlaced with roads (Budowski, 1988).

Wild plants can be seen as having special worth or, alternatively, as symbols of backwardness. Wild American ginseng *Panax quinquefolius* fetches ten times the price of cultivated material in East Asian markets and, conse- quently, it remains threatened in the wild even though it can be profitably farmed (Sheldon et al, 1997). In this case, the wild plants may well be physio- logically more efficacious, as is probably also the case with wild greens in Turkey, which are preferred over cultivated greens because they are regarded as more tasty and nutritious (Ertug, 2003). However, no similar functional justification can be claimed for the preference of some orchid lovers for wild- collected over cultivated orchids, valuing them for their rugged 'interesting' looks (IUCN and SSC, 1996). On the other side of the coin, there is a common belief in Eastern Africa that the use of wild foods is associated with poverty and low social status (Marshall, 1998; Asfaw and Tadesse, 2001). The Maasai do not hunt wild animals for food since they regard this as a degrading activ- ity best left to the poor and hunter-gatherers (Maundu et al, 2001).

Many plants are emblems of national or local identify. The maple *Acer* is a symbol of Canada, the cedar *Cedrus libani* of Lebanon and the daffodil *Narcissus pseudonarcissus* of Wales. Immigrants to Australia from Britain planted oak *Quercus robur* to remind them of their 'motherland'. The Gurage people, with a traditional territory to the south of Addis Ababa in Ethiopia plant symbolic specimens of *ensete* (*Ensete ventricosum*) in their gardens when they settle in the city. This relative of the banana has many uses for the Gurage, including for production of various starchy foods, such as *kocho*, a type of bread.

Local or national pride can be a major motivating force for conservation. Van Opstal (2001) has noted that a considerable number of plant species which have been accorded statutory protection in Europe are actually *not* threatened; he believes that they have been given protection for political reasons. BirdLife employees have found that local officials in Indonesia, such as provincial governors and district officers, have responded with interest and pride when informed that their territories support unique species or assemblages of birds (Jepson, 2001). National pride has been a major force behind the creation of new national parks in Indonesia (Jepson, 2001). The same is true of the Democratic Republic of Congo (formerly Zaire), where former President Mobutu greatly enlarged the protected area network, stating the need to guard his country's patrimony. To the average British person, an appeal to save *British* species would carry much more weight than one to 'just' save species. The conservation potential offered by appeals to local pride is a reason for hope that the plants of isolated oceanic islands can be saved despite the fact that so many of these floras are highly threatened. Islanders often take pride in the unique features of their island homes.

The ways in which people have looked at nature in the UK and the USA have been powerful forces behind the nature conservation movement globally, thanks to the extensive cultural, economic and political influence of these countries (Schama, 1995; Hamilton, 2001). In 19th-century Britain, the stresses and dirt of the new industrial cities resulted in a craving for the perceived lost peace of the countryside. This led eventually to the founding of the first British national parks, basically conceived as areas of old-fashioned agricultural land in which 'contaminating' modern developments were prohibited. In the USA, the new settlers deeply appreciated their 'freedom' as they broke loose of the religious intolerance and stifling class structures of Europe. However, as more land became 'civilized', there developed a need to retain a mythological 'wilderness' beyond the limits of the frontier. The first national parks in the USA, such as Yosemite, were conceived as monuments to wild, unspoilt nature – lands of mystery and spiritual uplift. Conservation in Africa has been influenced by both the British and American attitudes to nature. Many of the first national parks in Africa started life during the colonial era as 'game' reserves, founded on a British association between the hunting of certain animals and the ruling class – with the 'big five' (lion, elephant, buffalo, leopard and rhinoceros) replacing deer. Subsequently, attitudes associated with the USA seem to have become more dominant within the conservation movement in Africa, with a greater emphasis on wilderness.

Plants are closely associated with magic, religion and the supernatural world. There is a widespread belief in Africa that certain plants have the ability to cure, nourish and protect; some are used by diviners to control events by supernatural means (Cunningham, 1997a). The Loita Maasai in Kenya use a variety of plants in ceremonies and rituals. *Olorien* (*Olea europaea*) is the most common plant used in ceremonies, believed to bring good luck. A ceremony for blessing women is conducted under *oreteti*, a type of fig tree *Ficus thonningii* (Maundu et al, 2001). The colours and shapes of plants can carry meanings. Seeds that are two-coloured – with combinations of red, white or black (for example, the red-and-black seeds of *Abrus precatorius* and *Afzelia quanzensis*) – are widely believed in Africa to symbolize links with the supernatural world (Cunningham, 1996a, 1997a). The colour green – associated with leaves and vegetation – is commonly used in modern marketing to signify naturalness, goodness and health. The 'doctrine of signatures', associated with Paracelsus (1493–1541) and flourishing in European medical literature during the 16th and 17th centuries, holds that the shapes of plants signify their medicinal uses, a symbolism benevolently provided by God for his earthly children. Similar beliefs are common around the world. In the British Isles, the mistletoe *Viscum album* is a symbol of magic and love, while the yew *Taxus baccata*, which is almost invariably planted in graveyards, is a tree rich in associated folklore. Holly *Ilex aquifolium* and the Christmas tree – usually Norway spruce *Picea abies* – are strongly associated with the Christian festival of Christmas. Wild olive *Olea ferruginea* is commonly found in Islamic graveyards in temperate north Pakistan. Verses of the *Holy Quran* written with ink made from saffron (from the anthers of *Crocus sativus*) are believed to carry special power. Marigolds *Calendula officinalis* are widely used in Hindu ceremonies, while the grass *Cynodon dactylon* is given to the Hindu god Ganesh for good fortune as an essential part of marriage ceremonies. Buddha received enlightenment under a *papal* tree *Ficus religiosa*.

Nature conservation needs to be underpinned by basic values that appeal to the deepest levels of the human mind and spirit. Materialistic arguments – based upon utility or financial income – are important, but not enough. Human economies change and so do valuations of plants, so it would surely be foolish to base conservation purely on current economic values. In more traditional societies, conservation of key natural resources is promoted not only through practices based on rational arguments (as others might under-stand rationality), but also through religious and spiritual beliefs (see Chapter 10). Appeals to the more basic forces that are believed to influence human affairs reduce the chances of losing these vital resources through transient events.

Conservationists need to make efforts to find how plant conservation can be promoted today at the deeper levels of the mind in ways that make sense according to modern thinking. Plantlife International has had some success within the context of the UK through appealing to a basic British interest in wildflowers found in bluebell woods (containing *Endymion non-scriptus*) or traditional hay meadows. As yet, no one has yet come up with an image for plant conservation that is so universally appealing as the panda is for animals. The most likely candidate in the experience of the author Alan Hamilton is the (generalized) 'medicinal plant', which seems to carry resonance across many cultures.

NEEDS, MOTIVATIONS AND VALUES

Plant conservationists will be concerned with the needs, motivations and values relating to plants of the people with whom they work (Jepson, 2001; Campbell and Luckert, 2002). In Sweden, there is sufficient public interest in rare plants to be able to mount a monitoring programme for their conservation, involving local people across the country. Naturalists in the UK are also interested in rare plants, as well as in certain specific plant groups, such as orchids, and this has been instrumental in establishing many nature reserves. On the other hand, local people living near Bwindi Forest, Uganda, are interested in those forest plants that provide them with medicines and basketry materials (see Chapter 9). The moral is that, wherever they are working, plant conservationists need to approach the involvement of local people in conservation schemes, starting with those aspects of the plant world in which the local people are interested.

Exercises in valuing plants can make people reflect upon what is really important to them and help them to understand other people's perspectives. Valuation exercises can give useful information for guiding government policies. Although rural people in developing countries typically have many subsistence uses for wild plants and may also gather them for sale, these values are often under-appreciated by governments. The problem is that few of these plant resources or the goods derived from them enter formal markets; hence, they are under-recorded in official statistics (Shackleton et al, 2002). The International Institute for Environment and Development (IIED) has coined the phrase 'the hidden harvest' to draw attention to this problem (Guijt et al, 1995; Guijt, 1998; Shackleton et al, 2002).

The People and Plants Initiative has produced a book which describes a number of approaches to estimating the economic value of plant resources for rural households. Among the methods described are market research tools, non-market valuations and decision-making frameworks, such as cost–benefit analyses (Campbell and Luckert, 2002).

Various methods have been proposed for scoring or ranking the values that people attach to plants (see Chapter 12). A framework for listing and discussing values relating to nature conservation has been developed by Jepson, making using of the concept of the motivating value (see Box 7.1) (Jepson, 2001). A motivating value is a value that induces a tendency towards directed action. Jepson divides motivating values into two categories, intrinsic and extrinsic, the former based on an image of nature as having value in its own right (independent of human ends) and the latter concerned with material benefits. Jepson claims that the framework has universal validity.

Valuation exercises can throw light on whether (or not) values are shared between different individuals or groups. Such information can be useful when considering how resources should be managed. Plants can be valued differently by different people. Weeds in one culture are edible greens in another (Diáz-Betancourt et al, 1999). In Yunnan, China, different ethnic groups (minorities) place different values on tea *Camellia sinensis*, judging by the results of a valuation study (see Table 7.1). A valuation exercise in the East Usambara Mountains, Tanzania, revealed that there was much common ground concerning attitudes to forests on the parts of three major stakeholders – the Forest Department, tea estates and local communities – a measure of agreement that had not previously been properly recognized (Kessy, 1998). The exercise revealed that all three groups valued the forests, but for different reasons; crucially, none was in favour of forest depletion. The Forest Department valued the forests for their biodiversity, catchment values and (historically) as sources of timber. The tea estates saw the forests as useful for inducing a favourable climate for the growth of tea and as sources of fuelwood for drying tea in their factories. The local communities valued the forests for a wide range of material resources and their roles in inducing rainfall and maintaining stream flows, and used sacred groves for traditional rain-making ceremonies. At the time of the research, the approach of the Forest Department to forest management was said to be 'conventional' (rather authoritarian and distant from the people), reportedly tending to alienate the communities. John Kessy (1998), the researcher, considered that the central challenge for the Forest Department was to find ways to build cooperative alliances with the other two groups, taking account of their various values.

BOX 7.1 A FRAMEWORK FOR LISTING AND DISCUSSING VALUES, DESIGNED AS A FLEXIBLE TOOL FOR CONSERVATION PLANNING AT THE NATIONAL LEVEL

| | **Motivating values for conservation** | | | | | |
| | Intrinsic values | | | Extrinsic values | | |
	NH	WP	SC	GR	EH	SJ
Links with global biodiversity initiatives						
Links with general public sentiment						
Opportunities for successful conservation action						
Possible action responses						

Entries should be made in the boxes in the table according to information received from those consulted. Jepson provides these explanations for the motivations lying behind the various values:

- *NH = Natural history*: aesthetic and intellectual satisfaction which contemplation of nature brings. Monuments of nature are seen as helping to define cultural and natural identity.
- *WP = Wilderness preservation*: an ability to place human identity in the context of an 'other' (for example, non-human world or era). Contact with wild nature is seen as offering opportunities for personal development and an antidote to the stresses and alienation of modern life.
- *SC = Species compassion*: an assertion of humanity's special identity by extending compassion and respect to non-human life forms.
- *GR = Genetic reservoir*: recognition of the potential to derive useful compounds from the Earth's genetic library and ability to adjust to environmental change. It is seen as prudent to preserve biological diversity because species may be found to have previously unknown value in the future.
- *EH = Ecosystem health*: the desire to maintain healthy ecosystems to safeguard economic growth, quality livelihoods and social stability.
- *SJ = Social justice*: a desire to protect traditional peoples with nature-based cultures from exploitation and unmanageable change.

Source: Jepson, 2001

Table 7.1 *Comparison of traditional use of tea* Camellia sinensis *among five ethnic minorities in Yunnan Province, China*

Minority group	Social relations	Religious	Medicine	Financial income
Bulang	+++	++	++	+++
Wa	+	+++	+	–
Jinuo	+++	++	+	+
Hani	++	++	++	+
Dai	+++	–	++	–

Notes: +++ mentioned by over two-thirds of interviewees;
++ mentioned by one third to two-thirds of interviewees;
+ mentioned by less than one third of interviewees;
– not mentioned by interviewees.
Source: Chen Jin, personal communication, 2004

Numerous studies have revealed that people in many cultures value the intrinsic aspects of nature highly (Clarke et al, 1996; Aumeeruddy-Thomas et al, 1999). Three of the five top values placed on species of tree at two sites in Zimbabwe were non-material (see Table 7.2). Evidently, several types of value figure significantly in people's minds. Arthur Tansley was wise to give special stress to the non-economic values of nature when he appealed for a systematic programme of nature conservation in Britain. He argued that preservation of the countryside touches deep needs within the British for mental and spiritual refreshment (Tansley, 1946). Paul Jepson (2001) has criticized a tendency among international conservation non-governmental organizations (NGOs), operating in developing countries, to increasingly underplay the intrinsic values placed on nature. He writes that these NGOs have tended to emphasize economic arguments for conservation, downplaying intrinsic reasons, on the basis that these are imperialistic, sentimental, unprofessional and unscientific. Some of these NGOs receive significant funding from development agencies and this, Jepson believes, has tended to bias their public attitudes. Jepson has produced evidence showing that people involved in conservation in Indonesia (as an example of a developing country) are just as much motivated by intrinsic arguments for conservation as are their counterparts in Western countries. He suggests that the implementation of conservation strategies may flounder when notions of 'moral restraint' toward the environment are underemphasized (Jepson, 2001; Jepson and Canney, 2001).

Aside from asking their opinions, another way of gaining insight into the values that people place on plants is from studying how they behave towards them. For example, the types of crop grown can provide some idea of farmers' basic values. Studies have revealed that food security is a major driving force lying behind the choice of crops by subsistence farmers around the world (Teshome et al, 1999; Tuxill and Nabhan, 2001; Hamlin and Salick, 2003). A desire to minimize risk is a major reason why such farmers often grow many types and varieties of crops, rather than, for instance, trying to maximize crop

Table 7.2 *Values of woodland resources, identified by men, women and boys of Jinga and Matendeudze, Zimbabwe*

Order of importance	Value	Example of species valued for this purpose	Type of value
1	Water retention	*Ficus capensis*	Environmental
2	Rain-making ceremonies	*Ficus capensis*	Cultural
3	Poles	*Julbernardia globiflora*	Direct
4	Inheritance	*Combretum molle*	Cultural
5	Aesthetics	*Afzelia quanzensis*	Cultural
6	Prevention of soil erosion	*Combretum molle*	Environmental
7	Grazing	*Sclerocarya birrea*	Direct
8	Firewood	*Colophospermum mopane*	Direct
9	Wild fruits	*Adansonia digitata*	Direct
10	Camouflage/cover	*Afzelia quanzensis*	Cultural
11	Fibre	*Adansonia digitata*	Direct
12	Windbreaks	*Ficus* sp	Environmental
13	Shade	*Kirkia acuminata*	Environmental
14	Sacred places	*Ficus* sp	Cultural
15	Crafts	*Kirkia acuminata*	Direct
16	Medicines	*Adansonia digitata*	Direct
17	Fencing	*Combretum molle*	Direct
18	Seasonality indicator	*Afzelia quanzensis*	Environmental

Note: Direct = used for subsistence or sold.
Source: Clarke et al, 1996

production by concentrating on only a few high-yielding varieties. Modern crop varieties may give higher yields than traditional varieties in some years; but if they are extensively planted and harvests fail, the result could be catastrophic. Crop scientists were wrong when they predicted 30 years ago that traditional crop varieties would have almost completely disappeared by the end of the 20th century (Hawkes et al, 2001), though they have done so over extensive areas (see Chapter 1).

Hawaii is an unusual case in that all local varieties of kava *Piper methysticum*, sugarcane *Saccharum officinarum* and taro *Colocasia esculenta* are very similar to one another ecologically and genetically, so it seems unlikely that they have been selected for utilitarian reasons. Instead, they may have been selected largely for cultural reasons, including cosmological applications, aesthetic values and an interest in maintaining variability for its own sake (Meilleur, 1998).

Access to a diversity of wild plant resources can cushion poor people against external shocks and trends (Maheshwari, 1995; Nakashima, 1998; Dounias et al, 2000). Wild plants can provide many products; these products can be available at short notice and without financial charge; and the resources can be exploited without high levels of formal skills – all characteristics that can be a good match for the economic needs and social structures of poorer rural households (Campbell and Luckert, 2002). Wild plants can provide emergency foods at times of famine, as do the starchy trunks of the sago palm *Metroxylon* in Borneo, and the fruits of the trees *Maerua triphylla* in Tanzania and *Salvadora oleoides* in the Punjab.

Attempts to place monetary values on plant diversity need to be treated with caution. At one extreme, there are certain aspects of plant diversity that are so valuable that they are essentially beyond value – for example, a species such as maize *Zea mays* or a genus such as pine *Pinus*. If an average value for a plant species is considered a statistic worth calculating, then this would certainly be many hundreds of thousands of dollars annually. For the record:

- The average financial benefit from a new drug derived from plants has been calculated at approximately US$449 million (Mendelsohn and Balick, 1995). The total annual sales value of drugs (for example, taxol) derived from just one plant type – yew *Taxus* – was US$2.3 billion in 2000 (Laird and ten Kate, 2002). Probably the most expensive herbal medicine in the world is a preparation (*feng dou*) made in China from the orchid *he-jei-cao* (*Dendrobium candidum*); weight for weight, it is more valuable than gold. The preparation is used traditionally by singers of Chinese opera and more recently by officials of the governing party to restore their throats. The orchid is now extinct in China.

- The total global market value for 'genetic resource products' (principally based upon plants) was estimated in 1999 to be US$500 billion to $800 billion annually, with various sectors contributing as follows: major crops (US$300 billion to more than $450 billion); pharmaceuticals (US$75 billion to $150 billion); biotechnology (US$60 billion to $120 billion); botanical medicine (US$20 billion to $40 billion); horticulture (US$16 billion to $19 billion); cosmetics and personal care products (US$2 billion to $8 billion); and crop protection (US$0.6 billion to $3 billion) (ten Kate and Laird, 1999).

- The world market for herbal remedies in 1999 was calculated to be worth US$19.4 billion, with Europe in the lead (US$6.7 billion), followed by Asia (US$5.1 billion), North America (US$4 billion), Japan (US$2.2 billion) and finally the rest of the world (US$1.4 billion) (Laird and Pierce, 2002).

- Some studies have placed very high values on tropical forests as sources of non-timber forest products (NTFPs). It has been calculated that the harvest of NTFPs will provide a net revenue of US$341 per hectare per year in the case of one forest in Peru and US$564 per hectare per year for another in Belize (Peters et al, 1989; Sheldon et al, 1997). In the former case, there have been criticisms of the methodology, so the actual revenue may be lower (see, for example, Balick and Mendelsohn, 1992).

MATERIAL USES OF PLANTS

There are several ways in which useful plants or their products can be classified when grouped for purposes of analysis or discussion. Classifications are

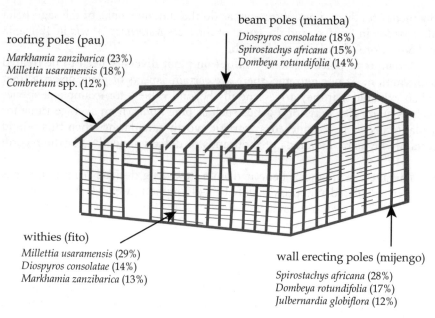

roofing poles (pau)
Markhamia zanzibarica (23%)
Millettia usaramensis (18%)
Combretum spp. (12%)

beam poles (miamba)
Diospyros consolatae (18%)
Spirostachys africana (15%)
Dombeya rotundifolia (14%)

withies (fito)
Millettia usaramensis (29%)
Diospyros consolatae (14%)
Markhamia zanzibarica (13%)

wall erecting poles (mijengo)
Spirostachys africana (28%)
Dombeya rotundifolia (17%)
Julbernardia globiflora (12%)

Figure 7.1 *Architectural design of a house near Kitalunghalo Forest Reserve, Tanzania*

Note: The types of wood used for different parts of the construction are indicated, with their frequencies of use.
Source: adapted from Luoga et al, 2000

sometimes made according to scientific or common names (for example, *Swietenia macrophylla*, or broadleaved mahogany), plant forms (such as climbers), parts used (such as fruits), product types (for example, sawn wood), end uses (such as wine) or management characters (for example, easy to collect, sold by children or gathered for home use) (Wong et al, 2001). Some principal types of use and plant-based products are listed in Box 1.3, and some notes follow later on a few of the categories of use. Figure 7.1 shows the species of tree used to construct different parts of houses at one locality in Tanzania. This is included here to illustrate how analyses can be made of species used for different purposes.

A distinction is sometimes made in studies of plant use between products made from wood and those made from non-woody materials. The latter can be quantitatively significant. A survey in tropical Mexico revealed that, out of a total of 579 types of products produced from plants, 82 per cent were made from non-woody materials and only 18 per cent from wood (Toledo, 1995). Two terms in common use for categories of products made from uncultivated plants are non-timber forest products (NTFPs) and non-wood forest products (NWFPs), excluding products made from timber and wood, respectively.

There are cultural differences in the ways that people classify the uses of plants. Western societies tend to make a clearer distinction between food and medicine than is typical of other societies. One study of four villages in Vietnam found that almost one third of the more than 40 types of wild plants used as vegetables were believed locally to have curative properties (Ogle et al, 2003). In conservation projects with a special focus on health, it is helpful to include medical doctors in project teams so that the equivalence of local and scientific ways of classifying human conditions and ailments can be assessed.

Some categories of plants or plant uses tend to be ignored by officials and conservationists. For example, woody climbers are sometimes common in tropical forests, but generally receive little attention from foresters. Only rattans (certain types of climbing palms) are often given special note – because the rattan industry is a major business, especially in Asia, where it is worth US$7 billion to $8 billion per year and employs 2 million people (Riley, 1997). Woody climbers have sometimes been systematically destroyed by foresters in efforts to liberate the growth of small timber trees, perhaps unaware of their value to local people. A study in Budongo Forest, Uganda, found that local people used 63 (44 per cent) of the 142 species of climbers in the forest. The purposes (in order of decreasing frequency) included medicine, home construction and food. Social differences in resource use emerged – for example, with craft-making using materials from climbers being mostly an occupation of relatively wealthy people – possibly, it was speculated, one of the reasons for their relative wealth (Eilu, 1999; Eilu and Bukenya-Ziraba, 2004). The climber

Figure 7.2 *Granary at a house on the margin of Bwindi Impenetrable National Park, Uganda, with researcher Onesimus Muhwezi*

Note: This basket is made of three plants, including *Loesneriella apocynoides* (a climber) and *Dodonaea viscosa* (a shrub).
Source: Alan Hamilton

omujega (*Loesneriella apocynoides*) is favoured for weaving granaries, tea baskets and stretchers in certain parishes around Bwindi Forest, Uganda, partly because of its long use life (see Figure 7.2). In the parish of Mpungu, 64 per cent of the granaries and 83 per cent of the tea baskets are made from this species (Cunningham, 1996b). Climbers are not only important in Asia and Africa. A survey of 20 craft shops in the Dominican Republic showed six species of climbers to be used for making baskets, the most important being *Hippocratea volubilis* (amount used: 217 tonnes per annum), *Smilax* (16 tonnes per annum) and *Trichostigma octandrum* (4 tonnes per annum) (Peguero et al, 2000).

A very large number of plant species are used for material purposes. Plants dominate over animals in nearly all societies in terms of the number of species used and in quantities. A study of 13 indigenous groups in the Mexican tropics found 82 per cent of products to be made with materials from plants, compared with 13 per cent from animals (Toledo, 1995). A word of caution is needed in interpreting statistics on plant use. In summary statistics, *all* species can be treated as equivalent; but there is a world of difference between a species that is regularly used by many people and one that has been tried once by one person. Also, a species that has *no* substitutes for a particular use is very different from one that can be readily replaced by others.

People frequently know uses for a large proportion of the species of trees that grow in their neighbourhoods. One study in miombo woodland in Tanzania found that local people had uses for 69 per cent of the woody species present (Luoga et al, 2000). A survey of four groups of forest-related people in Amazonia revealed that, on average, 66 per cent of the tree species present were used for one or more purposes, a figure that excludes the general categories of use of firewood and game attractants (Prance, 1995). Some examples of the numbers of species of plants used by people for specific purposes are given in Box 7.2.

Worldwide, many more wild than domesticated or cultivated species are used for material purposes. Fewer than 3000 plant species have been domesticated for food (Hawkes et al, 2001). A study of Kizilkaya, a village in central Anatolia, Turkey, revealed that local people gathered 100 species of wild plants for food (main types: 42 greens, 28 fruits), but grew 'only' about 50 varieties of crops (Ertug, 2003). Ninety-five per cent of rattan is wild-harvested (Riley, 1997). Very few species of medicinal plants are cultivated. Only 100 to 250 of the 10,000 species of medicinal plants used in China are cultivated and it is reported that more than 80 per cent of the 700,000 tonnes of medicinal plants used annually in China come from wild sources (Heywood, 2000). Only 130 to 140 of the 1200 to 1300 species of medicinal plants that are both traded in, and native to, Europe are derived predominantly from cultivation (Lange, 1998a). There is little or no cultivation of medicinal plants in large parts of the world – for example, Albania, Bangladesh, Pakistan and all of Africa (Marshall, 1998; Dold and Cocks, 2001; Schippmann et al, 2002; Hamilton, 2004).

The biggest use of wild plants globally, in terms of volume and time spent in collection, is as **fuel** (FAO, 1992; Dounias et al, 2000). Most of this fuel is

Box 7.2 Numbers of plant species used for various purposes

- 203 species of wild plants are eaten as foods in Ethiopia, representing 3 per cent of the flora (Asfaw and Tadesse, 2001).
- About 200 non-cultivated plant species, belonging to 139 genera and 62 families, are eaten in Uganda (Bukenya-Ziraba, 1999). This represents 4 per cent of the flora.
- 354 species of wild plants are eaten in southern Ecuador, representing 6 per cent of the flora (Van den Eynden et al, 2003).
- 800 to 1000 species are used as leafy vegetables in sub-Saharan Africa (Maundu et al, 2002). This represents around 3 per cent of the flora.
- 88 plant species are used to produce dyes in Ghana (Dounias et al, 2000).
- 123 plant species are used to produce dyes in Turkey (Dogan et al, 2003).
- Between 7 and 44 per cent of the floras of various parts of the world have been used medicinally (see Table 7.3).
- 299 (24 per cent) out of 1240 species of orchids found in China are used medicinally (Yan Zhi-Jian, 2004).
- Over 15,000 types of plants (some are varieties, not full species) are described in a major encyclopedia of ornamental garden plants (Brickell, 1999).

used for domestic cooking and heating, with other uses significant regionally – for instance, commercial brick-making, an activity that has grown greatly in scale in Africa over recent decades. The consumption of fuelwood in rural households averages around 1 cubic metre per person per annum, varying according to wood availability and climate (Kessy, 1998; Shackleton et al, 2002). Wood is often converted to **charcoal** before transportation for sale in urban markets. Seventy to eighty per cent of households in Dar es Salaam and Kampala use charcoal as fuel (Brigham et al, 1996; Cunningham, 1997a). The charcoal business is a significant source of employment for young men in Africa, with major impacts upon natural vegetation extending for distances of tens to hundreds of kilometres around some cities.

The collection of **fuelwood** for domestic use in developing countries is largely an activity of women, sometimes consuming large amounts of time and effort. Women of the Loita Maasai, Kenya, journey for fuelwood for distances of up to 8km from their homes, making three to four trips a week and carrying as much as 40 to 60kg per load (Maundu et al, 2001). The whole process of gathering and travelling takes up to two-thirds of their daylight hours. These women are selective by species, their most favoured being *olorien* (*Olea europaea*), valued because it burns brightly (helping to illuminate the homes), is sweet scented, produces relatively little smoke and burns even when fresh. Tradition decrees that women who use *olorien* win the endearment and respect of their husbands, a belief that is held so strongly that women will journey far for its collection, even when other species are available close by. The Loita

Maasai never use wood of certain species, such as *ololiondoi* (*Olea capensis*), probably their most sacred plant, and *osokonoi* (*Warburbia salutaris*), their most valued plant medicine. Species selection has also been noted elsewhere. Women living near the Kitalunghalo Forest Reserve in Tanzania most commonly collect *Combretum* and *Julbernardia globiflora*, but avoid *Bridelia cathartica* (because it produces a pungent smell when burned) and *Bridelia salicifolia* and *Dalbergia melanoxylon* (which are culturally taboo) (Luoga et al, 2000).

Plants are normally the major component of the human **diet**, with many types and parts of plants eaten. Wild plants are sometimes consumed as part of the regular diet (see Figure 7.3), but can be taken as luxuries or serve as emergency foods. There are over 300 species of wild plants with edible **fruits** in Cameroon (Dounias et al, 2000). Wild or semi-domesticated **leafy greens** are eaten on a regular basis in Africa, providing valuable sources of fibre, vitamins and minerals. The key plant families are the Amaranthaceae, Asteraceae, Brassicaceae, Cucurbitaceae and Fabaceae (Cunningham, 1997a; Kessy, 1998; Maundu et al, 1999; Marshall, 2001). The **underground storage organs** of over 100 species of wild plants are used for food in West and central Africa, most requiring detoxification before eating; wild yams *Dioscorea* are by far the most important (Dounias et al, 2000). **Fungi** are significant items in the diet in many parts of the world. For example, miombo woodland (found in South-central Africa) has an abundant and diverse fungal flora, with a mushroom season that extends from the start of the rains in November–December up to March–April. Journeys to the woodlands to collect mushrooms can be second in frequency only to those for collecting firewood (Clarke et al, 1996).

The use of plants in **medicine** represents by far the largest use of the natural world in terms of the number of species specifically targeted (Hamilton, 2004). Plants are used within the context of many medical systems, which can be classified into folk systems (transmitted by oral means) and scholarly systems of medicine (associated with literate societies, written pharmacopoeias and formal training – see Box 4.1). No one knows the exact number of medicinal plant species used globally; but a figure of about 50,000 species may not be too inaccurate, representing nearly 20 per cent of the world's vascular plant flora (see Table 7.3). The great majority of species of medicinal plants are used only in folk medicine, scholarly medicine employing relatively few: 500 to 600 commonly in traditional Chinese medicine (Pei Shengji, 2001); 1430 in Mongolian medicine (Pei Shengji, 2002); 1106 to 3600 in Tibetan medicine (Pei Shengji, 2001, 2002); 1250 to 1400 in Ayurveda (Dev, 1999); 342 in Unani (Shiva, 1996); and 328 in Siddha (Shiva, 1996). Some herbal remedies are widely known within communities and administered in home treatments, while others are prescribed only by specialist healers. Home treatments typically use single species or mixtures of only a few species, with more complex formulations of many species being associated with specialist healers (Lebbie and Guries, 1995; Dev, 1999; Cano and Volpato, 2004).

Many **pharmaceutical drugs** used in Western medicine are made from plants, or the properties of plants have inspired their development (Sheldon et

Figure 7.3 *Vegetables in preparation for dinner in Long Pasia, Sabah, Malaysia; these villagers eat a wide range of cultivated, semi-wild and wild leafy greens*

Source: Alan Hamilton

al, 1997). Among the early isolates were morphine from *Papaver somniferum* (1803), strychnine from *Strychnos nux-vomica* (1818), quinine from *Cinchona* (1821) and codeine from *Papaver somniferum* (1837). In the USA, 25 per cent of all prescriptions dispensed from community pharmacies between 1959 and 1980 contained plant extracts or active principles prepared from higher plants (Farnsworth et al, 1985). A study of the top 150 proprietary drugs used in the USA during 1993 found that 57 per cent of all prescriptions contained at least one major active compound that was either derived from or patterned after compounds from biological diversity (Grifo and Rosenthal, 1997). In some cases, chemicals extracted from plants are used directly as medicines – for example, reserpine (which lowers blood pressure), extracted from serpent-root *Rauvolfia serpentina*; ephedrin (a decongestant) from the shrub *Ephedra*; digitalin (used to treat irregularities in heart rhythm) from the foxglove

Table 7.3 *Numbers and percentages of medicinal plant species recorded for different countries and regions*

Country or region	Number of plant species		Percentage of flora that is medicinal	Source
	Total	Medicinal		
China	24,300 (angiosperms)	10,027 (angiosperms)	41	Xiao Pei-Gen and Peng Yong, 1998
India	17,000	7500	44	Shiva, 1996
Mexico	30,000	2237	7	Toledo, 1995
North America*	20,000	2572	13	Moerman, 1998
Zimbabwe	5000	500	10	Mavi, 1994
Tanzania (woodland site)	133 (trees only)	35 (trees only)	26 (percentage of tree flora)	Luoga et al, 2000
World	270,000	35,000–70,000	13–26	Farnsworth and Soejarto, 1991
World	270,000	52,885	20	Schippmann et al, 2002

Note: Figures are for vascular plants, unless stated.
* Native Americans and First Nations peoples.
Sources: the sizes of the floras (column 3) are from WWF and IUCN (1994–1997), except for the world figure (Baillie et al, 2004)

Digitalis; and vinblastine and vincristine (for treating childhood leukaemia and Hodgkin's disease) from the rosy periwinkle *Catharanthus roseus*. In other cases, chemicals are extracted from plants and used as chemical building blocks to create new compounds, which are then used as drugs. The drugs etoposide and teniposide (used for treating skin cancers and warts) are manufactured from epipodophyllotoxin, a chemical found in May apple *Podophyllum*. Progesterone (used as an oral contraceptive) is manufactured from diosgenin, found in certain species of yam *Dioscorea*. Some drugs, including a number that are synthesized today from inorganic sources, were inspired in their historical development by the physiological activities of chemicals found in plants. Aspirin, which is today synthesized inorganically, was inspired in its development by folk uses of the bark of willow *Salix* to treat fever. Although drugs extracted from or inspired by plants are used very extensively, the actual *number* of plant species that has contributed to Western medicine is rather few – only 95 vascular plant species, as reported in 1991 (Farnsworth and Soejarto, 1991). Apart from plants *sensu stricto*, fungi play major roles in Western medicine, notably through the use of genera such as *Penicillium* and *Streptomyces* as sources of antibiotics (developed during the 1940s).

Various plants yield **narcotics**, including coffee *Coffea*, tobacco *Nicotiana tabacum*, marijuana *Cannabis sativa*, opium *Papaver somniferum*, cocaine

Erythroxylum coca and cola *Cola*. The social context varies. Traditionally, the use of narcotic plants could be confined to shamans or rulers. Also traditionally, hallucinogenic substances are used more widely in Asia and the Americas than in Africa, where a principal substance used for this purpose is the bark of the shrub *Tabernanthe iboga* (Dounias et al, 2000). The root or stem bark of *olkiloriti* (*Acacia nilotica*) is taken as a decoction or in soup by the Loita Maasai before embarking on raids. It makes them energetic, aggressive and fearless (Maundu et al, 2001). The hallucinogen *ayahuasca* is prepared in the Amazon usually from the stem bark of the liana *Banisteriopsis*, which is boiled for several hours to produce a thick liquid. Its psychoactive effects are enhanced dramatically by the addition of a number of subsidiary plants, usually the leaves of the two shrubs *Psychotria viridis* and *P. carthaginensis* and a scandent liana *Diplopterys cabrerana*:

> *The initiation occurs usually through the mastery of the use of tobacco and* ayahuasca. *The personal disposition of the individual and his ability to stand the hard training and the dangers involved in shamanic induction will determine the degree of his development. He may continue to add new additives to the basic* ayahuasca *brew or consume other plant teachers to increase his knowledge and abilities. Each plant taken means entering a new dimension, where the initiate encounters beings who give him new powers to manipulate the environment, often through magic melodies or* icaros *and incantations. Each plant has its cluster of zoomorphic and anthropomorphic spirits with which it is associated. By establishing contact with these beings, the shaman acquires more knowledge and power.* (McKenna et al, 1995)

FUTURE USES OF PLANTS

One reason for conserving plants is to keep options open for the discovery of new uses. A similar argument holds for traditional botanical knowledge – if conserved, this knowledge remains available for the use of future generations. The medical is one field in which plants surely still have much more to offer. New medications are sorely needed to control those pathogens that are becoming genetically resistant to current treatments (Lewis, 2003).

The fact that few (approximately 100) plant species have contributed significantly to the discovery of pharmaceutical drugs should not be taken to mean that there will not be many more discoveries. Some idea of the potential is indicated by the fact that circa 50,000 plant species contribute to traditional medicine, with experimental evidence that many of them are efficacious (see Chapter 4).

Speculations about new uses for plants have tended to concentrate on pharmaceutical drugs; but actually there are several other relevant categories of use, among them fuel, clothing, construction, crop protection, horticulture, food, nutriceuticals, herbal medicine, personal care and cosmetics (Laird and ten Kate, 2002). The directions taken by research aimed at finding new products are influenced by questions of regulation and marketing, as well as by utility. One of the reasons why there has been more commercial interest in undertaking research towards new pharmaceutical drugs, rather than herbal medicines, is because they are easier to patent (Eldin and Dunford, 1999). Today, there is a growing interest in prospecting the natural world for genetic sequences, such as those that confer resistance to pests or allow crops to be salt tolerant, perhaps reducing research into phytochemicals in some quarters (Macilwain, 1998).

Although there is a degree of predictability about which types of plants are likely to yield new discoveries (see Chapter 9), history has shown that unexpected uses of species emerge periodically and sometimes even entirely new categories of use. Who would have guessed that the introduction of tulips to Holland in 1593 would have resulted in a 'tulip craze' (1634–1637), when single tulip bulbs were sometimes traded for entire estates? Who would have thought that obscure wild grasses in Turkey would one day be recognized as having immense value for breeding new varieties of wheat? Who would have imagined that an obscure temperate weed – thale cress *Arabidopsis thaliana* – would be chosen as the model plant for genetic analysis? Who, ten years ago, would have supposed that research on sugars in a resurrection plant *Selaginella lepidophylla* would lead to a way of preserving vaccines with the potential to benefit millions of people annually (Hawkes, 2004)?

8

The Patterns of Plants

Knowledge of the geography of plants is essential for deciding where to deploy resources available for conservation. It is even better if the *causes* of the patterns of plants are understood because this provides useful information for deciding the types of action that will be most helpful.

DETECTING PATTERNS OF PLANT SPECIES

Geography is a major consideration in plant conservation. Without some knowledge of their distributions, it is impossible to know if species are endangered (Kalema and Bukenya-Ziraba, 2005). Unless it is known where plant resources occur, sensible plans for their management cannot be constructed. The overall effectiveness of conservation depends upon judgements as to *where* best to concentrate the generally rather meagre resources available for conservation.

For many conservation purposes, it is useful to think of the patterns of plants that we see today as being formed of underlying natural patterns, modified to varying degrees by people. Natural patterns are those that *would* be present had humans never existed. Although there are considerable uncertainties in unravelling the contributions that people have made to plant geography, conserving some elements of the natural patterns of plants is a very significant purpose of *in situ* plant conservation.

The raw data used to map plant species are records of their occurrence, built up over time in files, publications and herbaria. Those engaged in floristic surveys should note carefully the localities where plants are found, using detailed maps and global positioning system (GPS) instruments, if available. Botanists who can identify many plants in the field and know how to identify the others using Floras, herbaria and referral to taxonomists form the hearts of systems of plant recording. The deposition of specimens in herbaria allows the identities of records to be confirmed at subsequent dates, if doubts arise about their authenticity.

Field botanists tend to visit more accessible places; therefore, records of plant species may not accurately reflect their true distributions. A more systematic coverage can be achieved by dividing a region into areas for recording purposes, noting how existing records are distributed and then arranging visits to places that seem under-studied for further recording. The areas used in recording schemes are often defined by political or administrative boundaries, with the advantages of easy recognition locally and a good chance of support by local botanists, proud of 'their' floras and happy to contribute their knowledge (Jonsell, 2003). Recording areas should be approximately equal in size, which can be achieved by subdividing larger political or administrative units into two or more smaller areas. This approach was used in 1874 by Watson, who subdivided the larger counties of the British Isles to create his 'vice-county' system for plant recording. Another way of encouraging systematic recording, also facilitating statistical analysis, is to divide a region into grid squares, such as the 10km x 10km cells used for mapping the flora of the British Isles (Preston et al, 2002).

Field surveys are often used in conservation projects to investigate the local distributions of plants. There can be several purposes. First, the results can contribute to knowledge of the conservation status of species. Second, the research can reveal something of the overall patterns of plant distribution in an area, which is useful for conservation planning (see Chapter 6). Third, information about the distributions and quantities of plant resources can be useful when considering how best to balance the conservation of plant diversity against the use of plant resources.

Sample plots are frequently used in field surveys of plants. These are usually small areas temporarily demarcated on the ground for recording the characteristics of plants and associated environmental variables. Several decisions must be made in the use of sample plots, such as on their numbers, locations, sizes, shapes and what to record. The answers to these questions will depend upon the precise purpose of the work, the ease of sampling, the time and other resources available, and advice from statisticians. Deciding *where* to place the sample plots is one of the most critical decisions. This will benefit from local knowledge (especially that of key local knowledge holders – see Chapter 12), evidence of patterns of vegetation apparent on aerial photographs or satellite images, and presumptions about the sorts of factors that are likely to influence the distribution of plant species. At the scale of local projects, some of the latter factors commonly include vegetation type, stages in vegetation succession, altitude, aspect and slope (Tuxill and Nabhan, 2001). Statisticians, especially those who have never participated in field work, can advocate *random sampling* for the disposition of samples; but totally random sampling can be difficult or impossible to achieve physically in many contexts (for example, in dense forest or on steep terrain). It is also time consuming and may cause a loss of interest in the work on the part of local collaborators. *Stratified random sampling* may be preferable. In this case, available knowledge is used to subdivide an area subjectively into sub-units, each of which is then sampled randomly. Both scientific and local knowledge can be used to determine how an area should be subdivided. The placing of sample plots at intervals along transect lines has been recommended for use in many conservation contexts in the tropics (Cunningham, 2001a).

Surveys of plant distributions are likely to be greatly improved if local people are involved. One advantage is that records can be made at all times of the year, which is useful because some species may only be conspicuous or identifiable during certain seasons (for example, due to periodic flowering). Local people with a special interest in plants are particularly valuable as collaborators. They will likely be mines of information about the history, uses and methods of managing local plants. Their knowledge can help ensure that new management plans are realistic.

The term parataxonomist has sometimes been applied for a local resident, typically living in a rural area in the tropics, who contributes to biodiversity inventories and assessments (Basset et al, 2004). Parataxonomists are trained by professional botanists to collect voucher specimens of local plants and sometimes, beyond that, sort them into morphological groups and add information about them to a database. An example of a project that has employed parataxonomists is Projek Etnobotani Kinabalu (PEK), with which the People and Plants Initiative has been involved (see case study at end of this chapter). The results of this project are so spectacular in terms of increasing knowledge of a flora that similar studies (modified methodologically to suit local conditions) should be tried in other parts of the world where the flora is poorly known scientifically.

Amateur botanists have contributed immensely to knowledge of the floras of certain countries, such as Australia, Sweden and the UK. A new edition of an atlas of the flora of Britain and Ireland – containing millions of records – is based on a partnership between government departments and voluntary organizations. Amateur botanists, some associated with societies such as the Botanical Society of the British Isles, have contributed many records (Preston et al, 2002). Amateur botanists, belonging to this tradition of natural history, are thin on the ground in most parts of the world; but it should not be presumed from this that there are no local people interested in plant diversity. Rural people with close dependencies upon local plant resources can be very knowledgeable about local plants – their types, uses and methods of management. Research has revealed that the Dai people of Xishuangbanna in China are (collectively) able to identify over 80 per cent of the plant species found in the forests and home gardens of their neighbourhoods (Wang Jinxiu et al, 2004). Traditional doctors are often the most botanically knowledgeable of local residents.

The use of two or more methods to investigate the patterns of plants at one locality can yield complementary information. An example is provided by two different approaches to studying the distribution of plant species for conservation purposes on the East Usambara Mountains in Tanzania. One survey involved the placing of small sample plots at regular altitudinal intervals along three transects on different aspects of the mountains – a type of approach with which many botanists will be familiar. The flora was carefully recorded in each plot and observations made on environmental factors, such as on the soil type present and evidence of human disturbance. This allowed a relatively high degree of ecological interpretation of the floristic results. The other survey was carried out using techniques commonly used by foresters for determining stocks of commercial timber, although extended in this case to include all species of trees (not just timber species). This survey involved a very large number of samples, spaced along transect lines and recorded rapidly. Only areas that were 'accessible' (for potential mechanical logging) were sampled, so that the sampling was less comprehensive in this respect than in the other survey. Figure 8.1 shows how floristic data from a sub-sample of plots of the second of these surveys were classified into forest types, based upon a numerical technique of classification. Labels are attached to the diagram to provide an interpretation of the forest types recognized (labelled A–J), in terms of altitude, degree of human disturbance and other features. In addition to revealing the distribution of forest types, the samples were also analysed in relation to the number of endemic species that they contained. More conservation weight was attributed to species that are totally confined to the East Usambaras (strict endemics) or nearly so confined (near endemics) than more widely distributed species. Altogether, the results of this work were used to recommend the establishment of a nature reserve in an area of the mountains where the forest remained relatively undisturbed and which contained the greatest concentration of endemic species. This area lies on the south-east side of the main range of the mountains, more or less contained within the Amani-Sigi Forest Reserve (as then existed). This is also the area of the East Usambara Mountains with the greatest value for delivering water supplies to the surrounding lowlands. The recommendation for the new nature reserve was accepted and the Amani Nature Reserve was created (Hamilton and Bensted Smith, 1989).

THE EVOLUTION OF PLANTS

Plant species are transitory phenomena, having origins, life histories and ends. A new species emerges – at a particular locality – when new genetic variation arises in the population of an existing species. There are various sources of genetic novelty (Van Rammsdonk, 1995):

- mutations at the level of the genes, chromosomes or whole genomes;
- the introduction of genetic material from other populations of the same species (introgression) or from other species (hybridization); or
- the transfer of genes from species that are essentially unrelated (horizontal gene transfer).

Speciation in plants can be gradual or sudden. The former occurs when a species becomes divided into two or more populations, which then evolve separately with time and eventually become sufficiently different to be judged distinct species. The latter can occur when polyploids are produced – that is, plants with multiple sets of chromosomes compared to their ancestors. Such multiplication of chromosomes can happen in a single generation. One way that polyploids can arise is through hybridization between two species.

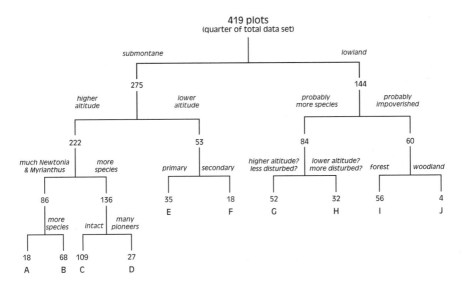

Figure 8.1 *TWINSPAN classification of 419 sample plots in the forests of the East Usambara Mountains, Tanzania*

Note: TWINSPAN (Two-Way Indicator Species Analysis) is a computer program used for classifying sample plots into categories, according to the species recorded within them. The classification is hierarchical, with all plots first divided into two groups (in this case, with 275 and 144 plots, respectively), then into four groups and so on. In this example, it makes sense to stop subdivision at the level indicated in the diagram, with nine forest types (labelled A–I) and one woodland type (J).
Source: Hamilton and Bensted Smith, 1989, based on an inventory carried out by the Forest Division, Tanzania, and the Finnish consultancy companies Silvestria and Finnmap

Commonly, the progenies of such hybridization are sterile because the chromosomes cannot pair up properly at meiosis; but if the number of chromosomes happens to multiply following hybridization, then the resulting plants can be fertile (because the chromosomes can now pair up). It is estimated that 30 to 80 per cent of higher plants have polyploid origins, many having arisen through hybridization between taxa that are not very closely related (Chase, 2001). Numerous crops, including wheat, are allopolyploids (polyploids between different species) (Arnold, 2004).

Species that have arisen relatively recently through polyploid events can have an exceptional ability to fertilize themselves – they do not need to cross-fertilize with other individuals to produce viable seeds or spores. This can give them a competitive advantage when there are extensive areas of habitat available for colonization, provided that they have produced a genotype that is ecologically well suited to the habitat. Such polyploids do not have the disadvantage of their related diploid (out-breeding) relatives. The success of polyploids as colonizers is shown by the fern genus *Asplenium* in relation to the colonization of northern Europe from southern refugia after the end of the last Ice Age. About half of the 50 European taxa of *Asplenium* are diploid and the other half are polyploid. The diploids, which are believed to represent the populations from which the polyploids have been derived, are more or less confined to the Mediterranean basin, where they grow in sites that are thought to have been temperate refugia during the Ice Ages. They are largely sexually reproducing species and have shown little ability to spread. In contrast, the polyploids are widely spread in continental Europe, including the British Isles and Scandinavia, having moved up from southern Europe after the end of the last Ice Age. They appear to be capable of self-fertilization (Vogel et al, 1999).

Over recent years, organelle studies and molecular analyses of genes and proteins have greatly advanced understanding of the origin of plants and how they are related to other forms of life (see Figure 8.2). These findings have challenged some established views about plant evolution. Most fungi are now known to belong to a single evolutionary group more closely related to animals than plants. The exceptions are members of the group of organisms known as the Oomycota – traditionally included in fungi, but actually more closely related to plants than animals. Various lineages of plants have split off the line leading from the first green plants to the flowering plants (angiosperms) – that is (in succession), the green algae, bryophytes, club mosses, ferns and gymnosperms. The classical division of the flowering plants into monocotyledons and dicotyledons is now known to be misleading from an evolutionary perspective. Rather, the monocots are a lineage that has emerged from *within* the dicots. The earliest divergent lines of the angiosperms contain, successively, *Amborella*, a shrub found in New Caledonia, the water lilies (Nymphaeales) and then a group of four small families: Austrobaileyaceae, Illicaceae, Schisandraceae and Trimeniaceae. The magnoliids (Laurales, Magnoliales, Piperales, Winterales) are confirmed as a group of angiosperms that diverged early from the main angiosperm body. The great majority of dicotyledons fall into a group known as the eudicots, which itself has two main divisions: asterids (containing Asteraceae, Lamiaceae, Rubiaceae and other families) and rosids (containing Brassicaceae, Fabaceae, Rosaceae and other families). Among the surprises of molecular studies are that whisk ferns *Psilotum* are related to the Ophioglossidae (adder's tongue and allies), rather than being relatively unaltered descendants of early vascular plants known from the fossil record, such as the psilotophytes. Three unusual genera of gymnosperms classified in the Gnetales (*Ephedra, Gnetum, Welwitschia*) belong to an evolutionary group (clade) that is close to the Pinaceae (*Abies, Pinus*, etc.). Together, these two groups are distinct from all other conifers (*Cupressus, Metasequoia, Taxus*, etc.). The Gnetales are not closely related to the flowering plants, as traditionally believed by botanists.

ENVIRONMENTAL AND HISTORICAL DETERMINANTS OF PLANT PATTERNS

Plants are unevenly distributed around the world (see Box 8.1). It can be useful for conservation purposes to distinguish between two broad sets of factors that determine the distribution of plant species – modern environment and historical. **Modern environmental factors** include such things as climate, soil

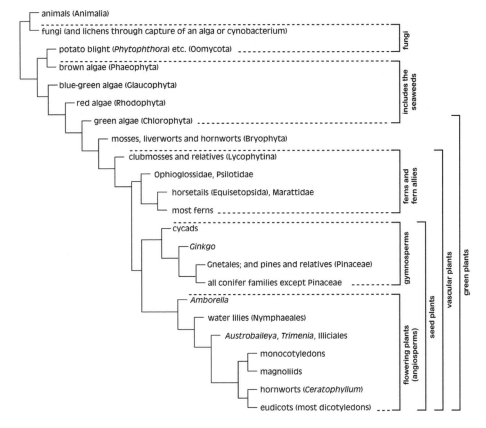

Figure 8.2 *Phylogenetic relationships of the main lineages of plants and some related groups*

Note: Groups connected by vertical lines on the left-hand side of the diagram are descended from a common ancestor. Plants are distinguished by having cells containing chloroplasts, a type of organelle capable of photosynthesis. Chloroplasts give plants the ability to form organic matter from inorganic materials and, hence, have allowed them to become the productive base of many ecosystems on Earth. Chloroplasts originated in evolution when a single-celled organism ingested a photosynthetic bacterium (a cynobacterium), the descendants of which (chloroplasts) were passed from cell to cell and species to species as plants have lived and diversified over time (Martin et al, 1998). Lichens originated on several occasions when fungi acquired algae or cynobacteria, most commonly of the green algal genus *Trebouxia*. There are some other types of organisms that are photosynthetic, apart from those shown in this figure. They include photosynthetic bacteria and a number of unicellular eukaryotes, such as *Euglena* and some dinoflagellates. The ancestors of these eukaryotes acquired their photosynthetic ability by ingesting single-celled green algae, already capable of photosynthesis from the possession of chloroplasts.

Sources: Stiller and Hall, 1997; Yin-Long et al, 1998; Mathews and Donaghue, 1999; Yin-Long Qlu et al, 1999; Baldauf et al, 2000; Bowe et al, 2000; Pryer et al, 2001; Ka Hou Chu et al, 2004; Schneider et al, 2004; Yajuan Liu and Hall, 2004

BOX 8.1 SOME NOTABLE GLOBAL PATTERNS
OF PLANT SPECIES

- About 64 per cent of the world's plant species occur in the tropics and sub-tropics. Of these, about 50 per cent occur in Latin America, 29 per cent in Asia and 21 per cent in Africa.
- The greatest concentration of succulent species is found in the Succulent Karoo of south-west Africa. Of the world's estimated 10,000 succulent species, 37 per cent occur in South Africa, mostly in the Succulent Karoo.
- Mediterranean ecosystems contain approximately 20 per cent of all plant species in only 2 per cent of the world's land area. The richest of these ecosystems is the Mediterranean itself (25,000 species), followed by the Cape Floristic Region (8550 species), south-west Australia (8000 species), California (4300 species) and central Chile (2400 species).
- Four large islands – Cuba, Hispaniola, Madagascar and New Caledonia – are ancient land masses in which many plants have evolved in isolation from the rest of the world over millions of year. Each has over 1000 endemic plant species.

Sources: Raven, 1987; Hilton-Taylor, 1994; Cowling et al, 1996

type and altitude (as a correlate of climate and associated factors), as well as small-scale processes such as gap-formation and succession (see Chapter 5). **Historical factors** refer to plant evolution (already mentioned) and the grander-scale processes described later in this chapter, including those related to major climatic change.

Climate is the modern environmental factor that is of supreme importance on the larger scale in imposing limits upon where plant species can grow. There are major floristic gradients associated with latitude and altitude, in both cases (very broadly) with decreasing numbers of species as these variables increase. **Soil factors** can be very significant for determining the distribution of species on the more local scale, with limestone and ultramafic rocks being particularly noteworthy for their distinctive floras, sometimes with many narrow endemics (May, 1998; Kiew, 1991). Repeated patterns of variation in vegetation and soils (**catenas**) are apparent in many parts of the world, with particular combinations of species on hilltops, hill slopes and in valleys. **Wetland vegetation** is distinctive, varying according to substrate, seasonality of flooding and degree of acidity.

The way in which plants are distributed in the world today is due to a long and complicated history, marked by the emergence of new species and then the vagrancies of whether they have managed to spread to new areas or have been subject to the local extinction of their populations. Changes in physical geography (for example, continental drift, mountain-building and changes in sea level and climate) have influenced patterns of plant geography. Certain places will have been especially favourable for the generation of new species in the past, just as certain places are today.

It is useful for the purposes of planning conservation strategies to know where modern centres of active speciation are situated. In South America, barriers to the movements of species *within* the northern Andes, combined with some stochastic dispersal, have created ideal conditions for active speciation. This hotbed of speciation is a priority for plant conservation. What is needed is a network of reserves along the eastern slopes of the Andes – dispersed so as to capture the diversity of plants and their habitats, along with speciation processes. In contrast, the Amazonian lowlands are dominated by old plant species with a wide distribution, so the precise siting of protected areas is less significant. A good strategy would be to establish reserves covering extensive areas, with the emphasis less on maintaining pristine habitats and more on sustainable use (Fjeldså and Lovett, 1997a, 1997b).

It is possible to divide the world geographically according to the distinctiveness of their floras. At the apex of such division are five floristic kingdoms – the Holarctic, Neotropical, Palaeotropical, Cape Floristic and Australasian kingdoms. This classification draws attention to the extraordinary flora of a small part of South Africa, where the Cape Floristic Kingdom is found (see Figure 8.3). It also illustrates the relative uniformity of the floras of north temperate regions (northern North America and northern Eurasia are both placed in the Holarctic Kingdom), compared with those of more equatorial and southerly lands. Despite the long historical separation of land masses in the southern hemisphere, some floristic similarities remain which are believed to owe their origin to ancient connections between the now widely dispersed southern continents. Such 'Gondwanaland' taxa – found in two or more of Australasia, South Africa and South America – include the families Proteaceae and Restionaceae, the monkey puzzle *Araucaria* and the southern beech *Nothofagus*.

There are some notable differences between the floras of tropical America, Africa and Asia. The latter two are relatively similar to one another in terms of shared families and genera, but there are still major contrasts. The family Dipterocarpaceae, so prominent in many forests in Southeast Asia, is absent from African rainforests (though a sub-family is sparsely represented in some African woodlands). Two characteristic plant families of tropical America, the Bromeliaceae and Cactaceae, are absent from the Old World, apart from the unusual epiphytic cactus *Rhipsalis* (found in Africa and Sri Lanka). It has been discovered through genetic research that there has been a considerable amount of transoceanic movement of plants. For instance, about 20 per cent of the Amazonian tree flora belong to groups of plants that are now known to have arrived in South America well after it became an island 100 million years ago (Pennington et al, 2004; Woodward et al, 2004). However, at the species level, there is little in common between the tropical forest floras of Africa and South America, the few exceptions including the forest trees *Carapa procera, Parinari excelsa* and *Symphonia globulifera* (Rejmánek, 1996).

Overall, the tropical flora of Africa is poorer in species than those of the other tropical continents, with an estimated 26,000 species compared with 90,000 (South America) and 50,000 (Southeast Asia) (Plana, 2004). It has

Figure 8.3 Brunia stokoei, *a species belonging to the family Bruniaceae,*
endemic to South Africa and concentrated in the south-west Cape –
illustrating the unique Cape Floristic Kingdom

Source: Alan Hamilton

been speculated that this may be due, in part, to extinctions in Africa at times
of aridity, Africa being more climatically susceptible in this respect. Palms
are noticeably much less of a feature of African forests than those of America
or Asia in terms of general abundance and numbers of species. There are
only about 50 species of palms in continental Africa, compared with 227
and 477 in just single countries on the other continents (Colombia and
Indonesia, respectively) (Johnson et al, 1996). The numbers of ferns and
orchids at just one site in Southeast Asia – Mount Kinabalu in Borneo – are
over 600 and 700 species, respectively, much higher than for any single locality in Africa.

   Long-term geographical isolation produces distinctive floras, as is evident
in Madagascar, New Zealand and many oceanic islands. Isolation *within* continents can have a similar effect, as illustrated by the floristic distinctiveness of
the Atlantic forests of eastern South America and the coastal and Eastern Arc
forests of Eastern Africa, both of which have been separated from the main
forest blocks of their continents (in the Amazon and Congo) by zones of drier
climate that have existed for several million years (see, for example, Burgess et
al, 1998). Floristic similarities between places that are today separated from
one another by water barriers are sometimes due to their former connection
by land bridges. Land connections during periods of low sea level during the
Ice Ages account for some of the similarities of the floras of Peninsular

Malaysia, Borneo and Sumatra (lying on the shallow Sunda Shelf), the floras of Australia and New Guinea, and the floras of North America and Eurasia.

The temperate forests of Europe are relatively impoverished in terms of numbers of plant species, compared with those of East Asia and North America. This is thought to be partly due to a greater number of extinctions of plants in Europe during the last few million years, a period of increasingly cold and perturbed climate in northern regions. North–south movement of species in Europe, responding to warm–cold climatic changes, was more impeded in Europe than on the other continents because of many east–west running mountain ranges, such as the Alps. Temperate genera that are known from the fossil record to have been present in Europe in the Miocene or Pliocene (24 to 1.8 million years ago), but no longer naturally occur there, include *Liquidambar*, *Magnolia*, *Sequoia* and *Taxodium*, all of which can be found today in East Asia, North America or both. Another cause of floristic change in Europe has been the steady desiccation of the climate of the Mediterranean region. This has been blamed for the disappearance of a type of evergreen forest known as *laurisilva* that was once found there. *Laurisilva* is characterized by an abundance of trees with leathery evergreen leaves (some belonging to the family Lauraceae) and can still be seen in the Canary Islands and Madeira in the Atlantic, where the climate must have remained relatively stable. The climate of the Mediterranean region was also much affected by an episode 5.5 to 4.5 million years ago when the Mediterranean Sea temporarily dried up, possibly due to tectonic events. This would have caused a much drier climate in Southern Europe at the time, no doubt resulting in extinctions.

ICE AGE INFLUENCES

Climatic changes associated with the Ice Ages have greatly influenced the distribution of vegetation types and plant species over the last 2.3 million years. The roughly 20 major climatic oscillations over this period became more pronounced 1 million years ago, when glaciations became more severe. Many parts of the world experienced major changes in climatic moistness during the Ice Ages, with a general association between glacial times in temperate regions and aridity in the tropics (Colvinvaux et al, 1996; Jolly et al, 1997; Hamilton et al, 2001; Maley, 2001).

During periods of unfavourable climate during the Ice Ages, many species became restricted to small areas of suitable habitat (refugia). Populations of a single species sometimes became isolated from one another in two or more refugia, resulting in their genetic isolation and therefore the possibility of their genetic differentiation. Refugia tended to be situated in the same places during successive periods of unfavourable climate. For some taxa, the populations in different refugia appear to be taxonomically identical, suggesting only recent separation. For other taxa, the populations in different refugia are taxonomically distinct – at varietal, subspecific or specific levels – and were possibly isolated from one another at earlier dates.

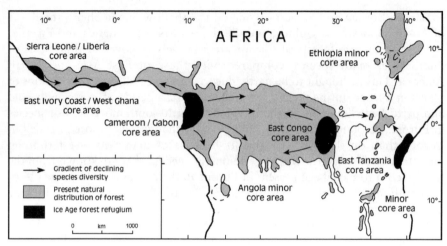

Figure 8.4 *Forest refugia in tropical Africa during the last Ice Age (maximum 18,000 years ago), showing routes of spread of forest species away from the refugia during the postglacial (since circa 13,000–10,000 years ago)*

Note: There are decreasing numbers of species away from the refugia, caused partly by differential rates of spread. Populations of species that occur both inside and outside the sites of former refugia will tend to be genetically more diverse within them.
Source: adapted from Hamilton and Taylor, 1991

The onset of a favourable climate at the end of the Ice Ages created the opportunity for species to spread out from their refugia. They did so at varied rates (Hamilton, 1974). With reference to the last Ice Age, in particular, this is why today we find gradients of increasing poverty in terms of numbers of species extending out from former refugia (see Figure 8.4). Identification of the sites of Ice Age refugia is important for conservation for several reasons. First, such refugia are often still centres of species richness. Second, species are likely to be genetically more diverse within the refugia than elsewhere (Hewitt, 1993, 1999; Petit et al, 2002). This means that saving populations of species within the sites of former refugia will be especially valuable for genetic conservation. Third, there is a good chance that the sites of past refugia will also be the sites of refugia during future Ice Ages, as are surely coming.

It might be supposed that Ice Age climatic change would have caused whole belts of vegetation to move latitudinally, resulting in many species no longer occurring at the sites where they survived the last Ice Age. Instead, it appears that many species survived near the sites of their Ice Age refugia by moving upslope (Davis, 1976; Hamilton, 1981; Hewitt, 1993; Hamilton et al, 2001).

Knowledge of floristic history during the period which extends from the height of the last Ice Age (18,000 years ago) to today is especially significant for understanding many details of the geography of plants at the regional scale. This period includes a time of major climatic transition from a glacial to a postglacial climate (circa 13,000 to 10,000 years ago). The last Ice Age saw a

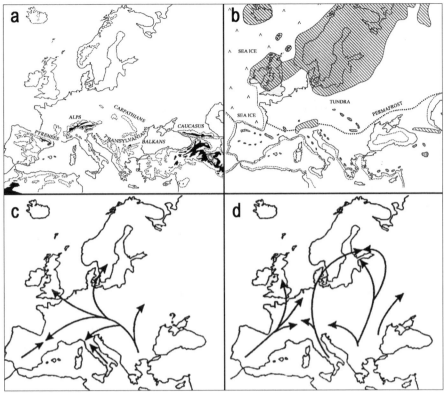

Figure 8.5 *Ice Age Europe and patterns of postglacial spread of plants:*
(a) physical geography of Europe, showing the presence of major
east–west running mountain chains towards the south (black = over 2000m;
dotted line = 1000m contour); (b) ice cover (hatched) and extent of
permafrost at the height of the last Ice Age, 18,000 years ago;
(c–d) proposed postglacial expansion routes of warmth-loving plants,
deduced from molecular data and fossil evidence: (c) beech (Fagus sylvatica);
(d) oak (Quercus robur and Q. petraea)

Note: As with Fagus, populations of many plant species in the Balkans were often more successful at colonization than those in the other southern peninsulas, being less restricted in their north-wards movement by east–west running mountains. For migration of *Quercus*, see also Cottrell et al (2002) and Kelleher et al (2004).
Sources: based on information in Hewitt, 1999, and Ferris et al, 1993

very large ice cap over much of northern North America and a smaller but still extensive ice cap over Northern Europe. Warmth-loving plants were confined to southerly localities, with some of the major refugia for temperate forest species lying in the southern Appalachian Mountains of North America and the southern peninsulas of Europe. In the case of species trapped in more than one southern peninsula in Europe, there could, in effect, be rivalry between the various isolated populations to colonize Northern Europe with the onset of warmer times (see Figure 8.5). The populations in the Balkans generally won, thanks to being less impeded in their northward spread by east–west running mountain barriers, such as the Pyrenees (Ferris et al, 1993; Hewitt, 1999).

During the last Ice Age, tropical Africa was more arid than now, with tropical forest largely confined to refugia in West Africa, Congo–Gabon, east Kivu (Congo) and near the coast in East Africa (see Figure 8.4). A study in Uganda has demonstrated that, once the influence of modern environmental factors has been discounted, the present distribution of forest tree species reveals patterns that are historically generated. Species have varied in their abilities to migrate away from refugia since the end of the last arid phase, about 12,500 to 10,000 years ago. This analysis has proved useful for conservation purposes because it has helped to identify the most important places for conservation of forest species in central Africa.

PEOPLE AND PLANT GEOGRAPHY

A division of floras into species that are naturally present (native or indigenous species) and those that have been introduced by people (introduced or alien species) is valuable for the purposes of plant conservation. People can introduce species to new areas intentionally or otherwise. There are often uncertainties as to whether species are native or not, with longer-established introduced species often being more difficult to identify than more recent immigrants. Longer-established introduced species also stand a better chance of being culturally accepted as 'truly belonging'. The authors of a recent atlas of the flora of the British Isles found it convenient to divide introduced species into three groups (Preston et al, 2002):

1   archaeophytes (species naturalized before AD 1500);
2   neophytes (species naturalized since AD 1500); and
3   casuals (species whose populations do not persist in the wild for periods longer than about five years).

Whether a species is native or not can have legal implications. Twinflower *Linnaea borealis* is an endangered species of pine forests in Britain. It has been given a Biodiversity Action Plan under the assumption that it is native, but it may be introduced (Welch, 2003). Another example is *Galeopsis segetum*, an arable weed listed as an extinct native in Britain. The fact that *Galeopsis* is regarded as a native means that a reintroduction programme might be developed (Rich, 2003).

Crop diversity is an aspect of plant diversity for which humans are directly responsible. Traditional agricultural systems are often rich in crop diversity (see Box 8.2). The great Russian geographer and geneticist N. I. Vavilov recognized eight regions in the world where crop diversity is especially high and where he believed many crops had originated: they are known as the Chinese; Indian; Inner-Asiatic; Asia Minor; Mediterranean; Ethiopian; South

---

BOX 8.2 EXAMPLES OF CROP DIVERSITY IN TRADITIONAL AGRICULTURAL SYSTEMS

- Fields in Malawi have an average of 13 varieties of beans *Phaseolus vulgaris*.
- There are 550 local varieties of beans *Phaseolus vulgaris* in Rwanda, with individual fields containing up to 30 components.
- There are 140 locally recognized types of cooking bananas *Musa* in Uganda.
- The average farmer in Cusco, Peru, grows 10 to 12 different landraces of potato *Solanum tuberosum*, representing up to seven different potato species.
- Research in 260 randomly selected fields in the north Shewa and south Welo regions, Ethiopia, revealed 60 landraces of sorghum *Sorghum bicolor* identified by the farmers. The number of landraces per field ranged from 1 to 24.
- There are 101 varieties of melons *Cucumis melo* in Xinjiang Province, China, of which 70 have local names.

Sources: Pickersgill, 1998; Teshome et al, 1999; Tuxill and Nabhan, 2001; Liu Wenjiang, personal communication, 2004

Mexican/Central American; and South American Andean Centres (Hawkes et al, 2001). Nowadays, it is recognized that crops can sometimes display diversity in places *outside* the places of their initial domestication. For example, the banana was originally domesticated in Southeast Asia, but has a secondary centre of diversity in Uganda. Traditional landraces of crops can be of special interest in plant conservation because of their usefulness to crop breeders (see Chapter 5).

Over time, people have introduced an increasing number of species to ever more places. The accumulative effects of introductions on local floras can be illustrated by the case of citrus in the Mediterranean. The only species present during Roman times was citron *Citrus medica*, a species especially appreciated for its medicinal uses. A further four species were introduced at the time of the Islamic Renaissance, reaching the Iberian peninsula between the tenth and 11th centuries. These included lemon *Citrus limon* and lime *C. autantiifolia*. The sweet orange *C. x aurantium* was introduced by the 15th or 16th centuries, and the mandarin *C. reticulata* and grapefruit *C. paradisi* between the 18th and 19th centuries (Ramón-Laca, 2003).

The European discovery of America (AD 1492) led to major intercontinental movements of crops and other plant resources (the Columbian interchange). Potatoes, maize, tomatoes and cassava are examples of crops introduced from America to other parts of the world, while wheat, coffee and sugarcane journeyed in the other direction.

Plants were very important elements in the economies of the global European empires (16th to 20th centuries) (James, 1998). Major commodities traded to Britain in the heyday of the British Empire included cotton *Gossypium* and tobacco *Nicotiana tabacum* from North America, sugar

Saccharum officinarum from the West Indies and tea *Camellia sinensis* from China (purchased with the profits from opium, *Papaver somniferum*, which was traded from India to China). The key that allowed Europeans to penetrate Africa was the use of quinine to treat malaria, extracted from the bark of *Cinchona* trees, native to South America.

Networks of botanical gardens, arboreta and experimental farms spread around the world from about AD 1750, mounting trials of crops and plantation trees to find new economic resources for the empires. This further increased the movements of plants. The development of the Wardian case (a mini-greenhouse) about 1835 by Nathaniel Ward made it easier and safer to transport plants by ship over long distances. Plant explorers penetrated deep into remote areas of the Himalayas and Andes to find new wonders for European gardens. Many trees planted for timber around the world are introduced, including such stalwarts of modern forestry as the pines *Pinus patula* and *P. radiata* from North America, silky oak *Grevillea robusta* and *Eucalyptus* from Australia, and Japanese cedar *Cryptomeria japonica* from Japan. The major timber trees Douglas fir *Pseudotsuga menziesii*, Sitka spruce *Picea sitchensis* and Japanese larch *Larix kaempferi* were introduced to Britain between 1827 and 1861.

A side effect of human introductions of plants has been the spread of invasive species (see Chapter 2). The problem has mounted in scale with the growth of human travel. Take the case of the Galapagos Islands, where the major threats facing native plants stem from introduced plants and animals. Until 1535, the Galapagos were uninhabited. By 1970, cargo boats were arriving once every three to six months. By 1980, there was a weekly flight to the islands. Today, large numbers of cruise ships visit the islands and there are daily flights to and within them (Davis-Merlen, 1998). In the case of Hawaii, the first arrivals were Polynesians, settling 3000 years ago. Prior to the arrival of Europeans (1778), the only 'weedy' plant species present were 7 intentional introductions (for example, *Dioscorea bulbifera*, the bitter yam), 28 accidental introductions and 16 native plants that were pre-adapted to the sunny conditions created by Polynesian agriculture. Today, the total number of weed species is 240, including some capable of invading native forest, especially if it has been disturbed (for example, *Albizia chinensis*, *Clidemia hirta* and *Funtumia elastica*) (Whistler, 1998).

CASE STUDY: PROJEK ETNOBOTANI KINABALU (AN ETHNOFLORISTIC INVENTORY IN MALAYSIA)

Projek Etnobotani Kinabalu (PEK) is a community-based ethnofloristic inventory that started in 1992 as a collaboration between Sabah Parks, members of Universiti Kebangsaan Malaysia (many of whom later transferred to Universiti Malaysia Sarawak) and the People and Plants Initiative. The results of this project (as of 1998) vividly illustrate how the involvement of local collectors and plant experts in a botanical survey can greatly add to documented infor-

Figure 8.6 *Mount Kinabalu in Sabah, Malaysia – possibly the richest site in the world for plants*

Source: Alan Hamilton

mation on a flora and to knowledge about the conservation status of its species.

Mount Kinabalu (4094m) in Sabah, Malaysia, has one of the richest floras in the world, with probably around 5000 vascular plant species in an area of less than 2000km² (see Figure 8.6). About half of Mount Kinabalu, including its summit, is protected by Kinabalu Park, which remains largely forested. The slopes outside the park display a complex patchwork of agricultural and forestry land under various conditions of occupancy and tenure, including by the Dusun, the indigenous people of this area.

Seventeen plant collectors from nine Dusun communities have worked part time for PEK. These communities are well distributed, so that the collections made under PEK provide a good picture of geographical variations in the flora. The results of PEK can be compared with those of traditional collecting by visiting botanists. More than 200 visiting botanists have collected plants on Mount Kinabalu over a period of 147 years, starting in 1851 with periods on the massif ranging from several days to several years. PEK collections are mostly from below 2000m, while visiting botanists have tended to collect at higher altitudes. The number of uniquely numbered plant specimens collected by PEK over a six-year period (1992–1998) was 9000, compared with about 30,000 specimens that had earlier been collected by conventional methods.

In six years (1992–1998), PEK increased the number of monocotyledons known from the mountain (excluding orchids, which PEK collectors were

instructed not to collect) by 7 families, 28 genera and 99 species, correspond-
ing to increases of 26 per cent, 21 per cent and 28 per cent, respectively. The
number of new taxa in a sample of dicotyledon families was 16 per cent. A
special effort was made to collect palms. Visiting botanists had earlier
(1856–1992) collected 372 specimens of palms on Mount Kinabalu, belonging
to 10 genera and 48 species and infra-specific taxa. In comparison, in six years
(1992–1998), PEK collected 404 specimens of palms belonging to 19 genera
and 74 species and infra-specific taxa. Altogether, the total number of palms
now known (based on 776 records) corresponds to 81 species and infra-
specific taxa in 19 genera. Even during the first four and a half years of PEK
(1992–1996), in merely 3.3 per cent of the period of time in which plant collectors had earlier been active, PEK was able to increase the number of collections
by 109 per cent, known palm genera by 90 per cent and known species and
infra-specific taxa by 54 per cent.

The productivity of the PEK collectors is attributed to their intimate
knowledge of local plants, localities and micro-environments, their abilities to
collect at a variety of locations throughout the year and their tendency to
collect all species – cultivated, semi-cultivated and wild. Most visiting botanists
have been from the US, Europe or Peninsular Malaysia and have often been
unfamiliar with the plants, ecology and geography of Mount Kinabalu. Many
have collected plants along established trails in accessible primary forest and
mature secondary forest, focusing on wild species.

In addition to greatly expanding knowledge of the flora, PEK has much
enriched information on the local names, uses and distribution of plants, and
on local systems of plant classification. The local names for only two species of
palms on Mount Kinabalu had earlier been recorded by collectors. An assessment of the conservation status of the palms of Sabah (the state containing
Mount Kinabalu) was published in 1991, shortly before the beginning of PEK
(Dransfield and Johnson, 1991). The authors of this assessment reported that
the conservation status of 113 out of 131 species of palms in Sabah was
unknown. Some tentative comments were made in the assessment about the
conservation status of a few species that grow on Mount Kinabalu, such as for
Arenga retroflorescens (classified as 'probably Endangered'), *Salacca clemensiana* ('Endangered') and *S. lophospatha* ('Extinct/Endangered in Sabah'). The
data now available from PEK provide a much clearer picture of palm distribution and conservation status on Mount Kinabalu. In general, many of the palm
species of Mount Kinabalu are believed from PEK data not to be Endangered.
However, of the three species mentioned above, only single specimens of each
of *Arenga retroflorescens* and *Salacca clemensiana* were recorded during the
PEK survey and *S. lophospatha* was not found, suggesting that the first two
may, indeed, be Endangered on Mount Kinabalu and the other possibly Extinct.

Source for case study: Martin et al, 2002

9

Plants and Places:
Choices, Priorities and Standards

Choosing where to work and which plants to work on are important matters in plant conservation. Resources available for conservation are scarce and should be used efficiently. Choices for individuals and organizations will depend upon their circumstances and the opportunities open to them. All efforts, large and small, will contribute to the cause, provided that all pull together in more or less the same direction. What is needed is to '*Think global, act local*' (Patrick Geddes, 1915), which means finding out what is practically possible, while keeping in mind the bigger picture.

PERSPECTIVES

Individuals in diverse walks of life and many types of social groups can play their parts in plant conservation. Their interests, remits and capabilities will differ and so, too, will the possibilities of what they can practically do. People and social groups should assess priorities for their own situations, taking account of the wider context. **International organizations**, surveying the global stage, may have the opportunity to select places for their operations based upon a grand perspective. **Governments** might like to know where to establish new protected areas or which species should be legally protected. **Forest managers** can contribute through identifying the types of vegetation and species under their care that deserve special attention. **Industries** using wild plant resources may seek to identify which of the species that they use require better management. Team members in **conservation projects** should know which plant species in their project areas merit special attention.

Existing patterns of effort in plant conservation reflect people's past choices. A good understanding of history is needed to know why protected areas are situated where they are or why certain species have become protected in law.

Likewise, the types of plants that people use are also partly a product of history. Some plants have been selected for use based on knowledge of their properties. For example, it is reported that certain families and genera of plants have been favoured across the tropics for woodcarving because people independently in different places have been interested in the same properties of wood. The species selected have wood that is even in texture, resistant to cracking and attractive in appearance. These plant families (and genera) include the Apocynaceae (*Alstonia, Holarrhena* and *Wrightia*), Ebenaceae (*Diospyros*), Fabaceae (*Dalbergia, Pericopsis* and *Pterocarpus*), Rutaceae (*Zanthoxylum*) and Santalaceae (*Santalum*) (Cunningham, personal communication, 2002; Cunningham et al, 2005). Certain families (and sometimes tribes within them) have similarly been selected for some of the major food crops. Among them are the Cucurbitaceae, Fabaceae (tribes Phaseoleae and Vicieae), Poaceae (tribe Triticeae) and Solanaceae. Again, this selection may be no accident, as is the fact that it was initially made in certain localized parts of the world. For example, some of the centres of domestication were in the mountains of the Asian subtropics, where the life-history strategies of certain wild plants may have pre-adapted them for domestication. The seeds of these weedy, fast-growing plants contain large reserves of food, so that they are able to germinate rapidly and the plants grow quickly once the rains have started. These species are capable of producing plump, nutritious seeds quickly – a feature noticed and exploited by the first farmers (Hawkes et al, 2001; Lenné and Wood, 1991).

Despite such examples, it is likely that some species have been selected for use and since retained for reasons partly of chance and human inertia. After all, if one species is known to produce 'the goods', why make the effort to replace it or introduce others – attitudes which may well become even more entrenched once species become commercialized and well known to markets. It is noteworthy that, out of a total of 3000 species of trees found in Peninsular Malaysia, only 400 have gained a degree of commercial recognition for their timber and only 30 are marketed in any quantity. In other parts of the world, the timber market is even more selective and the number of commercial species even fewer (Jennings et al, 2000, p502). One species, *Pterocarpus angolensis*, accounts for 85 per cent of furniture made from sawn wood sold in Tanzania (Clarke et al, 1996).

There are various methods for determining the values that people place on plants and places, and for the ranking or scoring of people's priorities (see Chapter 12). Valuation exercises are useful for helping people to better understand their own priorities, as well as those of others. Information about what really matters to people is useful when parties negotiate to resolve their conflicts. Very commonly, there is much ignorance between scientists, resource managers and local people about each other's concerns. For example, scientists can be concerned about the fate of certain species, which they classify as 'threatened'; but often resource managers and local people have no idea which species these are. Resource managers can be similarly ignorant of the priorities of local people. The managers of Bwindi Impenetrable National Park in Uganda, when it started to be managed as a strict reserve, had no idea that access to the bark of *nyakibazi* (*Rytigynia*) was extremely important to local people, sowing the seed for conflict (see case study at end of this chapter).

There are some matters so fundamental to people that they see no room for compromise (red line issues). However, normally, balances between different interests are possible and compromises can be agreed. An example of a red line issue for conservationists might be a particular area of habitat that has numerous threatened species and which is deemed irreplaceable, so that its protection as an inviolate nature reserve is regarded as absolutely essential. On the resource front, there are some species – especially those valued for cultural reasons – for which no alternatives are acceptable (Prance, 1995). On the other hand, there may be other areas of habitat or other resources where compromises or substitutes are possible.

Searches for the best balance between different interests are very common in practical conservation, inevitably resulting in debate and, often, controversy: What types of extractive uses are acceptable within this protected area? How much pesticide should be used in this agricultural system? What priority should be given to managing this pasture for livestock *versus* the production of medicinal plants? Jepson (2001) has argued that it is important that conservationists know about, and engage with, the values that other stakeholders place on nature. Through doing so, they stand a better chance of achieving socially credibility and, thus, their conservation goals. He believes that active engagement will maintain the public appeal of conservation and guard against it becoming an abstract scientific pursuit. It will promote public transparency and accountability in biodiversity policies and strategies, as formulated by public bodies and non-governmental organizations (NGOs).

Some techniques of categorization are based on multiple variables and complicated formulae. However, such a high degree of precision is generally unnecessary from the practical perspective. In the end, a simple assessment is often all that is needed. Judgements are best made by experienced people with access to all the information. Take the case of assigning a conservation status to a species for a Flora. Various types of evidence can impinge upon the question, such as information about changes in distribution and judgements about how the species is likely to cope with anticipated environmental trends. However, in the end, the writer of the Flora has to make a simple overall judgement (Preston, 2003).

It is striking that a threefold system of categorization has been used or proposed in several instances relevant to conservation:

- Luoga and colleagues (2000) recognized three categories of intensity of plant resource use in Tanzanian miombo woodland: major, minor and no use.
- Cunningham (1996b) divided plant resources used at Bwindi, Uganda, into three categories for the purposes of making recommendations for their management:
  1 used by many people; high impact (alternatives to wild collection needed);
  2 used by specialists only; low impact (controlled collection should be allowed); and
  3 impact uncertain (research and monitoring needed).

- Conservationists in Colombia have adopted an 'unofficial' tripartite scheme for classifying threatened plant species for practical purposes: threatened, not threatened and insufficient information.
- Menges and Gordon, as reported by Tuxill and Nabhan (2001), proposed that there should be three levels of effort for work on rare plants to make the best use of available resources:
  1 Visit the locations where a species of concern occurs, including other areas of suitable habitat, and note whether it has persisted, disappeared or spread (least intense). Only rough assessments of abundance are made.
  2 Make quantitative measurements of the abundance of the species and of the floristic composition of its habitat.
  3 Undertake demographic studies of the species by marking and counting the numbers of individuals of different age or size classes, as well as by recording other demographic variables (choosing the right variables requires an understanding of the life-history strategies of the species) (most intense).

Some schemes of categorization are intended to serve as international standards. Examples include the World Conservation Union (IUCN) Red List of threatened species and Important Plant Areas (IPAs) – see later in this chapter for more information on these. The designation of species or places in such schemes needs to be credible, which means that the reasons for designation must be rationally defensible (not necessarily the same thing as quantifiable). The definitions of the categories used by IUCN for listing threatened species have been revised twice (IUCN, 1994a, 2001b) in an effort to make the listing process less subjective.

A degree of formality is needed in the statement of the rules that govern how entities are placed in categories for schemes intended to be widely applicable. Use can be made of the concepts of principles, criteria and indicators, which are probably best known in conservation circles for their application to the setting of standards for sustainable forest management (Maini, 1993):

- A **principle** is a fundamental law or rule intended as a guide to action (for example, 'forests should be managed such that biological diversity is maintained').
- A **criterion** is a distinguishing characteristic of a thing by which it is judged (for example, plant diversity in a forest).
- An **indicator** is a variable that can be measured in relation to a specific criterion (for instance, population size of a threatened plant species).

A degree of stability in setting standards and lists of priorities is desirable. As de Klemm (1990) has pointed out in connection with the designation of plant species for legal protection, lists may become discredited if they are constantly changing. De Klemm has also noted, in the legislative context, that the length of a list is significant – a short list may look unprofessional, while a long list can be disheartening. A further consideration in advancing lists of priorities for widespread adoption is that some respect should be shown for earlier schemes. Jepson (2001) regarded new proposals for priority areas for conservation in Indonesia as unhelpful partly for this reason. Officials and scientists had been following an earlier scheme for some years and, he suggested, they might become confused and demoralized with the new advocacy (in any case, he regarded the earlier scheme as scientifically superior).

Plant conservationists should make it clear that, just because certain places or plants are highlighted for special attention, this does not necessarily mean that other places or plants are of no conservation consequence. Context is all important. For example, it is possible to pick out certain forests on the Eastern Arc Mountains of Kenya and Tanzania as being botanically more important than others, in the sense of having more plant species in total or more endemic species. This type of information is useful for potential projects that have not yet decided where to operate. However, *all* Eastern Arc forests are high priorities for conservation. All provide forest products for local use, help to regulate regional water supplies and are elements in a chain of ecological islands exhibiting fascinating patterns of evolution. For a conservationist living near one of the relatively minor sites (as judged in the overall context), helping to save the local forest patch could well be the most important contribution to conservation that he or she can make.

Despite all that is written about making priorities, opportunism is still important in plant conservation. If something practical can be achieved, however small, then it is often best to act rather than husband resources for a possibly bigger eventual gain. In conservation, as in other human affairs, sometimes conditions come together fortuitously that are conducive to success.

PLANTS, PLACES OR RESOURCES?

The weight given to different approaches to plant conservation will tend to vary between countries, related to their economies, cultures and the characteristics of their plants. Large-scale efforts to conserve threatened species, taking a species-by-species approach and integrating *in situ* and *ex situ* methods, is currently possible primarily in countries that are relatively wealthy or where there is significant popular interest in threatened or rare species for their own sakes. A programme to conserve the many hybrids of a genus that is of little current economic interest is only conceivable in a well-resourced country with a strong tradition of natural history. Such a programme has been proposed for montane willows *Salix* in Scotland (Tennant, 2004). An integrated *in situ/ex situ* approach on a species-by-species basis is also workable (and, indeed, much needed) on some oceanic islands, building upon and developing insular or national pride in the unique floras.

If a programme to conserve threatened species on a species-by-species basis is planned, then the approaches used, including the amount of effort devoted to conserving particular populations of plants, should be selected carefully. In a talk at the 16th International Botanical Congress in St Louis in 1999, William Bond pointed out that populations of a species may be subject to change due to factors *extrinsic* to the populations themselves, and that these factors will need to be addressed before detailed measures applied directly to the populations can succeed. Thus, populations of a species may die out anyway if the plant communities in which they live are unstable (for example, if they are stages in vegetation succession; see Chapter 5). Beyond that, there may be large-scale processes influencing the populations, such as changes in climate or nutrient enrichment, which may render efforts to conserve particular populations of species pointless, until they are combated.

Habitat conservation is *not* an alternative to species conservation because habitat conservation is essential anyway if species are to be conserved. At the same time, care is needed in categorizing habitats for conservation purposes. Habitats can be recognized at various levels of detail. The most detailed classifications, typically based upon detailed lists of the species they contain, may not be of much use for conservation purposes because of their historically arbitrary nature (see Chapter 5).

In many parts of the world, two top imperatives for plant conservation are to prevent destruction or degradation of important habitats and to find ways of engaging people's interest in conservation initiatives. Under such circumstances, the most important species for special attention in conservation projects are likely to include those wild species that are of special cultural or economic significance to local people, or else contribute substantially to the wider economy. Therefore, basic requirements in projects in these cases are to learn about the cultural and economic values that local people place on plants and to ascertain which plant resources are traded widely in the region. Assessments of resource sustainability, at various levels of detail, should often follow (see Chapter 11).

This approach of concentrating on culturally or economically significant species has been recommended for an initiative to conserve tropical dry forest in Mexico, one of the world's most threatened vegetation types (Peters et al, 2003). A species selected for special attention is *Bursera glabrifolia*, a slow-growing tree with several uses, including provision of raw material for the carving of small painted figurines known as *alebrijes*. The market for *alebrijes* has grown greatly since the 1980s, providing an income to many people but threatening the species through over-harvesting. Concentrating on improving the *in situ* management of *Bursera* is likely to enhance the value of tropical dry forest in the eyes of local people (in the face of competing land uses) and to augment their commitment to its long-term management.

PLANTS

There are various indicators that a species may be vulnerable (see Table 9.1 and Figure 9.1). Systems of scoring can be devised to allow the use of several indicators in combination, adding up to an overall score for vulnerability

Table 9.1 *Some indicators of vulnerability of plant species*

Criteria	Indicators of vulnerability
Geography	
Commonness	Small total population size; population greatly fragmented or confined to a single site; small area of occurrence or occupancy;[1] represented by few specimens in herbaria[2]
Change in commonness	Decline in range size or abundance; subject to extreme fluctuations in numbers
Biology	
Habit	Vulnerable to damage – for example, solitary palms with single stems
Rate of growth	Slow growing; long time to reach sexual maturity
Pollination	Dependent upon specialist pollinators
Seed dispersal	Dependent upon specialist seed dispersers
Habitat	
Habitat type	Suitable habitat type of restricted occurrence
Change in area of habitat type	Area of suitable habitat declining
Resilience of habitat type	Habitat type lacks resilience to change – for example, long recovery time if disturbed
Species that are harvested	
Whether or not in trade	Harvested for commercial trade
Extent of use	Popularity, as indicated by high expense or volume
Change in extent of use	Increasing amounts traded
Reactions to harvest	Inability to re-grow easily after harvest – for example, through re-sprouting
Part taken	Damage likely to be great – for example, if bark or roots are collected
How collected	Destructively harvested – for example, cutting down trees for their fruits
Irreplaceability	No substitutes available
Protection	
Statutory protection	No laws or regulations giving special protection; absent from protected areas
Customary protection	Not protected by customs, taboos or local rules; absent from sacred forests

Notes: 1 The **area of occurrence** of a species is that contained within the shortest continuous imaginary boundary which can be drawn to encompass all sites where it is known or believed to occur, excluding cases of vagrancy. The **area of occupancy** is the area, contained within its area of occurrence, that a species actually occupies, excluding cases of vagrancy (IUCN, 2001b). 2 Used by the Royal Botanic Gardens, Kew, Missouri Botanic Garden and New York Botanical Garden to provide a first-cut approximation of the conservation status of species in a world plant checklist.
Sources: adapted from various sources, including Boutin and Harper, 1991; Bond, 1994; Marshall, 1998; Cunningham, 2001a; Lama et al, 2001; Martin et al, 2002; IUCN, 2004

Figure 9.1 *An example of a threatened species – the Round Island bottle palm* Hyophorbe lagenicaulis *in Mauritius*

Note: This taxon is confined to this small island, with fewer than 15 individuals remaining in the wild. However, it is widely cultivated.
Source: Alan Hamilton

(Cunningham, 1997b). An application of this technique produced the following list for the ten most threatened medicinal plants in Turkey: *Acorus calamus, Ankyropetalum gypsophylloides, Ballota saxatalis* ssp *brachyodonta, Barlia robertiana, Gentiana lutea, Gypsophila arrostii* var *nebulosa, Lycopodium annotinum, Origanum minutiflorum, Paeonia mascula* and *Ruscus aculeatus* (Özhatay et al, 1997). *Barlia robertiana* is one of nearly 40 species of orchids used for making *salep*, a constituent of ice cream and other foods.

Sometimes *all* rare species have been treated as threatened (see, for example, Walter and Gillett, 1997). This is not unrealistic as a first approximation, given the extent of human influence in the modern world and the fact that tragedies can happen unexpectedly almost anywhere. However, a more detailed analysis of rarity can be useful. Three classes of rarity can be recognized. First, there are species that are habitat specialists and have always been rare. They have demonstrated their capacities to survive by their continuing existence. Their populations tend to show low levels of genetic diversity, a condition for which they may be genetically adapted. Second, there are species that occur at low population densities but are scattered over more or less extensive areas. These species are vulnerable if their habitats are reduced or fragmented, especially if they have features that add to their vulnerability (see Table 9.1). Third, there are species that are limited in their distributions for no obvious reasons of habitat. Most plant species classified as threatened in the USA fall into this category (Holsinger and Gottlieb, 1991; Huenneke, 1991; Karron, 1991).

A knowledge of the history of species is invaluable for assessing their vulnerability. An analysis of changes in the flora of Britain and Ireland between 1930–1960 and 1999 has shown that some of the greatest declines have been among weeds of arable fields and other plants of open habitats. These species flourished under the gentler agricultural practices of former times, but have become reduced with the spread of industrial agriculture, involving the improved screening of crop seed, herbicide application and generally more intensive farm management (Marren, 1999; Preston, 2003). Many of these species are archaeophytes (see Chapter 8).

A knowledge of the history of plant populations can be useful for assessing their vulnerability to loss of genetic diversity. Rapid reduction in the size of plant populations is very common today as a result of human influence. Populations that have shrunk recently in size are likely to be genetically more diverse than those that have long been small, as demonstrated for the fern *Dryopteris cristata* in Switzerland (Barrett and Kohn, 1991; Kozlowski et al, 2002). Therefore, taking action quickly to restore populations that have recently shrunk in size sometimes has the potential to save much of the original genetic diversity.

Some plants are elusive and may be present, though not readily seen. The cases of the starfruit *Damasonium alisma* and fen violet *Viola persicifolia* are described in Chapter 13. Another example is the Killarney fern *Trichomanes speciosum*, which was believed to have been rendered extinct over eastern England and much of continental Europe by fern hunters during the 19th century. However, it has been discovered quite recently that this fern has actually persisted in several localities from which it was thought to have been exterminated, having survived in its inconspicuous gametophytic form, which looks nothing like a fern to inexpert eyes (Moore, 1998).

If the objective of selecting priorities is to achieve practical results, then selection is best made by (or in collaboration with) those who have the potential to follow up with practical measures. An example of such involvement is

Table 9.2 *Four examples of the 22 species of high-altitude medicinal plants identified as of greatest conservation concern in the Himalayas of Nepal*

Amchi name	Botanical name	Part collected for medicine	Reasons for plant becoming rare	Recommendations for management
Bashaka	*Lagotis kunuwurensis*	Leaves, flowers, roots	Grows in a localized habitat; large quantities collected	Harvest only 50 per cent of plants; collect seeds and sow in loose soil
Honglen (*kutki* in Ayurveda)	*Neopicrorhiza scrophulariiflora*	Whole plant	Grows in a localized habitat; large trade demand	Enforce government ban on commercial collection; harvest 50 per cent of (mature) plants, leaving material to regenerate; initiate *in situ* cultivation
Tongzil serpo	*Corydalis megacalyx*	Whole plant	Grows only at high altitude (slow growing)	After harvesting, mix the soil to encourage the seeds to germinate (this species cannot be cultivated)
Upal ngonpo	*Meconopsis grandis*	Flowers	Naturally rare	Collect only some of the flowers, leaving others to set seed; disperse seeds; initiate cultivation

Source: developed by the Himalayan Amchi Association at a meeting in Kathmandu on 21–23 January 2004

provided by a project of the People and Plants Initiative in the Himalayas of Nepal. This case involves the identification of the most threatened species of medicinal plants by the *amchis* – traditional doctors following the Tibetan healing tradition known as *Sowarigpa*, a tradition closely linked to Buddhism and Bon. An initial step was to ask the *amchis* to bring specimens of species that they used medicinally to a workshop to confirm their scientific identities. Discussions then followed about how the *amchis* assessed the abundance of species and what aspects of abundance were of particular conservation concern to them. It turned out that the *amchis* conceive of abundance in terms of 'thick' and 'thin' plant populations and that their concern was with declines in abundance, not rarity in itself. An initial long list of species of concern was eventually whittled down to 22 priorities, and recommendations were made for their management (see Table 9.2). These recommendations are now being taken forward by the World Wide Fund for Nature (WWF) Programme Office for Nepal through their inclusion in a booklet to be distributed widely to *amchis* and resource managers.

Another example of an initiative that has identified priority plants, taking account of local perspectives, was one concerned with providing plant resources for the primary healthcare of the Highland Maya of Chiapas in Mexico. The Highland Maya recognize 20 major classes of health conditions (and more than 250 minor classes) and their treatments use over 1600 species of medicinal plants. Brent and Ann Berlin (2000) have isolated a set of 50 priority medicinal plants, targeting specific illnesses belonging to the eight most important classes of health condition. Initial laboratory bioassays have confirmed that almost all of these species show strong bioactivity. With one exception, the species are mutually exclusive in terms of their use to treat particular illnesses. Considered the basic core (*cuadro básico*) of the ethnopharmacopoeia of the Highland Maya, they are the species that are most widely known and which are regarded as providing the most effective treatments for the most significant health problems. They form the basic set of species used for developing gardens of medicinal plants, intended to provide easy access for treatments and to reduce collecting pressure on wild populations (Berlin and Berlin, 2000).

CAMP (Conservation Assessment and Management Plan) is the name given to a process for making rapid conservation assessments of species. The first step is to decide on a group of species for consideration – for example, medicinal plants of Himalayan Nepal. A group of 10 to 40 experts is then assembled for a workshop. The experts are selected so as to provide several perspectives, perhaps (for the Nepal case) field and herbarium botanists, research scientists, forest officers, botanic garden specialists, traditional doctors, herbal traders, policy-makers and members of NGOs. On the basis of discussions, the species are assigned to categories of threat and recommendations drawn up for conservation action and further research.

The most comprehensive and authoritative inventory of the global conservation status of plants is the IUCN Red List of Threatened Species (IUCN, 2004). Assignment of species to categories of conservation status are made on the basis of various criteria, which were revised in 1994 and again in 2001 (IUCN, 1994a, 2001b) (see Table 9.3). Species are regarded as threatened if they have been classified as Critically Endangered, Endangered or Vulnerable (2001 categories) (see Figure 9.2). A threatened species is defined as one facing a high risk of extinction in the near future – not a precisely defined period, but one that may be taken to extend from very soon to a few thousand years (for very long-lived species). All submissions of species for incorporation within IUCN Red List must be accompanied by an assessment, which has to include a minimum set of required information. Before inclusion, each assessment must be evaluated by at least two members of a Red List authority, normally members of a relevant taxonomic or geographic specialist group of the Species

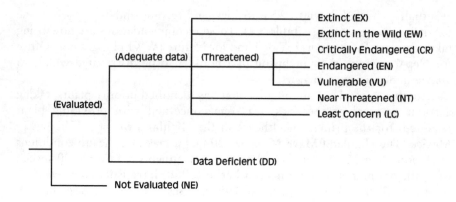

Figure 9.2 *Relationships between the IUCN 2001 Red List conservation categories*

Source: IUCN, 2001b

Survival Commission (SSC) of the IUCN (see Chapter 3). The 2001 IUCN categories of conservation status are:

- Categories of extinction:
  - *Extinct* (EX): species for which extensive surveys show there is no reasonable doubt that the last individual has died.
  - *Extinct in the Wild* (EW): species that survive only in cultivation, in captivity or as a naturalized population (or populations) well outside the past range.
- Categories of threatened species:
  - *Critically Endangered* (CR): species that are facing an extremely high risk of extinction in the wild (that is, when the best available evidence indicates that they meet any of the criteria A–E for Critically Endangered listed in Table 9.3).
  - *Endangered* (EN): species that are facing a very high risk of extinction in the wild (that is, when the best available evidence indicates that they meet any of the criteria A–E for Endangered listed in Table 9.3).
  - *Vulnerable* (VU): species that are facing a high risk of extinction in the wild (that is, when the best available evidence indicates that they meet any of the criteria A–E for Vulnerable listed in Table 9.3).
- Other categories:
  - *Near Threatened* (NT): species that do not qualify for Critically Endangered, Endangered or Vulnerable status now, but are close to qualifying or are likely to qualify for a threatened category in the near future.
  - *Least Concern* (LC): species that do not qualify for Critically Endangered, Endangered, Vulnerable or Near Threatened status. Widespread and abundant species are included in this category.

Table 9.3 *IUCN criteria (A–E) for assigning species to categories of threat (Critically Endangered, Endangered or Vulnerable)*

Use any of the criteria A–E	Critically Endangered	Endangered	Vulnerable
A Population reduction (declines measured over the longer of ten years or three generations)			
A1	≥ 90%	≥ 70%	≥ 50%
A2, A3 and A4	≥ 80%	≥ 50%	≥ 30%

A1 Population reduction observed, estimated, inferred or suspected in the past where the causes of the reduction are clearly reversible **AND** understood **AND** have ceased, based on and specifying any of the following:
- (a) direct observation;
- (b) an index of abundance appropriate to the taxon;
- (c) a decline in EOO, AOO and/or habitat quality;
- (d) actual or potential levels of exploitation;
- (e) effects of introduced taxa, hybridization, pathogens, pollutants, competitors or parasites.

A2 Population reduction observed, estimated, inferred or suspected in the past where the causes of reduction may not have ceased **OR** may not be understood **OR** may not be reversible, based on a–e under A1.

A3 Population reduction projected or suspected to be met in the future (up to a maximum of 100 years) based on b–e under A1.

A4 An observed, estimated, inferred, projected or suspected population reduction (up to a maximum of 100 years) where the time period must include both the past and the future, and where the causes of reduction may not have ceased **OR** may not be understood **OR** may not be reversible, based on a–e under A1.

	Critically Endangered	Endangered	Vulnerable
B Geographic range in the form of either B1 (extent of occurrence) and/or B2 (area of occupancy)			
B1 EOO	<100km^2	<5000km^2	<20,000km^2
B2 AOO	<10km^2	<500km^2	<2000km^2
AND at least two of the following:			
(a) Severely fragmented or number of locations;	= 1	≤5	≤10

(b) Continuing decline in any of (i) EOO; (ii) AOO; (iii) area, extent and/or quality of habitat; (iv) number of locations or sub-populations; (v) number of mature individuals;

(c) Extreme fluctuations in any of (i) EOO; (ii) AOO; (iii) number of locations or sub-populations; (iv) number of mature individuals.

	Critically Endangered	Endangered	Vulnerable
C Small population size and decline			
Number of mature individuals	<250	<2500	<10,000
AND C1 and/or C2:			
C1 An estimated continuing decline of at least: (up to a maximum of 100 years)	25% in 3 years or 1 generation	20% in 5 years or 2 generations	10% in 10 years or 3 generations

Table 9.3 *continued*

	Critically Endangered	Endangered	Vulnerable
C2 A continuing decline **AND** (a) and/or (b):			
(a) (i) number of mature individuals in each sub-population:	<50	<250	<1000
(ii) percentage individuals in one sub-population at least:	90%	95%	100%
(b) extreme fluctuations in the number of mature individuals.			
D Very small or restricted population **Either:**			
(a) number of mature individuals:	≤50	≤250	≤1000
AND/OR			
(b) restricted AOO.	not applicable	not applicable	AOO <20 km^2 or number of locations ≤5
E Quantitative analysis Indicating the probability of extinction in the wild to be:	≥50% in ten years or three generations (100 years max)	≥20% in 20 years or five generations (100 years max)	≥10% in 100 years

Note: AOO = area of occupancy; EOO = extent of occurrence
(see note under Table 9.1 for definitions).
Source: IUCN, 2001b

– *Data Deficient* (DD): species for which there is inadequate information to make a direct, or indirect, assessment of extinction risk based
on distribution and/or population status. A species in this category
may be well studied and its biology well known; but appropriate data
on abundance and/or distribution are lacking. Data Deficient is therefore not a category of threat.

The conservation status of species can be evaluated at more local levels, as
well as for the world as a whole. The national level is an obvious choice for
Red Listing, given that so much conservation work is country based (Oldfield,
1995). Red Listing at the country level introduces the problem of how to take
into account occurrences of species outside the countries of recording. After
all, a species may be threatened in one country, but not in another. A test in
Finland and Sweden has suggested a number of adjustments in application of
the IUCN Red List criteria at national level to deal with this problem. The
most notable recommendation is for a two-step process for species that are
not national endemics (further details in Gärdenfors, 2001):

1   their initial evaluation, as if they were endemic to the country, and
2   adjustments in these initial evaluations in the light of their wider distributions.

Despite the best efforts of botanists, information on the conservation status of many species remains poor, especially for the tropics and subtropics. Overall, the types, distributions and conservation status of plants are much less well known than those, for example, of birds, which is not surprising given that there are many more species of plants than birds (approximately 270,000 vascular plants *versus* approximately 9700 birds). Furthermore, there are probably many more amateur ornithologists who are scientifically competent to make accurate field records than there are competent amateur botanists. The percentages of species in various plant groups that have been globally assessed using the IUCN 1994 Red List criteria are: bryophytes < 1 per cent; gymnosperms 72 per cent; dicotyledons < 5 per cent and monocotyledons < 4 per cent (Hilton-Taylor, 2000). A few of the taxonomic groups and geographical areas that are better studied are included in Table 1.1.

Little is known about the conservation status of species of tropical plants based upon individual species-by-species assessments (White, 2001). This is partly because there are still areas of the world which are poorly explored botanically, such as New Guinea, Madagascar and parts of South America – in all of which there are likely to be substantial numbers of species awaiting discovery. A field survey of the palms of Madagascar during 1991 to 1994 resulted in an increase in the number of palm species known from the island from 116 to 176 (all but four of the total are endemics) (Johnson et al, 1996). However, as a geographical area, this is probably exceptional – it is most unlikely that many more species of palms await discovery in *continental* Africa, where the palm flora is certainly better known. However, even when fairly reliable lists of species exist for tropical regions (as is sometimes the case), generally very little is known about their conservation status. A review of the palms of Sabah in Malaysia revealed that a conservation status had been assigned to only 18 of the 131 species known to occur (Dransfield and Johnson, 1991). One way forward to obtain better data is to involve local people in floristic inventories and conservation assessments; but it is still uncertain how such local surveys can best be scaled up for assessments on the scales of countries or regions.

PLACES

Some indicators that can be used to identify places of special importance for plant conservation are listed in Table 9.4. The number of plant species, as a bald statistic, is a poor guide to the conservation importance of a site. Information on the indigenous status of the species and about the habitats in which they are found is needed for a proper evaluation. If this is not done, then suburban gardens in the UK and oceanic islands that are heavily invaded

Table 9.4 *Some indicators to identify places of special importance for plant conservation*

Criteria	Indicators
Species	
Number	Many native species
Geographic range	Many species of restricted geographic range
Threatened status	Many threatened species
Indicator species	Species indicative of habitat types of high conservation concern
Keystone species	Species playing key roles in ecosystems
Evolution	Site a centre of active species generation
Habitat type(s) present	
Conservation status	Habitat type of restricted occurrence
Ecosystem services	Habitat type important for providing ecosystem services (climatic moderation, water supplies, soil stability, etc.)
Resilience to change	Habitat type slow to recover from disturbance
Vulnerability to conversion	Land type or site attractive to people to convert for other purposes
Vulnerability to degradation	Habitat type yields products likely to be harvested unsustainably
Resources	
Material resources	High local importance rating for resources
Cultural significance	Significant cultural site
Genetic resources	Contains priority species or genetic diversity
Ecosystem change	
Greenhouse climatic change	Site will retain its plant conservation values based on predictions
Ice Age climatic change	Site will retain its plant conservation values based on predictions

Source: Alan Hamilton and Patrick Hamilton

by alien species might erroneously emerge as priority places for plant conservation based on their relative species richness. Information is also desirable on the processes responsible for the floristic characteristics of sites since this will assist in establishing the types of management that will be most useful.

The use of indicator species to signal sites of special conservation concern was pioneered in the UK in connection with the identification of ancient woodland (see Chapter 5). More recently, the presence of waxcaps (fungi in the genus *Hygrocybe*) has been found to be an excellent way of identifying and classifying unfertilized grasslands in the British Isles, some of which are of international conservation importance (Duckworth, 2002; Woods, 2002).

In general, more natural types of habitats are of greater plant conservation interest than those that are disturbed by people. For example, in tropical Africa, *secondary* forest (for example, as found on abandoned farms or where there has been heavy logging) contains many species that are widely distributed, while more *primary* forest contains more species of restricted distribution (narrow endemics). For instance, surveys in the Pande and Kiono forests of Tanzania have revealed that undisturbed forest is rich in plant species of restricted distribution (including some with very high site specificity), while forests heavily disturbed by people carry widely distributed species (Mwasumbi et al, 1994).

Despite this generalization, there are some secondary habitats of great conservation interest. Examples include semi-natural grasslands in parts of Europe and even some types of cultivated fields. Several weeds of cultivation have become rare in parts of Europe with the spread of intensive agriculture.

Local people can exploit a range of habitat types, though some will be more significant for certain purposes than others. In a study of the use of six floristically distinct forest types at Tambopata, Peru, the local *mestizo* people were asked to provide information on the values of woody species in sample plots placed in each of the forest types. The results revealed that very high percentages (average 94 per cent) of the woody stems in all forest types were useful to people, mostly for subsistence rather than commercial purposes. A weighting technique known as 'informant indexing' was used to distinguish *exceptionally* useful species from those with only *minor* uses. This demonstrated significant differences in the utility of the forest types. Mature forests on existing or former floodplains were found to be the most useful overall, mostly because of their values as sources of construction wood and food. Lower floodplain forest was the most valuable for botanical medicine. Extrapolating to other parts of Amazonia, the conclusions drawn from this study were that the maintenance of cultural autonomy requires that Amazonian people have access to *all* local forest types, that existing and former floodplain forests should be conservation priorities, and that conservationists should help communities to gain control of the resources in the floodplains (Phillips et al, 1994).

All plants are valuable as genetic resources, since they are either already known to be useful or potentially could be. However, because resources for conservation are limited, comparative valuation studies are needed to determine priorities. One category of plant resources that has, to date, been given high priority is that comprised of plants actually or potentially useful for crop breeding. For this reason, places that are centres of crop diversity or which contain wild crop relatives have sometimes been regarded as priorities for plant conservation. Within this context, it has sometimes been assumed that only primary centres of crop diversity are of much use as sources of genes for disease resistance, following the view that useful resistance evolves only through long association of wild species and pathogens. However, landraces in *secondary* centres of diversity are now known to also have value as sources of genes that provide resistance to disease (Hawkes et al, 2001).

The concept of Important Plant Areas (IPAs) has been promoted, notably by Plantlife International, as a tool for drawing attention to prime sites for plant conservation at scales appropriate to national-level planning and practical field-level management (Smart, 1999; Plant Europa, 2001; Anderson, 2002; Plantlife,

2004). Eventually, it is hoped, all countries will recognize IPAs – based upon their own national selection, but making reference to certain common criteria, ensuring reasonable comparability between countries. In principle, IPAs are intended to cover lower plants and fungi, as well as higher plants. An IPA designation is not an official conservation designation (for example, a protected area) in its own right and a wide variety of legal and other tools may be needed to protect them according to circumstances. Methods for the selection of IPAs are being developed with the intention that they will be sufficiently consistent, rigorous and transparent to convince botanists, authorities and lay people of their worth. To qualify as an IPA, a site has to satisfy one or more of the following criteria (further details are given in Table 9.5):

A   The site holds significant populations of one or more species that are of global or regional conservation concern.
B   The site has an exceptionally rich flora in a regional context in relation to its biogeographic zone.
C   The site is an outstanding example of a habitat or vegetation type of global or regional plant conservation and botanical importance.

The pilot country for IPAs has been Turkey, where work started in 1994, with a main project launched in 1998. Turkey is a country rich in plants, with about 9000 native vascular plant species (33 per cent endemic). During the pilot phase, a list of candidate sites was drawn up based upon the knowledge of local botanists and information in publications and on herbarium labels. This resulted in the identification of approximately 150 candidate sites, which were then visited to document their site characteristics, the diversity and quality of their vegetation, the presence and population sizes of rare species, and existing and potential threats. The main survey that then followed involved over 40 botanists from 17 universities. Eventually, the number of sites recognized was reduced to 130, covering 9 per cent of the Turkish land surface. These sites vary enormously in area (72ha to 1,250,000ha). It is thought likely that a further process is needed in order to determine which actions are most appropriate to ensure protection of the network and at what level they should be carried out (national, regional or local). The answers are expected to depend upon the type and level of threats at the sites themselves, as well as upon political opportunities for ensuring broader-scale protection (Byfield, 1995b, 2001).

The research carried out for the IPA survey in Turkey demonstrated great variation in the number of nationally rare taxa in different habitat categories. For example, there are 251 nationally rare species in a habitat type known as 'montane/forest', while there are only 4 nationally rare species in a habitat type termed 'raised bog'. The researchers emphasize that such figures must be appreciated within the context of the overall abundance of habitat types and their degrees of endangerment. Mountain tops harbour many endemic species; but there are many such sites and they are often not particularly threatened. In contrast, there is only one raised bog left in Turkey (Agaçbasi mire) and this is under great threat.

Table 9.5 *Selection criteria for Important Plant Areas (IPAs)*

Criterion	Description	Threshold	Notes
A (i) (threatened species)	Site contains *globally* threatened species	All sites known, thought or inferred to contain 5 per cent or more of the national population can be selected, or the five[1] 'best' sites, whichever is the most appropriate. (Populations must be viable or there is a hope that they can be returned to viability through conservation measures)	Species listed as *'threatened'*[3] on IUCN global Red Lists
A (ii) (threatened species)	Site contains regionally threatened species		Species listed as *'threatened'*[3] on regional IUCN Red Lists or regionally approved lists
A (iii) (threatened species)	Site contains *national endemic* species with demonstrable threat not covered by A (i) or A (ii)		Species listed as national endemics (on any recognized list or publication) and *'threatened'*[3] on national Red Lists
A (iv) (threatened species)	Site contains *near endemic/ restricted range* species with demonstrable threat not covered by A (i) or A (ii)		Species listed as near endemic/restricted range (on any recognized list or publication) and *'threatened'** on national Red Lists
B (botanical richness)	Site contains high *number of species* within a range of *defined habitat or vegetation types*	Up to 10 per cent of the national resource (area) of each habitat or vegetation type, or five[2] best sites, whichever is the most appropriate	*Species richness* can be based on a nationally created list of indicator species developed for each habitat or vegetation type. For example, characteristic species and/or endemic species and/or nationally rare and scarce species (where the endemic and rare and scarce species are numerous and/or are characteristic for the habitat). *Defined habitat or vegetation type* taken from or based upon a regionally accepted classification
C (threatened habitat or vegetation type)	Site contains threatened habitat or vegetation type	All sites known, thought or inferred to contain 5 per cent or more of the national resource (area) of priority threatened habitats can be selected, or a total of 20–60 per cent of the national resource, whichever is most appropriate	Threatened habitats or vegetation taken from a regionally recognized list

Note: 1 In exceptional cases – for example, where there are less than ten sites in the entire country or there are between five to ten large populations of a species – up to ten sites can be selected.
2 In exceptional cases – for example, where there are between five and ten exceptionally rich sites for a particular habitat – up to ten sites can be selected for each level 2 habitat type.
3 For criterion A, threatened species must be listed as *Critically Endangered (CR)*, *Endangered (EN)* or *Vulnerable (VU)* using the 1994 or 2001 IUCN criteria (IUCN, 1994a, 2001b), or *Extinct/Endangered (Ex/E)*, *Endangered (E)* or *Vulnerable (V)* using the pre-1994 IUCN categories.
Source: Plantlife, 2004

Some places that are important for plant conservation are sites of special value for conservation of other taxonomic groups apart from plants. There are initiatives, similar to that for IPAs, to identify important areas for other groups, such as birds (Important Bird Areas), fungi and butterflies (see, for example, Stattersfield et al, 1998; Van Sway and Warren, 2003). It has been proposed that places which are important for several taxonomic groups should be designated as Important Species Areas (ISAs) or Key Biodiversity Areas (KBAs), the latter covering additional aspects of biodiversity apart from just species.

SOME LARGE-SCALE PLANT PATTERNS RELEVANT TO CONSERVATION

Plant species are unevenly distributed on Earth. One analysis, based upon mapping plant distributions from numerous floras and other sources, resulted in identification of six places particularly rich in plant species: Chocó-Costa Rica; tropical eastern Andes; Atlantic Brazil; eastern Himalayas–Yunnan; northern Borneo; and New Guinea (Barthlott et al, 1996). The single province of Yunnan contains approximatley 17,000 species of seed plants, 56 per cent of the Chinese total (Zhang Qitai et al, 2003). Another study, associated with the NGO Conservation International (CI), has revealed that over 50 per cent of vascular plant species (and many vertebrates) are confined to just 34 hotspots, comprising only 2.3 per cent of the Earth's land surface (Myers et al, 2000; CI, 2005) (see Table 9.6).

A detailed study, the Centres of Plant Diversity project, identified many sites of prime importance for plant conservation on the global scale (WWF and IUCN, 1994–1997). Two hundred and thirty-four major sites and hundreds of minor ones were recognized (see Figure 9.3). The three-volume publication that resulted contains descriptions of all the sites, as well as regional overviews, making this one of the most complete surveys of global plant geography yet undertaken (WWF and IUCN, 1994–1997). The number of sites recognized per continent and per major habitat type (for example, rainforest) roughly corresponds to its relative species richness. On the local scale, strong reliance was placed upon the opinions of expert botanists, familiar with their areas – a reasonable approach (rather than relying upon statistics) given how little quantitative information on plant distributions is available for many regions (especially in the tropics). The principal criteria in selecting sites were that:

- the area is evidently species rich, even though the number of species present may not be accurately known; and
- the area is known to contain a large number of species endemic to it.

Table 9.6 *The global hotspots that harbour 50 per cent of all species of vascular plants*

North and Central America	Europe and Central Asia	Asia Pacific
California Floristic Province	Caucacus	East Melanesian islands
Caribbean islands	Irano-Anatolian	Himalaya
Madrean Pine-Oak	Mediterranean Basin	Indo-Burma
Woodlands	Mountains of Central	Japan
Mesoamerica	Asia	Mountains of south-west
		China
South America	**Africa**	New Caledonia
Atlantic Forest	Cape Floristic Province	New Zealand
Cerrado	Coastal forests of	The Philippines
Chilean Winter Rainfall-	Eastern Africa	Polynesia/Micronesia
Valdivian Forests	Eastern Afromontane	South-west Australia
Tumbes-Chocó-Magdalena	Guinean Forests of	Sundaland
Tropical Andes	West Africa	Wallacea
	Horn of Africa	Western Ghats/Sri Lanka
	Madagascar and the	
	Indian Ocean islands	
	Maputaland-Pondoland-	
	Albany	
	Succulent Karoo	

Source: Myers et al, 2000

Most mainland sites have (or are believed to have) in excess of 1000 vascular plant species, of which at least 100 are endemic either to the sites (strict endemics) or to the phytogeographical region in which the sites lie. In all cases the sites have at least some strict endemics. The Centres of Plant Diversity data were used extensively in the selection of the Global 200 sites chosen by WWF as priorities for ecoregion-based conservation (Olson and Dinerstein, 1998, 2002).

There are considerable overlaps in the sites identified as global hotspots for plants with those identified for other taxonomic groups or for biodiversity as a whole. For example, 25 places that were once selected by Conservation International as prime biodiversity sites on the global scale are reported to show 82 per cent overlap with the Centres of Plant Diversity sites, 92 per cent with WWF's Global 200 sites and 68 per cent with Birdlife International's Endemic Bird Area sites (Myers et al, 2000). A high degree of overlap between hotspots for plants and other organisms is to be expected because of functional dependencies (for example, some plants have specialist pollinators) and shared history (for example, related to Ice Age climatic change). However, not all plant-rich sites are also rich in other groups. A study of congruence in the distribution of endemic birds with endemics in other taxonomic groups in Amazonia found a good match *only* on the larger scale (Bush, 1994).

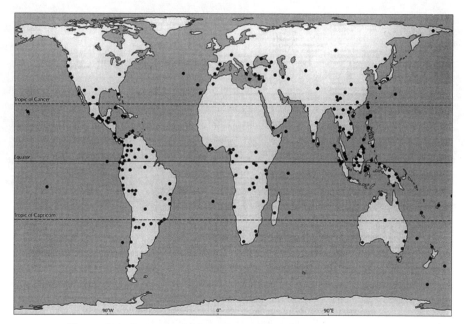

Figure 9.3 *Centres of Plant Diversity: These are prime sites for plant conservation on the global scale*

Note: The project was facilitated by WWF, IUCN and the Smithsonian Institution
Source: WWF and IUCN, 1994–1997

CASE STUDY: PRIORITIZING PLANT RESOURCES FOR COMMUNITY USE AT BWINDI, UGANDA

Bwindi Impenetrable National Park is one of the most valuable forests in Uganda for biodiversity conservation, including for plants. It stands on the front line of conservation (see Figure 9.4). It protects one of the catchments of the Nile, is home to half of the world's mountain gorillas, covers an area of only 321km² and is surrounded by one of the most densely populated parts of rural Africa (200 to 400 people per square kilometre). The forest was transferred from the status of a forest reserve to that of a national park in 1991 in recognition of its biodiversity value and to better control the extensive illegal exploitation of natural resources that had been occurring during preceding years of weak government. A survey during the late 1980s estimated that 61 per cent of the forest reserve had been heavily exploited for timber and the best hardwoods had been removed from a further 29 per cent of the reserve. Only 10 per cent remained intact (Butynski, 1984; Howard, 1991; Hamilton et al, 2000).

At first, there was little local support for the new park due to the rushed manner in which it was established and the lack of consultation with affected communities. Little attention was paid to local concerns about loss of access to resources and worries about increased crop-raiding by forest animals (Hamilton

Figure 9.4 *Boundary of Bwindi Impenetrable National Park, Uganda,
one of the most valuable forests in East Africa for biodiversity conservation,
surrounded by one of the most densely populated parts of rural Africa*

Source: Alan Hamilton

et al, 1990). Local people expressed negative sentiments about both the park
and the gorillas: 'When you mention the national park we want to vomit' and
'Gorillas should be put in cages and taken to zoos.' Sixteen fires, at least some
deliberately started or left uncontrolled, raged in the park during the dry season
that followed gazettement. Five per cent of the forest was burned and threats
were often made against the gorillas. Under the new status of Bwindi, the local
people enjoyed no rights in the park – not even, for instance, the right to collect
bark from the medicinal tree *nyakibazi* (*Rytigynia*), without which, they said,
'they would die' (Cunningham, 1996b). *Nyakibazi* is used to treat intestinal
parasites: 89 per cent of people around Bwindi are infested with whipworm
Trichuri and 34 per cent with roundworm *Ascaris* (Ashford et al, 1990). Yet, to
be caught collecting *nyakibazi* could result in a fine or imprisonment.

Several benefit-sharing schemes have been introduced at Bwindi, which
overall have contributed to a reduction in tensions between local people and
the park. These include the establishment of a trust fund by the Global
Environment Facility in 1996 and the sharing of revenues from gorilla tourism.
Of special relevance in the present context are agreements reached between
the park and some of the surrounding parishes, initially in 1994. These specify
the forest resources which the communities are allowed to collect in the forest,
including prescriptions about the quantities permitted and who may collect.
The initial agreements were between the park and forest societies established
in three pilot parishes, and in each case refer to a specific part of the forest (a

multiple-use zone). Permitted levels of off-take for plant resources were based upon assessments of the vulnerability of species to collection, considering their life forms, growth rates, parts used, degrees of endemicity, habitat specificity and whether or not the resources were commercially traded. These sustainability assessments were carried out by a team of scientists, park staff and local people, led by Dr Tony Cunningham. This work was undertaken on behalf of the development agency CARE International, whose staff (based in their local Development through Conservation project) later followed up the evaluation by facilitating the agreements between the park and the communities.

Plant resources collected in the forest were classified into three broad categories for the purposes of making recommendations about their use. One category was for resources that could be used with little conservation worry (at least so far as impacts on the species themselves were concerned). Attributes of resources placed in this group included originating from common or fast-growing species, and required in small quantities. At the other extreme were those resources that could clearly not be harvested from the forest sustainably because they originated from species that are rare or slow growing, or because they were required in large quantities. These resources included fuelwood, timber and stakes for climbing beans. In these cases, alternatives outside the forest were required – for example, through the growing of substitute resources on farms. The middle category included those plant resources about which judgements on sustainability were less clear. Tony Cunningham followed up this study with work under the People and Plants Initiative by mentoring many Ugandan postgraduate students studying these middle category resources in detail. The intention was to provide information for management.

Sources for case study: Cunningham, 1996b; Wild and Mutebi, 1996

10

Possession, Property and Protection

Questions of ownership and resource rights are central to plant conservation. Without a feeling of ownership – publicly recognized – there is little motivation to manage land or plants for long-term objectives. This chapter examines concepts of possession, property and protection, with special emphasis on traditional conservation, statutory law and protected areas.

TENURE AND RESOURCE RIGHTS

All societies which do not abuse their environments have customs and rules that govern how natural resources are used. Except in the most barren places in which no one is interested, a fundamental requirement for plant conservation is security of land tenure or the existence of acknowledged rights of use of natural resources (Tuxill and Nabhan, 2001). Without such assuredness, there is little motivation for good management.

Problems of habitat degradation and overexploitation of plant resources are found widely around the world as a result of unregulated open access to resources. The overgrazing of pastures by livestock and the depression of stocks of fuelwood – so widely encountered, especially in drier regions of the world – owe much to inadequate tenurial regimes. Even protected areas can be open-access areas, in effect, if authorities are weak and local people can collect the resources with impunity (in defiance of regulations or in ignorance). If resources are not available outside the protected areas due to overexploitation or privatization, then the protected areas can represent the last stocks of freely available resources, at least while they last.

Problems of inadequate tenure and resource rights are not restricted to developing countries. Take the case of Scotland, which many conservationists believe has an insufficient cover of native woodland. A problem here is that tenant farmers, who are many in Scotland, can be disinclined to plant trees because they have no guarantee that they will receive the benefits from doing so (Wightman, 1996):

> The UK government spends millions of pounds in its overseas aid budget to reform land tenure as part of wider development programmes. In Scotland, however, land reform is regarded as an

irrelevance, and feudal land tenure and the most concentrated pattern of private ownership in Europe is deemed to have no bearing on rural development policies and outcomes.

It can be much easier to offer advice on other people's problems than to deal practically with those at home!

Land and plants are key resources in rural societies, so questions of ownership and rights are political and can be highly charged (Tuxill and Nabhan, 2001). Conservationists need to be sensitive in raising these issues with communities. It may be best in projects to investigate them privately with individuals or small groups. Research into tenure and rights will reveal something of how communities function, including 'power relations' within them. Research may sometimes make matters explicit that are normally off the public agenda, perhaps creating tensions within communities that are unhelpful for solving conservation problems.

Some methods for the study of tenure and resource rights are given in Box 10.1. Numerous regimes of tenure and rights are found around the world. Frequently encountered concepts include **customary land**, which is land held under customary law, **communal land**, which is land held in common by communities, **state land**, which is land held by the government as a public resource (for example, many protected areas) and **private land**, which can belong to individuals, companies or other groups (de Klemm, 1990). **Freehold** refers to land over which individuals or groups have absolute and indefinite tenure – although, even so, there can still be laws and regulations limiting the types of permitted activities, including how plants can be treated. It is quite common for ownership and rights of different aspects of the same parcel of land to rest with different people (**pluralities**). For instance, the land may belong to one person, a house on it to another and the rights to harvest a fruit tree on the land to someone else. **Usufruct rights** are the recognized rights of certain individuals or groups to use specific resources.

Approaches taken by governments to questions of landownership and resource rights are strongly influenced by ideology. In some countries, such as Ireland, much land is privately owned and there is no legal limit to the extent of ownership. In more socialist-inclined countries, such as Tanzania, all rural land belongs to the state and may be either communally farmed or leased out to individuals or companies for fixed periods, often around 50 years. There can be limits to the sizes of individual landholdings, as in China (these are leased areas since all rural land in China belongs to the state). With regard to forested land, it is common for forests either to belong directly to the state (for example, as forest reserves) or there to be strict rules regarding the felling of trees. Mexico and Papua New Guinea are unusual in that most forested land is vested in communities (but with government restrictions about forest use and clearance in Mexico).

Tenure and resource rights commonly have statutory and customary elements. **Statutory law** is the law of the state, involving such matters as land titles, the leasing of land or other resources (for example, timber rights) by the

---

BOX 10.1 SOME APPROACHES AND METHODS FOR GAINING AN UNDERSTANDING OF REGIMES OF TENURE AND RIGHTS

- *Examining documents*:
  - analysis of documents and maps relating to land tenure and resource rights, including laws, inheritance documents, land titles, rental agreements and joint resource management agreements.

- *Analysing institutions and inheritance*:
  - studying governmental, village and other institutions to determine their interests and responsibilities;
  - finding out how rules are made and enforced by various types of institution;
  - studying how ownership or authority over land and resource rights is inherited.

- *Studying behaviour*:
  - seeing where *individual* people actually go and what they actually do – for example, in relation to collecting plant resources;
  - ascertaining the actual ways in which people behave in relation to plant resources in their *institutional* roles (for example, as members of government departments, village institutions, commercial companies or other groups).

- *Noting signs of ownership*:
  - fences, hedges and boundary plants;
  - signifiers or protectors of property, including marks, signs and charms.

- *Recording knowledge and opinions*:
  - recording views of individuals and groups on their knowledge of local tenure and rights, how these have evolved with time and their attitudes towards them;
  - studying variations in knowledge between social groups.

- *Analysing costs and benefits*:
  - who does the hard work and who receives the rewards?

---

state, agreements with communities on the management of land or resources (for example, joint forest management; see Chapter 11), and the legal recognition of customary tenure or rights. **Customary law** refers to long-standing rules and customs regulating land tenure and resource rights, often with roots that pre-date modern states.

Traditional understandings of land tenure in Africa are based upon kinship rights, with the rights to settle and cultivate land vested in clan or tribal heads. The saying in Buganda is '*Omwami tafuga ttaka, afuga bantu*' ('A chief does not rule land, he rules people'). Land in Africa was traditionally not seen as an absolute property of individual landowners, as with modern freehold. However, claims to the continued holding of usufruct rights could be greatly strengthened through forest clearance, long occupation or, in particular, the presence of burial sites (West, 1972; Almquist, 2001).

Although customary law sometimes remains influential in Africa, major changes in tenure and rights were introduced during the colonial period, notably the concepts of individually owned land and state-owned land. Areas perceived as not being 'beneficially occupied' by local people were commonly assigned to the state (for example, as Crown land) (West, 1972; Twaddle, 1993). The earlier colonial years were confusing to many Africans who did not appreciate hard English-style concepts of private ownership and who sometimes signed away their land without realizing what they were doing. There were similar problems in India and Ireland, where peasants who were unable to produce documents to the land could be treated to summary rent increases or evictions (Schama, 1995). There is a trend towards stricter privatization of land in Africa (whether legally registered or otherwise), resulting in some local people losing access to natural resources (Maundu et al, 2001).

Contrasts between customary and statutory approaches to rights over resources are apparent in a comparison of communal (Shona) and resettlement areas in Zimbabwe. Traditional lineage-based institutions remain important in the Shona areas. Authority is vested in chiefs, whose duties include presiding over rain-making ceremonies, at which they intercede with the ancestral spirits on behalf of their people. In contrast, the resettlement areas are socially mixed and follow a more democratic model. In this case, responsibility for ensuring proper woodland and agricultural management rests with resettlement officers and agricultural extension workers (both state employees), along with village development committees (VIDCOs). In both the Shona and resettlement areas, the rights to land and woodland resources are assigned to men, with access to these resources by women being contingent upon their relationships with men. However, widows are treated differently in the two cases. In communal areas, women lose their rights when their husbands die and are evicted, while in resettlement areas the trend is for the rights of deceased husbands to be reassigned to widows (Goebel, 2003).

An historical perspective is invaluable for understanding regimes of tenure and rights. The story of landownership in Mexico illustrates how modern arrangements are rooted in the past and how they have been influenced by events. One of the products of the Mexican Revolution (1910–1920) was the creation of *ejidos*, which are communally registered lands belonging to villages – or at least male villagers, since the registered members of *ejidos* (*ejidatarios*) have been almost exclusively male. By 1980, about 80 per cent of forested land in Mexico belonged either to *ejidos* or to *comunidades* (communal landholdings associated with indigenous communities). There were certain limitations on the rights of *ejidatarios* – for example, they could not raise capital by mortgaging their land and the rights to timber on their land were assigned to outside companies by the government, though with stumpage fees paid to the *ejidos*. *Ejidatarios* began to take more direct control of their forest resources from the 1980s. Rules governing agricultural land in *ejidos* changed in 1992 when, under the influence of the North American Free Trade Agreement (NAFTA), the number of registered *ejidatarios* in an *ejido* became frozen and agricultural plots could be privately registered, subdivided and sold.

Another way to try and understand regimes of tenure and rights is through analysing the journeys that people take through life, as they take opportunities and suffer restrictions in relation to natural resources. The landscape can be seen as a mental map with a social overlay, not just as a bald physical reality. For those who have long lived in an area, much of the social geography of resources is accepted with little thought – 'this field belongs to someone; here is a path that I can use; these are wild fruits that I can pick'. The interactions of people with nature can then be analysed in relation to statutory and customary rules – according to how the rules are *supposed* to operate, how they *actually* operate and how individuals *understand* them. The legal position, as evidenced by land registration and titles, can lag far behind the position on the ground (West, 1972). Confusion over land and resources is very common – for example, with complex and confusing laws, disputes between government agencies over which is responsible for particular protected areas, differences in the perceptions of parties about where property boundaries lie, and divergence between boundaries as shown on maps and realities on the ground. There can be disagreements as to whether customary or statutory rules should be followed in the inheritance of land (West, 1972; Tibatemwa-Ekirikubinza, 1999).

One advantage of taking a 'mental journey' approach is that it readily allows the inclusion of illegal activities into analyses. Illegality can be rife in relation to natural resources. Indeed, it can be so common that knowledge of how 'alternative social networks' (corrupt networks) work can be more useful for unscrupulous individuals wishing to gain access to land or plant resources than dealing legally with official channels. Illegality or semi-legality can be a major impediment to good management of plant resources, as so often seen with the issuing of timber licences. However, on occasion it can yield conservation benefits. In Soviet days, the exchange of germplasm of fruit and nut trees between smallholders in Uzbekistan was illegal; but this traditional practice continued anyway, contributing to the survival of genetic diversity (Eyzaguirre, 2002).

CUSTOMARY CONSERVATION

There are many customary beliefs and practices that are supportive of plant conservation. In so far as people engage in such practices for conscious logical reasons (rather than just because *that's how things are*), some explanations given by people are down to earth and probably widely comprehensible across cultures, while others can seem less rational or even superstitious, being based upon understandings peculiar to the societies. Conformity to customary practices is encouraged in societies by rewards and punishments, the latter ranging from social disapproval, fines, ostracism or, in very serious cases, expulsion from society (Tsouvalis, 2000). An example of customary conservation that is probably readily comprehensible to all is that of the Khumbu Sherpa of Nepal, where the villagers appoint guardians (*nawas*) over communal resources to serve on a rotational basis. The guardians regulate the movement of herds and cutting in the forests, with their authority backed by measures of enforcement that include a system of fines (Nakashima, 1998). A less rational belief, at least to cultural outsiders, might seem to be one common in Ireland, affording protection to isolated trees of hawthorn *Crataegus monogyna*, known as 'fairy trees'.

People of many cultures view the natural world, or aspects of it, as sacred or otherwise worthy of respect. According to a poster exhibited by Jan Salick and her colleagues at the 2004 meeting of the International Society of Ethnobiology in Canterbury, England, Tibetans consider the whole environment as sacred, without a strong polarization into the secular and sacred domains. However, certain springs, groves, trees and mountain passes are seen as especially worthy of conservation. They are auspicious places for rituals to accrue merit, honour the past and placate the deities.

Mountains are often held to be special. The Kikuyu, Embu and Meru traditionally view Mount Kenya as the House of God, while the Dusun in Borneo believe that the upper slopes of Mount Kinabalu are the resting place of the deceased and are thus taboo to the living (Martin et al, 2002; Morimoto and Omari, 2002). The Romantic Movement in Europe (circa 1790–1850) saw the growth of a somewhat similar interest in wild mountains, the seashore and ancient groves of gnarled trees – places to find one's soul and seek spiritual refreshment. This was one of the foundations of the modern conservation movement.

Small areas of semi-natural vegetation are widely conserved around the world as sacred or holy forests (for example, in Africa; see Lebbie and Guries, 1995; Luoga et al, 2000; Yaa, 2001) (see Figures 10.1, 10.3 and 10.4). Occasionally, other types of semi-natural vegetation apart from forests are represented, as in the UK, where many churchyards carry semi-natural grassland that can be of high conservation value. Sacred forests are commonly used as burial sites and occasionally provide communities with special resources, such as medicinal plants. Where natural vegetation has largely been removed from landscapes, as in northern Ethiopia (see case study at the end of this chapter), sacred forests can stand out as patches of darker greenery set amidst the more pallid shades of agricultural or pastoral land. In some cases, sacred forests are valuable for conservation of traditional crop varieties, as with pears *Pyrus* in patches of disputed land (*serai*) administered by the Islamic ecclesiastical establishment in the Swat valley of Pakistan (Habib Ahmad, personal communication, 2000).

Many species or varieties of plants receive customary protection or encouragement. Indeed, there are some species that are unknown or barely known in the wild, such as the medicinal trees *Eucommia ulmoides* and *Ginkgo biloba*. They have only survived into modern times because they have been cared for through cultivation (He Shan-an et al, 1997). Some clans have totemic ancestors represented by certain types of plants (though more commonly animals), which are then afforded special protection. Protected tree species are common in Africa. *Sclerocarya birrea*, *Sterculia africana* and *S. appendiculata* are examples of species valued for traditional worship in one area of Tanzania (Luoga et al, 2000).

Figure 10.1 *Forest patch conserved to protect a source of holy water,*
Anchucho Church, Ethiopia

Source: Alan Hamilton

There are many traditional beliefs and practices that restrict the amounts of plant resources that may be collected or encourage the regeneration or regrowth of plants after harvest. Villagers in Garhwal Himalaya of Uttaranchal, India, consider alpine meadows to be the home of God. They only harvest medicinal plants during the festival of Nanda Ashtami. Collection starts when selected villagers (*jagaryas*) travel to the alpine areas to collect flowers to offer to the goddess Nanda. Only then can general collection commence. If medicinal plants are collected prior to the ritual, then disasters such as droughts may follow. The festival falls in the last fortnight of September or first fortnight of October – the time of seed set for most of the high-altitude medicinal plants – so these beliefs and practices can be seen as cultural mechanisms to avoid unnecessary depredation of the plants (Dhar et al, 2002).

It is normal for individuals or groups of people to protect certain specimens of plants in their localities. This is partly a matter of maintaining what has become familiar and accepted as part of reality. New homeowners can find it easier to fell trees in their gardens when they begin occupancy before attachment has developed. Some plants are protected for obvious practical reasons, such as trees providing shade in tropical villages. People often want to maintain certain specimens of trees for their aesthetic or mystic qualities, related to their size, shape or form. Who would dare fell the tallest tree in the land? The concept of 'the heritage tree' (an outstanding specimen) is increasingly being promoted and is expected to be a useful devise for promoting plant conservation in many countries.

There are many customary conservation practices associated with traditional agriculture. 'Wild' plants and local varieties of crops can be nurtured, and forest patches retained, as parts of overall systems of production (see Chapter 6). The following example demonstrates how traditional beliefs and practices can be undermined by modernity, but how (more hopefully) total environmental disaster can be averted once the values and wisdom of traditional ways are recognized. The Dai are a long-established people practising irrigated cultivation of rice in the plains of Xishuangbanna, the main area for tropical forest in China. These plains are surrounded by hills, once largely covered by forest and home to various other nationalities, practising shifting agriculture and living off forest produce. Until recently, Dai society was feudal, with the management of the complex irrigation system under the control of lords. Villages and households were assigned tasks to maintain the system and share its water, with severe punishments for those who failed to act their parts. Water played then – and continues to play – a major role in Dai culture. It is not seen just as a material commodity (Gao Lishi, 1998):

> *The Dais need water not only to cultivate paddy fields and other crops, but also to raise livestock and poultry like buffaloes, geese and ducks. They also need water to worship their gods and hold ceremonies... Hence, the Dais have always worshipped water and they have tried their best to protect water resources as they have protected their life. In the past they not only looked upon their Village-God Forests and Meng-God Forests as the dwelling places of their ancestors' spirits, but also as their water resources and no felling was allowed... Their music, dance, literature and art were closely related to water.*

The Dais have traditionally believed that their fates in this world are dependent upon the water goddess Nantuolani. No killing of animals, collection of plants, felling or cultivation were allowed in the watershed forests, which once extended over 100,000ha.

The founding of the People's Republic of China in 1949 resulted in many transformations at Xishuangbanna. The emphasis was now on economic growth with little consideration given to the ecological consequences. Religion was held to be superstition and was banned: 'Fight against the evils, ask forests for grain! Topple old idols and emancipate the mind!' Some traditional tools of irrigation, such as wooden dams, bamboo dykes and earth-banked canals were replaced with structures in concrete, seen as a more 'progressive' material. The area of forest rapidly declined, largely because of a big expansion in plantations of rubber *Hevea brasiliensis*. Many Han Chinese were brought in to work in the plantations, contributing to an increase in the population of the Xishuangbanna Prefecture from 230,000 individuals during 1953 to 817,000 in 1995. At one time the entire tropical forest area of Xishuangbanna was destined for replacement by plantations. The forest area declined from 66 to 70 per cent (during the early 1950s) to 30 per cent (1980) (Gao Lishi, 1998).

Fortunately, the excesses of this unbridled drive for economic expansion and 'scientifically based production' became curbed from the mid 1980s, when more consideration started to be given to traditional knowledge, the ecology of the area and the biodiversity values of the forest. A seminal event was the Second National Symposium on Rubber and Tropical Economic Crops at which Professor Pei Shengji managed to argue successfully against the whole-sale replacement of tropical forest in Xishuangbanna with rubber plantations. He could do so because he came well armed with evidence from research of the value of produce from the forest to local people (and potentially the nation) and the essential role of the forest in maintaining the irrigation system.

STATUTORY LAW

Government policies, laws and regulations are critical to plant conservation, given the primacy of the state in modern systems of governance. They may be formulated under various jurisdictions, according to how the state is structured (see Chapter 3).

Some countries have regulations that protect individual specimens of plants. In the UK, local planning authorities can place Tree Preservation Orders (TPOs) on single trees (or woodlands) that make it an offence to fell, cut or otherwise damage them. The aim is to protect trees that make a significant impact upon their local surroundings. The penalty for deliberate destruction or damage to a tree can be a fine of up to £20,000 (US$38,000).

Some types of plants can be declared undesirable and measures may be instigated for their control. In the UK under the 1959 Weeds Act, landowners are required to control five named species of arable weeds, such as ragwort *Senecio jacobaea*, which is poisonous to livestock. Under the 1981 Wildlife and Countryside Act, it is an offence to plant or grow Japanese knotweed *Reynoutria japonica*. Wild species can be targeted if they are seen as enemies of agriculture. Barberry *Berberis vulgaris*, though possibly not a native plant, but introduced in antiquity as a medicinal, is now rare in Britain following measures taken for its eradication after it was discovered to be an alternate host for black rust fungus *Puccinia graminis*, which attacks wheat *Triticum*. Plants that are desirable at one time can become undesirable at others. Henry VIII of England (1509–1547) required farmers to plant hemp *Cannabis sativa* to provide fibre for ropes for his navy. The cultivation of *Cannabis* was *banned* in 1961 as a result of its narcotic properties. However, hemp is increasingly being grown under license in the UK, once again as a source of fibre. Varieties are used that have little of the narcotic chemical (Mabey, 1997).

Many states have laws and regulations to protect wild plants. In the UK, no wild plants may be uprooted (except by landowners on their own land), although the aerial parts of most species can be freely picked. There is a list of specially protected species which no one, not even landowners, is allowed to uproot or pick, except under licence. As of 1995, this list included 107 species of vascular plants, 33 bryophytes, 26 lichens and 2 stoneworts Characeae. Legal protection of wild plant species has not proved easy to enforce in the UK (or anywhere else) and few legal cases have been brought (de Klemm, 1990; Palmer, 1995).

Some countries go further and require or encourage species recovery plans for threatened species. The USA was the first country to pass a law stipulating that recovery plans must be prepared and implemented for endangered species as a provision of the 1973 Endangered Species Act. By 2004, 609 of the 749 species of American plants protected under this act had their own recovery plans. The concept of the recovery plan is gaining ground in Europe, where Finland and Spain now have similar legislation. The UK has adopted a less legalistic approach, but Species Action Plans have been prepared for some threatened species. They are implemented by various statutory and voluntary bodies, such as Plantlife International.

While endangered species should receive legal protection, there are merits in allowing people to collect some wild flowers for their pleasure or edification. Many field botanists became interested and knowledgeable about plants through making plant collections in their youth. De Klemm (1990) has recommended a 'one bunch' (small handful) rule for non-endangered species. He believes that it would be helpful if a simple rule like this could be widely applied across several jurisdictions to increase the chance that it will become widely known and respected.

Certain habitat types, or particular areas of habitat, are sometimes protected by law, aside from any measures that apply specifically to protected areas (see 'Protected areas'). Tanzania prohibits the clearance of natural forest for the establishment of plantations. In Denmark, there are prohibitions against making alterations to a number of natural features and processes – for example, to the states of most rivers, streams, peat bogs and salt marshes, and to grasslands and heathlands if larger than a quarter of a hectare (de Klemm, 1990). Some legal protection is afforded in the UK to sites designated as Sites of Special Scientific Interest (SSSIs) in Britain or Areas of Special Scientific Interest (ASSIs) in Northern Ireland. There are over 4000 SSSIs in England, covering 8 per cent of the land area. They have been established to conserve the native flora and other worthy natural features. Owners who intend to make changes in these areas are obliged to give official notification of their intentions. Permission for the proposed developments can be refused.

A range of laws and regulations can govern access to, and use of, certain plant resources. Forest clearance and the felling of timber trees are controlled by law in many countries, even on private land. In the UK, an application must be made to the Forestry Commission to fell trees if the yield of wood is more than five cubic metres per three-month calendar period. In the North West Frontier Province of Pakistan, there is a form of land tenure known as *Guzara* (covering 53 per cent of forested land), whereby the land is privately owned; but licences are required from the Forest Department for the felling of timber trees. Occasionally, non-timber products are subject to regulation. In Turkey, the commercial collection of certain wild bulbs, such as those of the snowdrop

Galanthus, is allowed only under licence. Amounts are subject to annual quotas.

Plant resources can be the focus of a range of regulations. These can cover the quantities permitted for collection, the methods of collection allowed and the fees that must be paid. The payment of fees can be required at different stages along market chains, including for harvest (for example, stumpage payments or royalties in the case of timber trees), transport, sale or export. Policies relating to quotas and licences can significantly influence how well the plants are managed. There may also be issues of justice, relating to the ways in which the benefits and costs associated with resource management, harvest and use are distributed through society. If fees are set too low, there may be little incentive for proper management, operations can be wasteful and profits to merchants may be considered excessive (Brigham et al, 1996). On the other hand, if fees are set too high, more people may be tempted to avoid payment and illegality is therefore encouraged.

A range of incentives can be offered to landowners to encourage them to undertake measures in favour of conservation. The Forestry Commission in Britain (the Forest Service in Northern Ireland) runs a Woodland Grant Scheme that offers grants to create new woodlands and to encourage the improved management of existing woods. In 1988, the European Union (EU) introduced a Set Aside Scheme as a response to the excessive production of food in the EU. By 1996, farmers were required to set aside 10 per cent of their cropping areas to qualify for payments. How far set-aside land is valuable for plant conservation is debatable. One study in Scotland found benefits to plant diversity from set aside to be minimal and concluded that a better approach, if plant conservation is an objective, would be to sow appropriate seed mixtures (Christal et al, 2001). In 2005, many schemes to encourage more environmentally friendly farming in the UK were brought together under a new Agri-Environmental Scheme (with two levels of standard).

Despite all of their laws and regulations, in practice some governments have rather little direct control over the management of resources. Many are financially weak and have few field staff. In some cases, natural resource management has been further undermined by laws that remove authority for natural resource management from traditional institutions, but fail to provide effective substitutes. One way forward for governments struggling to improve resource management is to draw upon local support by granting legal recognition to traditional systems of resource management. However, thought is needed in doing so, considering that societies and environments are not the same as in the past. Adaptations are needed to suit modern times. There have been moves towards giving more authority to local institutions in several countries, such as India and Nepal (with regard to forest management), and Kenya (with regard to sacred groves and other sites of local cultural interest) (Morimoto and Omari, 2002; Pant, 2002). A return to traditional communal management of pastures has been recommended for China, although government ideology and a rigid belief in the individual household responsibility unit have proved obstacles to taking this course (Richard, 2003).

PROTECTED AREAS

Protected areas form one of the most useful tools for conserving plants. Their value for conservation is widely appreciated around the world, judging by the great expansion in their numbers and aerial coverage over recent years (MacKinnon, 1997; Jepson et al, 2001). A protected area, by definition, must have biological conservation as one of its management objectives (CBD, 1992; IUCN, 1994b). However, it is relatively rare for places to receive protection solely for conservation of biological diversity. Common additional objectives include protection of timber and water resources, the safeguarding of wilderness or valued landscapes, recreation and tourism. Some protected areas are in places where traditional agriculture is practised. In these cases, protected area management may be able to assist with the conservation of crop diversity – within or near the protected areas – as well as with the diversity of wild plants (Tuxill and Nabhan, 2001).

Few protected areas have been established specifically for plant conservation. However, more are being founded, such as in Spain, where many small areas have recently been registered for the conservation of species of restricted distribution (see Chapter 6). The first national park in tropical Africa, proposed specifically for the protection of plant diversity, is an area of 135km^2 on the Kitulo Plateau in the Southern Highlands of Tanzania. The special interest is grassland with an exceptionally rich herbaceous flora (WCS, 2002).

There are numerous categories of protected areas – for example, 45 named types in Australia alone and over 1388 terms applied worldwide (Green and Paine, 1997). The World Conservation Union (IUCN) has devised an international system of classification for protected areas to facilitate comparison between countries (see Box 10.2). There is an unresolved question as to how to treat forest reserves, which are generally excluded from the IUCN system. Their exclusion is not surprising in many instances, given the low level of protection that forest resources often receive. However, some forest reserves are better protected and contain very significant levels of biological diversity. Furthermore, forestry departments in many countries are giving increasing attention to the conservation of biodiversity as a management objective (Hamilton et al, 2003; Burgess et al, in prep).

Protected areas are the responsibility of various government agencies, each with its traditions. National parks often fall under wildlife departments, which can have special interests in animals and tourism, while forest reserves belong to forestry departments, typically most interested in timber. Regulations covering national parks tend to be stricter than those for forest reserves, with fewer activities permitted. Because of this, conservationists have sometimes advocated the transfer of certain forest reserves to national park status; but this is not necessarily beneficial for conservation. So many national parks have been declared in some countries that it is possible that the public will come to believe that too much land has been alienated from productive use. Some forest reserves in Ghana have been converted to wildlife conservation areas over

---

Box 10.2 The World Conservation Union (IUCN)
Protected Area Categories

The IUCN definition of a protected area is 'an area of land and/or sea especially dedicated to the protection and maintenance of biological diversity, and of natural and associated cultural resources, and managed through legal or other effective means':

- category I: *strict nature reserve/wilderness area*, managed mainly for science or wilderness protection;
- category II: *national park*, managed mainly for ecosystem protection and recreation;
- category III: *natural monument*, managed mainly for conservation of specific natural features;
- category IV: *habitat/species management area*, managed mainly for conservation through management intervention;
- category V: *protected landscape/seascape*, managed mainly for conservation of a landscape/seascape or for recreation;
- category VI: *managed resource protected area*, managed mainly for the sustainable use of natural resources.

Source: adapted from IUCN publicity literature

---

recent years under the argument that the Forestry Department does not adequately conserve wildlife. However, the real issue may be understaffing and it might have been better to incorporate wildlife management within the activities of the Forestry Department (Yaa, 2001).

Plant conservationists can often usefully forge alliances with other parties with an interest in the same areas. Some countries have indigenous reserves, established to protect indigenous people from erosive cultural and economic influences. These are particularly relevant to nature conservation when people's lives remain closely tied to nature. Colombia has 400 indigenous reserves, covering 24 per cent of its territory. The Parque Nacional Indi Wasi, established in Colombia in 2002, is the first national park in Latin America with the *dual* objectives of conservation of biodiversity and indigenous culture – and, possibly, the first anywhere. Indigenous reserves are unknown in most of Africa, where generally the concept of indigenous people is more difficult to apply than in Latin America. However, there has been a call for the establishment of indigenous reserves for 'pygmy' people in central Africa, threatened as they can be by loss of habitat and cultural erosion through the incursions of land-hungry settlers (Peterson, 2001).

Archaeologists can be natural allies for plant conservationists, with a shared interest in 'heritage'. Harappa is an ancient city of the Indus Valley Civilization in the Punjab, Pakistan, dating to the third millennium BC. The archaeological reserve at Harappa is only 1.5km² in size; but, even so, it is of great value for plant conservation because it carries one of the very few remain-

Figure 10.2 *Tsering Youdon of Pragya and Elizabeth Radford of Plantlife International at a community* in situ *plant conservation site with the endangered medicinal orchid* Dactylorhiza hatagirea, *Khangsar, Himachal Pradesh, India. A community-based conservation project of Pragya.*

Source: Alan Hamilton

ing tracts of native thorn forest remaining in the region (Khan, 1994a). Military reserves, such as training areas, can also be excellent sites for endangered species and habitats.

The great majority of protected areas are state owned; but protected areas are sometimes declared by communities and individuals. Private reserves are being increasingly established in Colombia, Italy, New Zealand, South Africa, Spain and Venezuela, among other countries (Figure 10.2). Their value for conservation will often be enhanced if they receive a degree of recognition from the state since this may help to guard against their loss – for instance, if they change owners.

There are mixed indications about the effectiveness of protected areas. On the positive side, a survey of 93 protected areas in 22 tropical countries, undertaken to test the hypothesis that parks are an effective means of protecting tropical biodiversity, found that the majority have been successful at stopping land clearance and, to lesser degrees, at mitigating logging, hunting, fire and grazing. The level of effectiveness of reserves was found to correlate with that of their basic management, such as enforcement, boundary demarcation and direct compensation to local communities. A conclusion from this research was a prediction that even modest increases in funding would directly increase the abilities of protected areas to protect tropical biodiversity (Bruner et al, 2001).

Protected areas are faced with many challenges. For example, in Southeast Asia, there are numerous reports of conflicts between conservation and development, and some protected areas that exist in law have never actually been established on the ground and have become severely degraded (MacKinnon, 1997; Jepson, 2001). A lack of resources is a fundamental problem in many developing countries, leading to understaffing, few management plans and a high turnover rate among senior management.

War, civil unrest and political anarchy are common in parts of the world, creating testing conditions for protected areas. In some cases, as in Angola, Mozambique and Liberia, civil war has led to a breakdown in the management of protected areas and degradation of habitats (Bandeira et al, 1994; Huntley and Matos, 1994; Appleton, 1997). A loss of the authority of the central government in Indonesia since 1998 has led to rampant illegality in protected areas in Sumatra and Kalimantan, with numerous unlicensed logging operations and extensive burning of forest (Jepson et al, 2001). In contrast, a period of weak government in Uganda during 1971 to 1986 failed to result in massive loss of forest reserves to agriculture. According to an analysis, the reserves had been established for sufficiently long (mostly 30 to 40 years) to be accepted as realities by local people and their continuing existence provided psychological reassurance as symbols of orderliness and normality in an unstable and sometimes very dangerous world. Where encroachments did occur, this was often due to illegality on the part of government leaders, not ordinary people (Hamilton, 1984). A similar acceptance of protected areas has been reported from Ghana, where a survey of four villages near forest reserves in Ghana found that 90 per cent of residents considered that the establishment of the reserves had been useful and contended that there would be no primary forest left if this had not been done (Yaa, 2001).

Almost by definition, protected areas will commonly result in the curtailing of certain activities that could produce short-term benefits. There are many such benefits that may be forgone, such as conversion to agriculture, the harvesting of timber and the mining of minerals, to mention but a few. Protected areas are likely to prove most defensible (individually and as networks) if it can be demonstrated that they are rationally conceived in terms of biodiversity conservation, serve a range of interests across society and there is transparency about how benefits from their existence are distributed. Since protected areas are public assets, there should be a measure of democratic accountability in how decisions about their management are made. Problems can arise if benefits from protected areas are thought to be distributed inequitably (Vandergeest, 1996; Naughton-Treves and Weber, 2001; Noss, 2001; Yaa, 2001).

There has been debate in conservation circles about whether and how local people should be involved in the management and use of protected areas. These questions become pressing when local people are economically dependent upon the resources of the protected areas. Agencies responsible for protected areas may naturally be wary of giving local people rights to use resources within protected areas if they see that the land outside the protected areas has been denuded of the same resources and fear that the same fate will happen next within them. On the other hand, there are advantages to involv-

---

BOX 10.3 SOME BENEFITS OF INVOLVING LOCAL PEOPLE IN THE MANAGEMENT OF PROTECTED AREAS

Benefits of community involvement in protected areas include the following:

- impracticality of locking local people out of protected areas;
- maintenance of cultural ties between people and nature;
- continuation of long-standing practices that support conservation;
- access to local knowledge relevant to resource management;
- local support for management available at times of political conflict or anarchy;
- management continuity (senior staff of protected areas are sometimes moved frequently);
- less credibility in the charge that protected areas are against human rights.

Sources: Naughton-Treves, 2001; Peterson, 2001; Tuxill and Nabhan, 2001

---

ing local people in the management of protected areas (see Box 10.3); in any case, the question of whether or not to involve local people in managing protected areas is not really a policy option that can simply be accepted or rejected. Community involvement in protected areas is a reality in many cases – whether legal or illegal – and the real question is how this can be developed and built upon. Attempting authoritarian suppression could invoke violent resistance and be politically suicidal for governments in cases where local populations depend closely upon the resources of protected areas. The management of protected areas has to be practical in local social and political terms, as well as make sense biologically and ecologically (Adams and Hulme, 2001; Wilshusen et al, 2002).

One approach to involving local people is collaborative resource management (see Chapter 11). Another is to divide protected areas into zones designated for different purposes, along the lines of the biosphere reserve model (see Figure 6.3). Local people might then be allowed to collect designated resources in one zone, while another is treated as a strict nature reserve. Whatever arrangements are developed, they should be based on a good understanding of how local people relate to local nature. For example, it might be that forests in a protected area owe their origin and continuing existence to the beliefs and practices of local people. Maybe some are holy forests, receiving traditional protection from local people, while others are stages in systems of shifting agriculture. The conservation of these forests could well be imperilled if local people are excluded from the protected areas because the processes that have been responsible for creating and nurturing them have then been removed.

Integrated conservation and development projects (ICDPs) became popular with agencies and international non-governmental organizations (NGOs) during the 1980s as a way of reconciling the conservation of biological diversity in protected areas with the economic needs of local people. One of the major strategies attempted was to reduce the dependency of local communities upon the resources of protected areas by providing economic alternatives. However, many first generation ICDPs failed either to enhance conservation or to generate benefits for conservation that persisted after the external funds became exhausted. Their weaknesses have included (Wells et al, 1999; Worah et al, 1999):

- ineffective linkages between conservation objectives and developmental activities;
- failure to appreciate that effective enforcement of laws and regulations is a basic requirement for successful conservation;
- insufficient knowledge bases, leading to poorly informed decision-making;
- limited focus and scale, and, thus, failure to deal with external threats;
- lack of consensus among stakeholders on objectives, roles and responsibilities, leading to continued conflicts of interest and confusion over priorities.

A second generation of ICDPs is benefiting from the lessons of these early experiences. Conditions for success are now seen to include high-level political and administrative support, close communication with local governments, a degree of local flexibility in planning, being able to respond to local needs as they arise, and adoption of adaptive approaches to project development and resource management (McShane, 1999; Wells et al, 1999). Sejal Worah and her associates have made many other suggestions for improvements and, beyond that, have now widened the coverage of the 'ICDP approach' to include *all* attempts to achieve conservation of natural resources in conjunction with sustainable human development. They have introduced the concept of the integrated conservation and development (ICD) initiative, no longer seen as applicable only to protected areas. Some of the key elements of the ICD approach are (Worah et al, 1999):

- participatory situation analysis (this is the gathering and analysis of information, using participatory rural appraisal, root-cause analysis, action research and other approaches; all key stakeholders should be included in the analysis);
- stakeholder negotiations and agreements;
- development of partnerships for implementation;
- the building of capacity for change (this involves the development of necessary attitudes, knowledge and skills among all concerned groups);
- working on a wide scale, including at the policy level;
- assessing impacts and sharing lessons.

CASE STUDY:
CHURCH FORESTS IN ETHIOPIA

There are numerous protected forests associated with the churches and monasteries of the Christian Ethiopian Orthodox Tewahedo Church. They range in size from 0.5ha to greater than 50ha, church forests typically being smaller (generally less than 1ha) than those of monasteries. The heartland of the Orthodox Tewahedo Church is the northern Ethiopian plateau, an area heavily farmed and with very little natural forest remaining outside church land. *Eucalyptus*, quite recently introduced from Australia, is the main tree in the farming landscape. The small church forests, often on hilltops, can be visible at great distance as little islands of darker green. There are some 34,000 churches and just over 1000 monasteries of the Orthodox Tewahedo Church in Ethiopia, 25 to 50 per cent of the churches having some forest.

The churches have a typical architecture consisting of two concentric courtyards (*atsed*) surrounding a central church building (see Figures 10.3 and 10.4). The inner courtyard is mostly covered by a low grass-dominated sward with scattered large trees towards its periphery. Ornamental and fragrant plants, such as *Artemisia afra*, *Canna*, *Datura* and rue *Ruta graveolens*, are commonly planted immediately around the church building to provide beauty and fragrant materials for church use. The inner courtyard is generally surrounded by a palisade of upright split timbers, preferably of juniper *Juniperus excelsa*. Small houses are sometimes found towards the periphery of the inner courtyard for the shelter of pilgrims and other religious purposes. The outer courtyard, which may be demarcated by a wall, fence or hedge, is typically forested. Both the inner and outer courtyards are used for religious purposes, such as for services, ceremonies and meditation. Both contain graves, especially the outer courtyard. All churches have access to nearby sources of holy water (*tasabel*), which variously may be springs, rivers or lakes. Springs supplying holy water are protected by small patches of consecrated forest (*beWigz yetekebere den*), preserved to maintain water supplies.

The churches are fundamentally embedded within local communities, from which church officials, including priests, are appointed. Many priests are farmers. Monasteries differ from churches, not only in typically having larger areas of associated forest, but in that many of their monks and nuns may not be drawn from local communities. Churches and monasteries have considerable degrees of autonomy, though the Orthodox Tewahedo Church has a hierarchy with bishops headed by a patriarch.

Species composition in the forests varies with altitude (1400–3000m) and from the wetter south of Ethiopia to the drier north. Many of the trees are characteristic elements of drier montane forests in Eastern Africa, such as *Acacia abyssinica*, *Allophylus abyssinicus*, *Celtis africana*, *Croton macrostachyus*, *Euphorbia candelabrum*, *Juniperus excelsa*, *Nuxia congesta*, *Olea europaea*, *Pittosporum viridiflorum* and *Podocarpus falcatus*. Tree heathers *Erica* are found at higher altitude. The ground vegetation beneath the

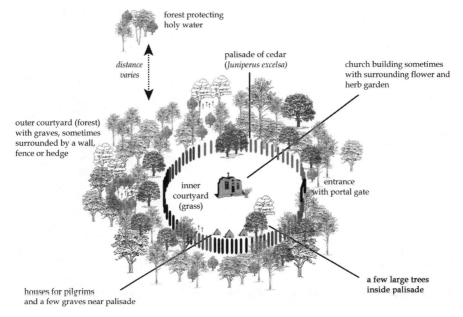

forest protecting
holy water

*distance
varies*

palisade of cedar
(*Juniperus excelsa*)

church building sometimes
with surrounding flower and
herb garden

outer courtyard (forest)
with graves, sometimes
surrounded by a wall,
fence or hedge

inner
courtyard
(grass)

entrance
with portal gate

houses for pilgrims
and a few graves near palisade

a few large trees
inside palisade

Figure 10.3 *Architecture of a church of the Ethiopian Orthodox
Tewahedo Church, showing holy forest*

Source: Alan Hamilton

trees varies from being very open to dense, and commonly contains many
Acanthaceae and sometimes *Ocimum lamifolium*, important as a source of
traditional medicine. Fresh burials in the forests are commonly planted with
trees or shrubs, especially the succulent *Euphorbia tirucalli*. The number of
species of trees and shrubs recorded in 39 forests sampled by a project of the
Ethiopian Wildlife and Natural History Society (EWNHS) and the University
of Wales at Bangor (UWB) through rapid biodiversity assessment ranged
between 21 and 68 (average 41). Many of the 233 forest species recorded by
the EWNHS/UWB survey proved to be consistently rare.

Members of the Orthodox Tewahedo Church are keen to preserve their
forests, which they value for theological and aesthetic reasons, as well as for
supplies of certain products. Believers view nature as a gift of God for human
welfare, and church forests as sacred places which people should respect and
maintain. The forests are symbolic of the presence of angels surrounding and
guarding each church. Trees are traditionally cut only for the purposes of the
church – for instance, when repairs are needed to church buildings. Some
species have special religious significance. *Juniperus* is used for making
palisades and is said to be always present in the forests. *Olea* is favoured for
making crosses and other religious artefacts. *Artemisia afra* and rue yield
fragrance. *Premna schimperi* is used symbolically in marriage ceremonies.

The Orthodox Tewahedo Church has a Church Development Agency,
concerned with forest conservation, among other matters. Altogether, the

Figure 10.4 *Church building and holy forest at the monastery of Mitak Abune Teklehayimanot, Ethiopia, altitude 2600m.* Juniperus excelsa *on right.*

Source: Pierre Binggeli

forests represent an invaluable network of habitats, protected by religious beliefs and customs, and strongly supported by church members. However, challenges remain. There is no legal basis for ownership of the forests by the church and this can become an issue if communities contain members of other branches of Christianity or of other religions. Some non-Orthodox villagers may wish to use the resources in the forests, which they can see as secular property. This is more of an issue in southern than northern Ethiopia, which is more homogeneously Orthodox. Another issue is that the populations of many species of plants are very small in the forests in which they are found; but as yet there has been little research into whether and how they are connected through biological or social mechanisms.

Sources for case study: Dessalegn, 1998; Bekele et al, 2001; Teklehaimanot and Healey, 2001; P. Binggeli, personal communication, 2002; and personal observations

11

Approaches to *In Situ* Conservation

Achieving effective conservation of plants in diverse local contexts is a major challenge for our time. This chapter outlines some approaches to *in situ* conservation. Complementary information is given elsewhere, such as on techniques of plant and land management (see Chapter 6), issues of land tenure and resource rights (see Chapter 10), working with communities (see Chapter 12) and trade in wild plants (see Chapter 14).

THE LOCAL: THE HEART OF PLANT CONSERVATION

The local should be at the heart of plant conservation. This is where plants actually live and where people interact directly with them, and everywhere can be local. It will often be necessary in plant conservation to give attention to the details of managing plants; therefore, the physical relationships between people and plants must be a matter of primary consideration. Of course, there are actions away from the local that can be taken in favour of plant conservation – for example, the improvement of conservation information systems, the persuasion of consumers to purchase 'green' products or the passage of enabling laws; but 'distant' measures will only be useful if they feed back to the local level, influencing the actual state of plants.

The approaches described here are certainly not the only ways in which *in situ* plant conservation can be pursued. In any case, they should be applied creatively, adapted to local conditions. A combination of approaches, interacting synergistically, will likely often prove most effective in the context of projects. *In situ* plant conservation is a poorly developed science and practitioners have much to gain from sharing experiences and determining which approaches and methods work best in defined contexts. Approaches and methods should not be followed blindly, as if they were magic formulae; rather, each region, village and project area should be approached as a unique case (Tuxill and Nabhan, 2001). *In situ* conservation is an art (as well as a science) and the personal skills of practitioners can be honed through apprenticeship, experience and thought.

ASSESSING THE SUSTAINABILITY OF RESOURCE SUPPLY

Introducing the issues

Many people around the world harvest uncultivated plants to support their livelihoods, using them directly, giving them to others (thus strengthening their social relationships) or selling them for income. The plants harvested may be managed or unmanaged, and may stand on land under various tenurial regimes. For present purposes, these uncultivated plants are referred to as 'wild'.

Biological sustainability of wild plant resources is defined (for the purpose of discussion in this chapter) as the use of plant resources at levels of harvesting and in such ways that the plants are able to continue to supply the products required indefinitely (Wong et al, 2001). The key question addressed for any given plant resource is whether supplies will be sufficient to meet demands within a defined system, such as a particular geographic area (Peters, 1994; Cunningham, 2001a). Sustainability analyses may reveal that the level of harvesting is unsustainable. If so, two possible next steps are to try and improve the management of the resource or to promote alternatives.

Evaluations of sustainability provide insights into possible *future* conditions. They can be fraught with uncertainties (Menges, 2000). There are many reasons why the demands on a resource at a particular locality can unpredictably surge, decline or even disappear. All sorts of changes in wider political, economic and social systems can influence what happens locally. There can also be various complexities and uncertainties relating to other demands on the species of interest, or to biological or ecological factors (see Box 11.1).

Despite the many complexities and uncertainties that can be associated with sustainability assessments, they are still valuable. On the *conceptual* front, an environmentalist would argue that there is great value in people striving for greater sustainability in resource use, according to their abilities. Taken together, their many small efforts will add up to a powerful force for conservation. The main *caveat* is that it is important for people to try and identify those things that are conservation priorities, within the contexts of their situations (see Chapter 9). On the *practical* side, the uncertainties of analyses of sustainability suggest that recommendations about harvesting quotas or other suggestions for management should err on the side of caution.

There are a number of reasons why it can be worth devoting considerable time to sustainability assessments. First, they provide opportunities to engage local residents and other stakeholders (for example, resource managers, traders and industry) in initiatives in plant conservation, based upon aspects of plant diversity that interest them. They can lay the foundations of collaborative partnerships for subsequent conservation action. Second, the research provides opportunities for members of project teams to learn about the social, economic and cultural aspects of resource use and management in their project areas, all important in the search for solutions to plant conservation problems. Third, quantified information, as can result from this type of research, may be persua-

---

BOX 11.1 SOME POTENTIAL BIOLOGICAL AND ECOLOGICAL COMPLEXITIES AND UNCERTAINTIES IN SUSTAINABILITY ASSESSMENTS

- The species of interest may be used to supply other resources apart from that which is the focus of research.
- Collection of the resource may have significant knock-on impacts on the wider ecosystem, perhaps lowering its overall value for conserving other species or influencing supplies of ecosystem services.
- Modes of harvesting may cause serious genetic erosion – for example, if better-formed specimens are chosen for collection.
- Vegetation at a site may be in an unsteady state for reasons quite unconnected with the harvesting of the resource of interest. For example, if a species is not regenerating adequately, this may be *not* because of overharvesting, but rather because the type of vegetation in which it grows represents a successional stage that will naturally give way to other vegetation types (with different species) with the passage of time (see Chapter 5).

Source: Alan Hamilton and Patrick Hamilton

---

sive to officials and policy-makers. Fourth, the fact that the researchers are prepared to put effort into collecting detailed data can persuade other parties that they are serious in their intentions and worthy of respect as partners in development (Peters et al, 2003).

A sustainability assessment must be carried out with reference to a defined system, which may be demarcated in various ways according to the interests of those concerned. The simplest case is where a well-defined social unit, such as a family, village or even a whole nation, obtains a plant product from a single well-defined area, such as a single private woodland, community reserve or country. Resource managers, such as forestry officers, are likely to define their reference areas according to the places for which they are responsible. Approaches taken will vary according to whether the intention is to produce a continuous supply of a product from a particular area of land or whether rotational harvesting is intended. An agency, such as a forest department, may decide that the best way to provide a continuing flow of a product from its forests is to rotate production around its reserves (or compartments within reserves), each managed for occasional rather than continuous harvest.

There are two types of assessment relating to the supply of a resource (see Figure 11.1). One is *baseline* research, concerned with making an initial estimate of the amount of the resource (see 'Quantitative assessments'). The other is *monitoring* research, which is carried out as a part of routine management (see 'Strengthening resource management', p196).

Research into sustainability can be undertaken at various levels of detail. The most detailed analyses are based upon quantification of a range of variables relating to supply and demand. They are time consuming, with some

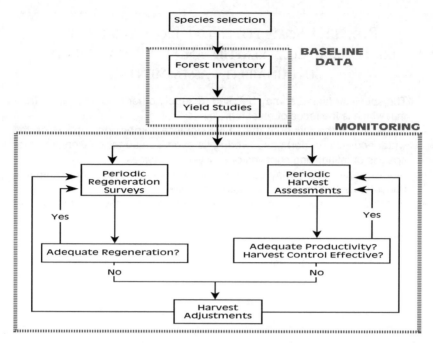

Source: adapted from Peters, 1994

Figure 11.1 *Flow chart of basic strategy for establishing sustainable harvest of non-wood plant resources*

methods requiring special expertise (Peters, 1994). More simply, there are a number of indicators, revealed through potentially less arduous research, that suggest that a resource is being over-harvested. They include:

- a '**retreating resource frontier**' – for example, in the case of ebony *Dalbergia melanoxylon*, used by the woodcarving industry in Kenya, supplies of wood used to be obtained from Kenya, but now largely come from Tanzania, thanks to exhaustion of Kenyan stocks (see Chapter 14);
- **higher prices** of goods over time;
- use of **less desirable specimens** of plants over time as more desirable specimens become eliminated – for example, smaller sizes of timber trees;
- use of **less easily harvested parts of plants** – for example, in the case of collection of *olorien* (*Olea europaea*) by the Loita Maasai for fuelwood, collection has sometimes shifted from dead branches to live branches, then to live trunks, stumps and, finally, roots (requiring excavation) (Maundu et al, 2001);
- **substitution** by other species – for example, ebony *Dalbergia melanoxylon* became largely replaced by *muhuhu* (*Brachylaena huillensis*) and later by neem *Azadirachta indica* in the case of the coastal woodcarving industry in Kenya (see case study at end of Chapter 14).

Quantitative assessments

Mathematical comparisons of supply and demand require that the two sides of the resource equation are **expressed in the same units**, such as kilograms of materials produced by the plants per year (supply) and kilograms of materials used by the people per year (demand). Measurements and observations can be made at many points along the chains leading from the plants on the ground to end uses (described in detail by Cunningham, 2001a). Harvesters, traders, manufacturers and end-users all have their own **units for quantities** (such as bundles, baskets or bags in the case of materials sold to traders by harvesters). The names and definitions of these units should be recorded and their scientific equivalents determined. Elements along these chains that are commonly studied in assessments of sustainability include:

- the plants or parts of the plants that are sought by harvesters;
- the materials that are physically removed from the sites of collection;
- (for traded items) the goods that are traded or used as raw materials by manufacturers; and
- the end products.

Living plants, end products and the various goods and materials in between vary in their quantities of water because water is lost after harvest. Thus, **measurements of water content** are needed all along the chain to allow **standardization of weights to dry weight equivalents**.

The collection of data in detailed studies of biological sustainability will often continue simultaneously on several fronts, influenced by logistics and opportunities. However, it is probably best to conceptualize work as starting on the **demand side of the sustainability equation** since the supply side cannot be tackled without having at least some idea of the types of plant materials required and the places where they are harvested.

Observations and measurements relating to demand can be made in many ways, such as joining harvesters in their work, undertaking household surveys or finding out what is traded or used by manufacturers (for methods, see Chapters 12 and 14). When studying collection, note should be made about *what* is selected for harvest – in terms of parts of plants, their sizes and shapes, and other features. *How* plants are harvested is significant, since this can greatly influence sustainability – for example, *positively* if harvesting is done carefully to give plants a good chance to re-grow or regenerate, or *negatively* if, for example, trees are felled just to collect their fruits, which can be a common practice (Peters, 1994).

Quantitative data on various aspects of demand will need to be converted into terms that allow them to be compared with supply-side data. Thus, 'harvester units' (for example, a bundle of cut stems) will eventually need to be referred back to the equivalent size of area from which the stems in the bundle have been derived (Cunningham, 2001a). Calculations of demand require that time is considered, which should be investigated through

Figure 11.2 *Exposed roots of the climber* Decalepis hamiltonii
*at an experimental site to determine community-based management
systems for the sutainable harvesting of medicinal plants;*
Savana Durga, near Bangalore, India

Note: Research is by a village Task Team, backed up by FRLHT (see Box 11.7) and the
Environmental Change Institute, University of Oxford.
Source: Alan Hamilton

gathering information on the **amounts of materials harvested, traded or used
per unit time.** For example, as a very rough approximation, the total amount
of wild edible leafy greens consumed in a village *per year*, can be calculated
from determining the amount consumed in an average household per week,
multiplied by the number of households in the village and by the number of
weeks in a year.

Various detailed methods have been devised to estimate the **supply side of
the sustainability equation** (Peters, 1994; Cunningham, 2001a). These involve
making measurements and observations in sample plots placed in vegetation
containing the species of interest. Decisions on what to sample within the plots
should be based upon knowledge of what the harvesters are collecting and
how they collect it (Figure 11.2). Some of the main parameters recommended
for recording are given in Box 11.2. These data may be used to calculate:

- the density of the species yielding the resource;
- the amount of the resource present, including partitioned according to size
  and age;
- the impacts of harvesting; and
- the rate of production of the resource.

BOX 11.2 SOME MEASUREMENTS THAT CAN BE MADE IN SAMPLE PLOTS TO HELP UNDERSTAND THE SUPPLY SIDE OF A PLANT RESOURCE

- Locate, count and measure the size of all plants of the species of interest, living and dead (sub-plots may be used for smaller size classes).
- Collect evidence of the ages of plant, where this can be found (for example, from local informants, growth rings and periodic measurements in permanent plots).

For plants that have been harvested:

- Record damage levels (use standard damage scales).
- Determine when harvesting occurred (for example, from local informants, states of decay of stumps and growth rings on re-sprouts).

Source: adapted from Cunningham, 2001a, which contains information on standard damage scales

---

The **yield of a resource per individual plant** is best determined by taking a few sample plants of different sizes and measuring the amounts of the resource that they provide. For example, the amount of charcoal produced from trees of different sizes can be determined by combusting a number of sample trees of various sizes and measuring the amount of charcoal that each produces. Once established, regression relationships can be used to calculate resource yields from plants according to size.

There are various methods of determining the **ages of plants** (Cunningham, 2001a). They include examining tree rings or other types of seasonal growth marks produced by plants, data from permanent plots and asking local people, who may know, for example, when particular specimens of plants became established. Once the relationship between plant size and age has been established, then regression lines can be calculated between the ages and sizes of plants, and plant size may subsequently be used to provide a surrogate measure of age. Determining the ages of plants can be difficult, especially in places of equable climate where there is little seasonality in plant growth.

One of the key decisions in studying the supply side in sustainability assessments is deciding **where to place the sample plots** (see Chapter 8). Sampling is possible at many different scales and levels of detail. For example, a forestry department might wish to sample on the national scale in order to gain some idea of overall timber resources in its country, while, at the other extreme, the mapping of individual timber trees in a forest has sometimes been undertaken to decide on detailed strategies for their harvesting.

Information from local people can be extremely useful for formulating sampling strategies. Many plant resources occur patchily and local people will often know where the patches are, as well as other useful information about them, such as their ownership. Sample plots may then be placed randomly in these patches, checking in allegedly barren areas to confirm the accuracy of local information. Transects can be useful – for example, extending along lines of likely declining intensity of use, as from villages outwards into forests. Where the boundary of resource extraction is expanding, a good approach is to place some samples in areas that have already been harvested, others in places where harvesting is currently under way and others again in places where the resource is yet to be targeted (Cunningham, 2001a).

STRENGTHENING RESOURCE MANAGEMENT

In situ plant conservation requires **managers**, and therefore an early step for a project concerned with plant conservation is to identify who, if anyone, owns the plants. Weak management can have a variety of causes, including a lack of interest or knowledge on the part of owners, or unclear or disputed ownership. Projects may sometimes be able to improve management by suggesting who *should* be involved in resource management. There will often be roles for others in resource management, apart from the recognized owners – for example, governments in the case of private land and communities in the case of protected areas (see 'Collaborating in resource management', p200, and Chapter 10).

The **boundaries of the management system** need to be determined. This could involve establishing and mapping the boundaries of a forest reserve or farm, or, in the case of scattered landholdings, establishing a comprehensive list of all the reserves in a system of forest reserves or all the scattered fields of a dispersed farm. Frequently there is uncertainty, ignorance or dispute about where the boundaries of landholdings lie – matters that may require attention in projects. **Boundary marking** – for example, with planted trees – can make a significant contribution to conservation.

Geography will be a major consideration as projects become engaged in more detailed planning – for example, perhaps involving the demarcation of a management area into **zones** for different purposes. Local perceptions of physical and social geography can be important considerations in placing boundaries on the local scale. In the case of a protected area, the zones might serve variously for nature conservation, low-impact extractive use and more intensive exploitation (see Figure 6.3). Another basis for recognition of zones can be compartments for rotational use, such as for livestock grazing or the harvesting of wild plants.

There must be some **vision** of what strengthened management is designed to achieve in terms of the future condition of habitats and plants. To contribute to plant conservation, the vision of a project must make some reference to this as an aim. Specific management objectives can then be formulated to achieve this vision (see Table 12.1 in Chapter 12). Visions should be informed by scientific perspectives on conservation, as well as the values held by local stakeholders (see Chapter 9).

Management systems vary in their **degrees of formality**. Most management of land and plants worldwide is based upon unwritten systems, in which managers follow established practices or adopt new ones according to a mixture of tradition, received advice, experimentation and, perhaps, government directives. Formal (written) plans are useful for managers who are responsible for complicated holdings or where several parties are involved and require coordination. Formal management plans should contain (Walters, 1986, 1997; Walters and Holling, 1990; Tuxill and Nabhan, 2001):

- statements of visions and management purposes, goals and objectives;
- activities that have been prescribed, including monitoring;
- assignment of responsibilities; and
- arrangements for periodic review of the plans.

Management plans should be prepared by those actively involved in their management, and should be understandable and practical for them, as well as capable of adjustment. Some suggestions for creating a management plan are given in Box 11.3. It has been recommended that plans should *not* be produced by teams of consultants coming up with long, detailed (and expensively produced) documents, as has sometimes been the case with protected areas. Such plans tend not to work, because they have inadequate local buy-in and insufficient weight is given to on-site flexibility and local decision-making (Wells et al, 1999).

There may be an interest in achieving a certain **recognized standard** in resource management – for example, for the marketing of products labelled as being sustainably produced (see Chapter 14). Various categories of standards have been applied to land or resource management, for both natural habitats and cultivated land, and to one or more of: product quality; ecological sustainability; and social responsibility (Pierce and Laird, 2003):

- (for an area) **ecologically responsible forest management standards** that assess water and soil conservation, wildlife habitat, forest management planning and harvest activities (for instance, Forest Stewardship Council, or FSC, certification);
- (for an area) **organic standards** that assess pesticide-free and inorganic fertilizer-free agricultural practices and are occasionally applied to agroforestry and forestry operations (for instance, the Soil Association Organic Standard);
- (for harvesters) **wildcrafter standards** that outline best harvest principles for gatherers;
- (for all those involved in trade systems) **fair trade certification**, which ensures the equitable sharing of benefits among producers, workers' rights and decent working conditions.

There are many uncertainties in resource management. Ignorance abounds about how ecosystems function and about the future vagaries of man and

Box 11.3 Steps in creating a management plan

Step 1

Establish **conservation priorities** (species, habitats, ecological features), including from perspectives of local stakeholders.

Steps 2 to 5

Formulate **management purposes for each conservation priority**, taking each in turn:

- Step 2: translate conservation priorities into **management goals** (desired general situations and/or accomplishments to achieve over time).
- Step 3: list **threats and opportunities external to local management** related to management goals (for example, anticipated development of external markets for resources).
- Step 4: list **organizations** that are relevant to achieving management goals, as well as their **strengths and weaknesses**. These organizations may include community institutions, government agencies and private companies. Aspects for consideration include available human, material and financial resources and (for government agencies) credibility, especially with local residents. This process should result in the highlighting of areas where improvements are needed.
- Step 5: formulate **management objectives** that minimize threats and maximize opportunities and strengths. They should be clearly defined, practical and short-term accomplishments designed to achieve parts of more general goals. They should be assessable against criteria, which should be:
  - specific – describes a single key result to be accomplished;
  - measurable – readily monitored and quantifiable, if possible;
  - result centred, not activity centred;
  - concise – to the point and understandable;
  - realistic and substantial – neither too hard, nor too easy to accomplish;
  - consistent with other objectives; and
  - time bound – where possible, specify a date for accomplishment.

Step 6

List **management activities in order to achieve management objectives**. These are specific activities to be carried out that will address management objectives directly. They should keep within the framework of threats, opportunities, strengths and weaknesses that has already been identified. There should be agreement on the *roles* of different people, their *responsibilities* for activities and *timetables* for their accomplishment. Assessments will be needed of the availability of resources necessary to accomplish the tasks. Subsequent meetings by those assigned responsibilities can flesh out more details.

Step 7

Compile this and other planning exercises into a **management plan**. A management plan provides a record of analyses that have been undertaken and agreements reached, as listed under steps 1 to 6. It serves as a reference to see whether actions have been undertaken.

Source: adapted from Tuxill and Nabhan, 2001

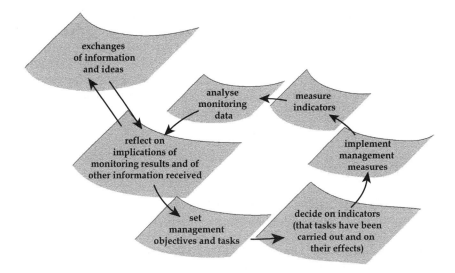

Figure 11.3 *Stages in adaptive management*

Note: Notice the similarities to stages in applied conservation research (see Figure 4.2).
Source: Alan Hamilton and Patrick Hamilton

nature, and there are uncertainties about the actual (contrasted with the intended) effects of human interventions. For this reason, an **adaptive approach** to natural resource management is recommended, incorporating stages of monitoring, reflection and learning as parts of normal management procedures (see Figure 11.3). There is nothing particularly new about adaptive management. It is normal for people who manage land and plants, whether foresters, farmers or gardeners, to decide what they should do in future, taking into account past experience and new information.

Research is an essential feature of resource management. Two types can be distinguished: *baseline research* (carried out before developing management plans) and *monitoring research* (a routine part of management). Figure 11.1 illustrates the two types with reference to questions about the biological sustainability of a plant resource. Monitoring provides periodic 'snapshots' of the state of a system, allowing changes in management procedures to be made – for example, adjusting quotas for harvesting or taking steps to encourage increased regeneration of a species. Formal monitoring systems should include procedures to check whether the prescribed management tasks have been undertaken, as well as to record the state of the system itself.

Formal monitoring uses measurements or observations made on **indicators**. These should be selected in collaboration with those who will undertake the monitoring, taking into account their interests, duties and technical competencies. In the case of protected areas, different indicators might be considered for three different groups of people – resource managers, resource users and professional scientists. The indicators selected for *resource managers* could, perhaps, refer to the states of certain species that are priorities for conservation or resource use (if harvesting is allowed), the licences that have been issued for resource harvesting, illegalities detected and income received. Indicators selected for *resource users* might refer to the quantities of resources that they have harvested, with their locations and dates. Indicators selected for *scientists* might relate to overview surveys of the states of habitats, species and resources, perhaps carried out at intervals of a few years, as well as detailed studies of species of special concern.

COLLABORATING IN RESOURCE MANAGEMENT

Protected areas exist within cultural, social, economic and political contexts. Their effectiveness for biological conservation will depend, in part, upon how they relate to these aspects of their environments. Relationships with local people can be especially critical to their success, particularly when there are strong livelihood dependencies upon the resources of the protected areas or local people see the protected areas as parts of their traditional territories.

Joint forest management and **participatory resource management** are terms used for certain forms of collaboration between protected areas and local people. Nepal and India have been pioneering countries in these developments. In the case of Nepal, the Sixth Five-Year Plan (1981–1985) emphasized community participation in the management, conservation and utilization of forest resources – a major change from an earlier authoritarian approach. Communities were now able to gain control of certain types of forests, provided that they could show that they had the capacity to manage them. The fundamental organizing concept was the **forest user group**. In the case of India, a new National Forest Policy was declared in 1988, encouraging the involvement of local people in managing degraded forests. By 1999, 10,000 Forest Protection Committees had been formed, institutionalizing community involvement in around 1.5 million hectares of government-owned forests (Aumeeruddy-Thomas et al, 1999).

Collaborative management should be implemented under a **supportive legal framework**, which may require modifications to pre-existing laws and regulations governing protected areas. However, on its own, an enabling legal environment is insufficient to bring about new arrangements. What can be useful are trial projects to demonstrate how, in practice, collaborative management can be achieved and training programmes for the staff of protected areas and local people.

The terms of a **collaborative management agreement** between a protected area and local people might refer to:

- an obligation on each party to educate the other in its culture – for example, through the running of educational programmes by the protected area and the welcoming of protected area staff at cultural events by the community;

- the rights of local people to visit sites of cultural importance within the protected area and engage in related rituals and ceremonies;
- the rights of local people to harvest specified resources in the protected area on agreed terms in relation to who, where, when and how;
- the duty of the community to undertake specified management tasks in the protected area – for example, the scattering of seeds of plants that they have harvested or help with the control of wildfires;
- the duty of the protected area to assist residents in developing alternative ways to support their livelihoods in cases when the continuing use of resources in the protected area is inadvisable or prohibited (see 'Finding alternatives to wild harvest');
- the respective roles of the protected area and the community in the monitoring system of the protected area;
- prescribed benefits and penalties for adhering to or breaking the terms of the agreement;
- mechanisms for settling disputes and reviewing the agreement.

Collaborative agreements are reached between parties. Inclusiveness and legitimacy are two matters that should be considered when deciding the best **institutional arrangements** for agreements. On the community side, this means ensuring due representation of those members of communities who have strong economic or cultural ties to the protected areas – that is, frequently people who are politically marginalized in local social systems. It can be landless people, indigenous people and women who have the greatest dependencies upon protected areas; yet, these can all be groups which have little influence on how decisions are reached within communities. If protected areas make agreements with communities 'over the heads' of those local people who depend upon the resources of the protected areas, then the agreements may prove unworkable.

Sometimes there are **existing community institutions** that can serve as social foundations for new arrangements. The collaborative agreement reached at Bwindi Impenetrable National Park in Uganda was based, in part, upon the pre-existing institutions of the *abataka* (responsible adults of a local area) and the *ebibina bya engozi* (stretcher-bearer societies, responsible for transporting the sick to clinics) (see Chapter 9) (Cunningham, 1996b).

Trust often needs to be built between protected areas and communities before collaborative agreements can be reached (Aumeeruddy-Thomas et al, 1999). Sometimes, there can be histories of conflict between protected areas and local people, with illegalities and even violence. Non-governmental organizations (NGOs) may be able to play pivotal roles in building trust (Tuxill and Nabhan, 2001). One approach is for an NGO to organize a series of meetings and workshops – at first separately with the two 'sides' and then bringing them together, when the time is right, for joint studies and discussions (see Box 11.4 and case study at end of Chapter 9) (Wild and Mutebi, 1996).

Participatory research (see Chapter 4), involving professional researchers working alongside members of local communities and government agencies,

Box 11.4 Project steps towards collaborative management in West Bengal, India, and Bwindi Impenetrable National Park, Uganda

India

Purpose: Resolve conflicts between the Forest Department and villagers, resulting in agreements on the use of certain resources – especially non-timber forest products (NTFPs) – in forest reserves:

- Raise awareness of foresters through training: 'Listen and be sensitive; learn from villagers.'
- Understand the position of local communities – for example, learning about amounts of resources used, their distribution and problems of access.
- Identify institutions in both government and communities relevant to resource management.
- Ensure mutual recognition of government and community institutions.
- Identify similarities in the aims of these institutions.
- Establish good quality information about resources that are sources of conflict.
- Promote discussions between agencies and villagers to build a broad-based consensus, rather than individual goals.
- Address questions about particular issues, with the listing of options and suggestions for resource management, together with practical ways in which management measures can be implemented.
- Develop and monitor a plan of action.

Uganda

Purpose: Establish defined areas (multiple-use zones) in the marginal areas of a national park where local villagers are allowed to collect certain plants in designated ways, and provide substitutes outside of the park for resources that cannot be harvested sustainably within it:

- Assess the resources through:
    - rapid vulnerability assessments, which examine biological, social and economic factors;
    - participatory resource appraisal through workshops with communities (resources identified and ranked; ground mapping of forests and villages; resource flows);
    - interviews with nominated resource users;
    - quantification of individual products and demands per household;
    - surveys of marketplaces;
    - forest surveys with resource users to locate and estimate the abundance of resources and to locate potential sites where harvesting might be allowed (estimates based upon measurements taken in participatory ways with resource users);
    - participatory harvesting trials.

> - Select resources and areas that communities might be permitted to harvest.
> - Establish joint workshops between communities and the park to build trust and to jointly assess the resources.
> - Develop joint workshops to negotiate and agree upon the conditions for multiple use, the identification of community management structures and the nomination of specialist resources users who will be allowed to harvest nominated resources.
> - Establish Forest Societies.
> - Sign a joint forest management agreement.
>
> *Source:* Aumeeruddy-Thomas et al, 1999

has been used as a standard technique by the People and Plants Initiative in the context of protected areas. It has generally proved an excellent way of forging relationships between communities and agencies, and identifying priorities for conservation attention. A symposium on experiences in participatory assessment, monitoring and evaluation, organized at the UK Department for International Development (DFID) by the European Tropical Forest Forum and the Environmental Change Institute of Oxford University in 2002 resulted in the listing of these advantages of a participatory research approach:

- Local people can provide their knowledge on species and habitats, resulting in short cuts to scientific assessments.
- Research can concentrate on matters important to resource managers and/or local needs.
- The process is inclusive and the recommendations resulting from the research are more likely to be accepted.
- From the perspective of *scientists*, local people (once trained) can collect data faster and cheaper than they could themselves.
- From the perspective of *local people*, they can contribute their views and embark upon negotiations during the process of the research, develop motivations for conservation and learn about scientific approaches.

Lessons learned about participatory research approaches at this symposium included the following:

- Keep the research simple.
- Ensure proper institutional anchorage.
- Give equal priority to biological and socio-economic data.
- Develop mechanisms to integrate information within decision-making.
- Take time (needed to develop a sense of ownership of the work).

It was concluded that a participatory approach is not appropriate in all cases. At the same meeting, Finn Danielson presented conclusions from his experience in developing systems of resource monitoring by local people in a protected area in the Philippines. He advocated that there should be little delay between the gathering of monitoring data and its translation (if necessary) into management actions since this helps to retain people's attention and deals effectively with problems as they arise. He has found that useful methods for gathering data include focus group discussions (village meetings at three-monthly intervals, at which village groups present their findings to the staff of the protected area), transect walks, recording information based upon regular patrols and fixed-point photography at intervals of selected hillsides.

FINDING ALTERNATIVES TO WILD HARVEST

There is a long history of substituting wild-collected resources with alterna-tives, stretching back to the time of the first cultivation, and this is a process that continues widely around the world today. Until quite recently, houses at Kerinci, Sumatra, in Indonesia were constructed of local hardwoods, such as dipterocarps; but, as the supply of these has dwindled – with clearance of natural forest close to homes and restrictions placed on forest use – so people have switched to fast-growing trees such as *surian* (*Toona sinensis*), planted in agroforestry systems. Today, *surian* is seen as perfectly adequate for house construction (Aumeeruddy-Thomas, 1994).

Conservationists may wish to catalyse the adoption of alternatives to the harvesting of wild resources in order to save species or to avoid a habitat becoming so degraded that it is virtually useless for biological conservation or, indeed, anything else. In some cases, the management of resources may be so complex or expensive to achieve in practice that the provision of substitutes to wild-collected resources can become an attractive alternative.

One way of providing an alternative is to produce the same resource from cultivated plants – of the same or other species. If species' substitution is impossible – which may be the case with species of exceptional cultural value or efficacy – then domestication of the species may be needed. However, the domestication of a species can be a laborious task, with no guarantee of success.

Switzerland provides examples of the substitution of plant resources at country level, in the case of wood. Forests in Switzerland were subject to major devastation during the late 19th century as the country became caught up in the Industrial Revolution and forests were cut extensively for fuel and timber. Relief was achieved through a combination of replacement of fuelwood by coal and of local timber by tropical timber, backed up by appropriate legisla-tion. The forests then regenerated and today stocks of trees in Switzerland are continually increasing. A comparative analysis with the history of forests in Nepal raises many questions as to whether resource substitution to ease pressure on the forests will be possible in that country too (Küchi, 1996/1997).

Projects can play significant roles in stimulating the uptake of alternatives – for example, through facilitating research for the selection of alternatives, mounting demonstration projects to showcase new technologies or providing incentives to kick-start local buy-in (such as meeting part of the costs of new nurseries). Steps may be needed to ensure that those who currently harvest wild plants are included among the beneficiaries. Harvesters of wild plant resources often lack land or otherwise have access to relatively few resources to maintain their livelihoods. If cultivation to produce a resource is introduced as an alternative to wild collection without other measures, then some of the current harvesters of wild plants may fail to benefit and will continue to collect wild plants, just as before (Sheldon et al, 1997).

The development of alternatives – and, indeed, confirmation that alternatives are needed in the first place – requires a participatory approach. Conservationists may believe that a wild resource is being over-harvested; but resource users may not have the same opinion or may regard it as a low priority for project attention.

Local knowledge can be important in identifying alternatives. For example, there may be individuals within communities who already cultivate plants as alternatives to wild collection and who may be prepared to demonstrate to other people how this can be done.

Promoting an alternative to collecting a wild plant resource can have the potential to produce major benefits for conservation if undertaken *within the context of a protected area*. The fact that a protected area is present can help to give anchorage to an analysis of the conservation impacts of harvesting a resource – that is, taking account of issues relating to biodiversity conservation more widely, not just relating to the species providing the resource considered in isolation. If it is decided that an alternative needs to be developed for a wild-collected resource, then promoting the alternative can form part of an overall agreement between communities and the protected area. In this way, assistance provided to communities to develop an alternative resource can help to build protected area–community relationships and thus assist in the conservation of biodiversity, generally, in the protected area.

A *partial* solution to dealing with the over-harvesting of a resource can be to raise the efficiency of its use. The use of many materials collected from wild plants can be wasteful. For example, it is widely reported that many medicinal plants collected for sale are improperly dried and have to be discarded. The connection between increased efficiency of use and reduced collection is most likely to be made when plants are used for subsistence purposes and the amounts needed per household remain fairly constant over time. This is the case, for example, with fuelwood use at Ayubia in Pakistan, a site where the introduction of fuel-efficient stoves to about 500 households by a project of the People and Plants Initiative can be expected to result in some reduction in pressure on the forest (see case study at end of Chapter 6). These stoves reduce fuelwood use by 40 to 50 per cent (Aumeeruddy-Thomas et al, 2004). However, raising the efficiency of resource use may not always reduce pressure on resources. If wild-collected resources are sold, then increased efficiency in post-harvest processing might sometimes be expected to actually *increase* the amount of harvesting because now more money can be made and the demand may rise.

BUILDING CULTURAL SUPPORT FOR CONSERVATION

There has been massive cultural change globally over the last few hundred years. Before then (and often much more recently), almost everybody lived in rural societies organized along clan or tribal lines, or were peasants in pre-industrial states. There was close dependency upon local natural resources. Around the world, people belonged to a great diversity of cultures and inter-acted with a great diversity of local plants. Today, the world is rapidly becoming culturally globalized. These developments are associated with many other changes in human societies, such as a massive expansion of monetary economies, reduction in subsistence-based economies and local systems of exchange, mass migration, urbanization, the widespread introduction of formal schooling (based upon common national curricula), greater specialization in occupations and leisure activities, and the intrusion of the radio, television and DVD even in remote villages.

Local interest in the diversity of local plants is an important foundation for plant conservation. Thus, the major decline in traditional knowledge of plants, as is now occurring, is of major conservation concern, especially when it is not replaced by new types of interest in local plant diversity – for example, associated with 'whole plant' scientific botany or the hobby of natural history. Some ways that can be used to promote cultures favourable to plant conservation are listed in Box 11.5. Initiatives that encourage 'learning through doing' are particularly useful since it is through doing (contrasted with book learn-ing) that much deep-rooted knowledge is acquired. It will be through the continuing use of local plants that much cultural knowledge of plants will be transmitted, which is why the continuing use of wild plant resources, rather than substitution with alternatives, can be desirable for conservation, provided it is sustainable.

A project in Kenya, aimed at conservation of local diversity of the bottle-gourd *Lagenaria siceraria* has used several of these tools. They include the building of a resource centre, the holding of seed fairs, the documentation of indigenous knowledge through reports, videos and demonstrations in farmers' fields, story-telling by elders, and events to share myths, songs, riddles, poems and dances. The experience of this project was that greatest effectiveness was achieved when project personnel had frequent interactions with the society and a good understanding of its culture, and full use was made of traditional social gatherings and traditional means of communication (Morimoto, 2002).

Many people worldwide are faced with trying to find a reasonable balance between feeling rooted in local society and local place, while benefiting from national and global culture. As has been written of Indonesia: 'Everybody in Indonesia is looking for a strengthened local identity within a new decentral-ized Indonesia' (Jepson, 2001). Of course, people coming from more traditional backgrounds vary in their response to Western influences, ranging from open-hearted welcoming, scepticism of the West, shame in their own cultures or even outright rejection of their own cultural heritage (Ruddle,

---

**BOX 11.5 SOME WAYS TO PROMOTE LOCAL CULTURES
FAVOURABLE TO PLANT CONSERVATION**

- Respect local knowledge and customs relating to plants. This can give people more confidence in their cultures and provide firm foundations for dealing with conservation issues.
- Record and return local knowledge about plants to communities (for example, through publications, drama, exhibitions in community centres and at cultural gatherings).
- Establish ethnobotanical resource centres, such as collections of local products made from plants, libraries with reports on local plant use, community herbaria, and ethnobotanical or medicinal plant gardens.
- Form alliances with local and indigenous cultural groups based upon a linked interest in heritage (botanical, cultural).
- Form alliances with religious groups based upon common interests, such as thankfulness for creation and duties for its care.
- Incorporate studies of local plant diversity and local ethnobotanical knowledge within school curricula.
- Encourage schools to develop extra-curricular activities, such as nature clubs, ethnobotanical gardens and the interviewing of parents and elders by pupils about their knowledge of plants.
- Make scientific information about local plants available. Draw attention to information relevant to local issues in conservation and sustainable development, as well as any distinctive or unique features (potential sources of new pride in local nature).

Sources: adapted from Martin, 1995; McShane, 1999; Jepson, 2001; Tuxill and Nabhan, 2001; Long Chunlin and Pei Shengji, 2003; see also 'Linking livelihood support and plant conservation' in this chapter

---

2000). Plant conservationists have an intrinsic interest in helping people to retain and develop knowledge of plants associated with their traditional cultures. Through such cultural promotion, conservationists can play a part in giving people more confidence in their cultures and can thus contribute to helping them face the challenges of modern times.

Robbie Mathew, an elder of the Eeyou (Cree) First Nations peoples of the James Bay region in Canada is concerned with the social disorientation that affects some youth of his tribe. There are unacceptable levels of depression and juvenile delinquency. He believes that Eeyou boys and girls need a fresh start, and that this depends upon them knowing where they have come from – as a basis for deciding where they want to go. He has found that the passing on of traditional Eeyou knowledge in today's world is no easy task. Robbie has started a bush school, where traditional knowledge is transmitted on the land by doing things, and through legends and story-telling. Youths attend for three to four months. Away from town life, they learn how to live according to another rhythm and set of values. They are taught survival skills and how

to live off the land. The curriculum covers traditional hunting, trapping territories, fishing and Eeyou spirituality. The boys and girls are received as
members of one family in an atmosphere of caring and sharing. They start to
learn about respect for the land, other people and themselves. They learn to be
proud of who they are (Mathew, 2000).

LINKING LIVELIHOOD SUPPORT AND PLANT CONSERVATION

Most people around the world are interested in development, which can constitute either an opportunity or a threat to plant conservation. Three common
views of development are with reference to *quality of life* (physical, mental,
spiritual), *security of livelihoods* and *financial income*. Some types of development that can be supportive of plant conservation are listed in Box 11.6. Apart
from these, it may be possible to achieve benefits for both conservation and
development through making the use of plant resources more sustainable or
through promoting alternatives to the unsustainable harvesting of wild plant
resources (see earlier sections in this chapter).

The Working for Water Programme in South Africa is an excellent example
of a scheme that links economic development closely to plant conservation.
Run by the Department of Water Affairs and Forestry in collaboration with
the Western Cape Nature Conservation Boards and other partners, the
programme operates throughout South Africa to clear invasive plants from
catchment areas. Invasive plants are an extremely serious conservation
problem in South Africa, especially in the Cape region. Clearance of invasives
has advantages for both conservation and development. It creates habitat for
native species, enhances water supplies and provides an income (through
employment and fuelwood sales). Some of the invasive plants, such as *Acacia
cyclops*, are thirsty water users. Using a two-stage clearance process, 30,675ha
were cleared during the working year 2002–2003 in the 'initial clearance'
phase and 91,939ha in 'follow-up clearance'. The programme aims to create
jobs for 18,000 people per annum (Richardson et al, 1996; DWAF, 2002).

The use of medicinal plants is by far the biggest use of biodiversity on a
species-by-species basis. Since healthcare is a prime concern of people worldwide, in principle there should be many opportunities to develop linkages
between the strengthening of local healthcare and the conservation of local
plant diversity (Figure 11.4). Some examples of actual initiatives are given in
Box 11.7.

It would seem sensible for countries to develop their national health
services drawing upon all available traditions of medicine and their associated
practitioners. In fact, there is a growing trend towards official recognition of
traditional, largely plant-based, medicine, opening up the hope that governments will start to take serious action with respect to the conservation of
medicinal plants. It has been estimated that 70 to 80 per cent of people worldwide rely chiefly upon traditional medicine to meet their primary healthcare

---

BOX 11.6 TYPES OF DEVELOPMENT THAT CAN BE SUPPORTIVE OF PLANT CONSERVATION

Development that encourages plant conservation includes the following:

- retention or enhancement of *environmental services*, such as those providing more equable local climates, more reliable and cleaner water supplies, greater control over flooding and landslides, and more efficient treatment of effluent (see Chapters 1 and 6); typically, these services can be best safeguarded by retaining or restoring more natural types of vegetation, so there can be a direct link with conservation of plant diversity;
- development of *forestry and agriculture* using local species and varieties;
- use of a diversity of plants (especially local plants) for *health* provision; the benefits include nutritional (for example, use of local leafy greens), medicinal (for example, use of local medicinal plants) and psychological (for instance, beautiful flowers, gardening as therapy, calming landscapes for convalescence) uses;
- maintenance or expansion of *local industries* based upon local plant diversity – for example, craft production, horticultural businesses and botanical products (such as medicines, local paper and food supplements);
- *guaranteed or premium payments* for products that are produced in environmentally friendly ways (see Chapter 14);
- *receipt of income from visitors* interested in local plant diversity or associated aspects of local culture – for example, through providing guidance on nature walks, sales of guidebooks, home stays and attendance at cultural events;
- *employment* of local plant experts by protected area agencies;
- benefits received from the use of plant diversity or associated local knowledge by *outside interests*; such uses may include biodiversity prospecting (see Chapter 4) and ecosystem services exported externally (for example, water supplies generated by areas of natural vegetation and exported to cities);
- *trust funds* established for developmental activities linked to conservation.

Source: Alan Hamilton and Patrick Hamilton

---

needs (Farnsworth and Soejarto, 1991). Traditional medicine is increasingly gaining official recognition. In 1952, Mao Tse-Tsung pronounced that 'Chinese medicine is part of our cultural heritage; we must work hard to study it and develop it for the health of the people.' Today, a variety of medical traditions – including Dai, Mongolian, Tibetan and Uigur medicine – as well as Traditional Chinese Medicine (TCM) – are officially recognized by the Chinese government. There are 36 universities in China teaching traditional medicine (Pei Shengji, 2002). In 1967, a society was formed in Japan to support Kampo; it now has 10,000 members and 200 preparations have been cleared for prescription by the National Institute of Health as well as by insurance companies. In 1970, the Indian Parliament passed the Indian Central Council Act,

Figure 11.4 *Local healer in her medicinal plant garden, near Savana Durga,
India (a development initiative supported by FRLHT – see Box 11.7)*

Source: Alan Hamilton

establishing a Central Council for Ayurveda. This was followed by the accreditation of courses at colleges and universities. In Africa, the African Union has
declared 2001–2010 as the Decade for African Traditional Medicine. The
medical establishment in the UK, as embodied in the British Medical
Association (BMA), has traditionally opposed 'alternative medicine'; but in
1988 the Royal Society of Medicine recommended bridge-building with other
traditions.

   Botanical tourism has great potential in some parts of the world. The
spectacular display of spring flowers in the Namaqualand Uplands of the
Succulent Karoo in South Africa is already a well-known attraction (SKEP,
2003). A *Rafflesia* Conservation Incentive Scheme was introduced by Sabah
Parks in Malaysia during 1994 for sites on private land containing *Rafflesia*
(see cover photograph). There are more than 13 species of *Rafflesia* in Sabah.
The idea was to generate an incentive to landowners to retain forest through
charging a fee to visitors who wished to see this parasitic plant, which has the
largest flowers of any plant worldwide. A total of 83 sites for *Rafflesia* had
been documented by 1996 in Sabah, 44 of them lying outside protected areas.
It was known that a further 20 sites outside protected areas had been destroyed
over the period of 1988 to 1996, mostly due to forest clearance for shifting
cultivation. As a result, the need for improved conservation was clear. The fee
charged to view *Rafflesia* was set at MYR5 (5 Malaysian ringgit; approximately US$2) per person for foreign tourists (but less for local visitors), with

BOX 11.7 EXAMPLES OF INITIATIVES CONCERNED WITH HEALTHCARE, MEDICINAL PLANTS AND CONSERVATION

- In 1993, the Foundation for Revitalization of Local Health Traditions (FRLHT) initiated an *in situ* conservation programme for medicinal plants in collaboration with the Indian Ministry of Environment and Forests, the forest departments of five Indian states and a number of forest communities. A network of 55 forest sites (Medicinal Plant Conservation Areas, or MPCAs), each approximately 200ha to 500ha in size, was established. The sites, located in various vegetation types and altitudinal zones, are managed jointly by forest departments and local communities. The MPCAs are intended to serve as *in situ* genebanks for medicinal plants. In total, they now capture more than 50 per cent of the wild medicinal plant species diversity of the region. The MPCAs are also used as sources of germplasm for developing cultivation systems by local communities and for breeding new crop varieties. Planting material supplied by the MPCAs has led to the establishment of nearly 20 district-level ethnobotanical gardens and more than 150,000 home medicinal plant gardens.

- A project with the Highland Maya in Chiapas, Mexico, encourages the establishment of community-based medicinal plant gardens for the production of low-cost herbal remedies, using culturally appropriate horticultural procedures. During 2002, the gardens contained 150 species of medicinal plants, very few of which had previously been in cultivation locally. In return for technical and scientific advice, each community is expected to maintain their gardens in good condition, record and update information on the species planted, and provide systematic instruction to younger Maya on the traditional knowledge of each species. Pilot research on cultivation techniques has been undertaken in collaboration with El Colegio de la Frontera Sur. Ethnobotanical surveys have documented local variations in the way in which medicines are prepared, which has proved useful for identifying recommended standard procedures (Berlin and Berlin, 2000).

- The Instituto Etnobiologia, founded by German Zuluaga in 2001, strives to promote healthcare with conservation in Colombia. Both indigenous groups and *mestizo* communities are involved in its programme. In the case of the indigenous people of the Andean piedmont, the institute tries to strengthen the autonomy and authority of community organizations – including an umbrella group, the Union of Indigenous Doctors of the Colombian Amazon (UMIYAC). The methods used include the encouragement of apprenticeships with traditional doctors, the full use of ceremonial houses, the designation of community protected areas and the establishment of medicinal plant gardens. The philosophy of the programme stresses the synergy possible through working at the *interface* between culture, healthcare and conservation.

- A project led by Nat Quansah at Manongarive in Madagascar has tried to strengthen community support for forest conservation through developing integrated medicine. Integrated medicine involves drawing upon both traditional and Western medicine as medically, economically and culturally appropriate. Selected traditional treatments have been tested in a laboratory at the University of Antananarivo in order to establish their efficacy, toxicity and optimal dosage. Western-trained doctors and traditional medical practitioners work side by side in an integrated healthcare clinic constructed by the project. The project has tried to encourage community pride in local culture and its links to biodiversity, and to identify and tackle local conservation issues relating to the use of medicinal plants (Baranga, 1999).

- The Traditional Medicine for the Islands (TRAMIL) network in the Caribbean and Central America links public and private research organizations, nongovernmental organizations (NGOs) concerned with biodiversity conservation and public health, public health agencies and local communities to work together in an interdisciplinary programme of research on the ethnopharmacology and traditional health practices of communities. The TRAMIL programme has three main goals:

1 revalorization of cultural traditions using medicinal plants;
2 provision of a scientific basis for the rational application of traditional health practices using medicinal plants, based upon criteria for safety and efficacy; and
3 identification of significant interactions between biodiversity of medicinal plants, local people and their tropical rainforest environment as a basis for conservation.

The results of research into the efficacy and toxicity of frequently used and widely known remedies are returned directly to communities through workshops, theatrical presentations and educational pamphlets, and are also published in a regional pharmacopoeia.

Source: Alan Hamilton and Patrick Hamilton (see also case study below)

an additional MYR10 (US$4) charged for each camera used. Pilot studies suggested that each site could yield MYR200–8000 (US$80–$3200) annually, depending partly upon the number of blooms, a level of income which compares favourably with the annual income from cultivating hill rice (MYR250 = US$100 per hectare per year) (Nais and Wilcok, 1998).

CASE STUDY: *AMCHI* MEDICINE AND CONSERVATION OF MEDICINAL PLANTS IN NEPAL

The Himalayas are a major source of medicinal plants, used by many hundreds of millions of people in India, China and elsewhere. Virtually all of the plants are collected in the wild. There is an urgent need to move towards greater sustainability in the supply of medicinal plants from the Himalayas through improvements in the management of wild populations and the development of cultivation.

Shey Phoksundo National Park (SPNP) is a remote, very large and newly established national park covering part of the traditional Dolpo region of

Figure 11.5 *Inuaguration of a traditional healthcare centre based on Tibetan medicine at Dolpo, Nepal*

Source: Yildiz Aumeeruddy-Thomas

Himalayan Nepal. Isolated until recently, when an airstrip was build nearby, SPNP still carries substantial supplies of medicinal plants. It is famous for snow leopard. Apart from its value for nature conservation, SPNP is an important area for Tibetan culture, with many Buddhist and Bon monasteries. About 50 *amchis* (practitioners of Tibetan medicine) live in the scattered settlements of the park and its buffer zone. The villagers cultivate barley *Hordeum vulgare* and buckwheat *Fagopyrum* and graze yak in high pastures.

In 1997, the World Wide Fund for Nature (WWF) Programme Office for Nepal started an integrated conservation and development project (ICDP), the Northern Mountains Conservation Project (NMCP), at SPNP to assist the authorities in developing a management plan for the park, ensuring that community interests are incorporated. Within the context of this project, the People and Plants Initiative (PPI) worked with WWF-Nepal to develop a project focusing on the conservation of medicinal plants.

A team was formed for the PPI project, consisting of botanists, sociologists and other external specialists, as well as community members and park staff. *Amchis* have been closely involved in advising the project. The *amchis* here, as elsewhere in Tibetan communities across the Himalayas, are extremely knowledge about local plants and traditionally have held considerable influence over natural resource management – for instance, to prevent overgrazing by yak. Much of the first year of the project was devoted to a scoping survey to gain an overall picture of the status of medicinal plants and related community concerns. This survey confirmed that commercial trade in medicinal plants had been increasing. It also revealed that the *amchis* are responsible for virtually all professional healthcare in the SPNP (there are almost no 'modern' health services in this region). Tibetan medicine, unlike Ayurveda, has failed to gain official recognition from the government of Nepal. The *amchis* at SPNP are concerned about future supplies of medicinal plants (they use both locally growing plants and others brought up from the lowlands) and the future of their profession, including the training of apprentices.

The results of the scoping survey were fed back to the community, partly through booklets written in Tibetan and Nepali. Following discussions, the project team and the community decided that the project should concentrate in its next phase on two related themes – the management of pastures for sustainable harvesting of medicinal plants and the strengthening of Tibetan medicine. These are linked, in principle, through the dependency of local medicine upon local plants and the fact that many *amchis* have traditionally had large measures of responsibility for natural resource management.

Surveys were then undertaken to identify the medicinal plants most vulnerable to over-harvest and to identify and promote methods of more sustainable harvest (including through rotational collection) (see Figure 4.3). Medicinal Plants Management Committees have been developed at village level. The strengthening of Tibetan medicine has been partly through establishing two traditional healthcare centres (see Figure 11.5). These serve multiple roles, including provision of healthcare (they have resident *amchis*), the training of apprentices, the provision of education in primary healthcare for women and the monitoring of medicinal plants for conservation purposes. Cultivation trials of a few selected medicinal plants were initiated in 2002 in the buffer zone of the park.

In 1996, *amchis* in Nepal formed an association, the Himalayan Amchi Association (HAA), for communication between themselves and to promote their causes. The WWF project has worked closely with the HAA, recognizing that, where they occur, the *amchis* are the most knowledgeable people about plants in the Himalayas and have an intrinsic interest in their conservation. In 2004, WWF was able to assist HAA in organizing an international *amchi* conference (the first ever held) to discuss the status of Tibetan medicine in various countries (Bhutan, China, India, Mongolia and Nepal) and to seek ways of strengthening their profession, including conservation of medicinal plants.

12

Projects with Communities

In situ plant conservation will almost invariably involve local residents, given that local people are in such a strong position to immediately influence the fates of plants. Projects are relatively short-term initiatives aimed (in the present context) at improving the state of people–plant relationships from the conservation perspective. An element of institutionalization in new patterns of behaviour will increase the likelihood that benefits will continue after projects end.

THE USEFULNESS OF APPLIED ETHNOBOTANY

It is widely recognized today that local residents must normally be involved in initiatives in *in situ* plant conservation. The relevant academic discipline, dealing with relationships between people and plants, is ethnobotany. Any social group can be the subject of (or partner in) ethnobotanical research, although in this chapter the emphasis is on one group in particular – the rural community. Applied ethnobotany is the branch of ethnobotany which is concerned with its applications to conservation and sustainable development. Applied ethnobotany is a participatory and problem-orientated subject that is evolving rapidly. Research in applied ethnobotany is both a means of gathering information and a social process, involving the building of relationships. Various types of participation between researchers and local people are possible and appropriate according to circumstances (Cunningham, 2001a). Often, members of communities will be involved in helping to define the objectives of the research, the approaches and methods used, the gathering and interpretation of the data, and the dissemination of the results. The involvement of local collaborators throughout the work means that there is a greater chance of buy-in and a better chance that practical actions will follow from the research (Davis, 1995; Tuxill and Nabhan, 2001).

Usually some local people have particularly close ties to local plant diversity; it is especially important to involve them in projects. Among them might be people who regard certain areas of natural habitats as parts of their territories, or who are involved in the harvesting of wild plant resources, or who practise traditional agriculture. Some users of wild plant resources may have special knowledge (**specialist resource users**) – for instance, in the art of using plant materials for constructing houses or concerning information about which wild plants are edible. However, the knowledge of 'ordinary' people about plants should not be neglected in plant conservation projects. For example, in the field of medicine, it is easy to become mesmerized by the fascinating knowledge and practices of some traditional doctors; but it is housewives and mothers who are responsible for much everyday treatment of health problems in communities, not shamans (Kothari, 2003).

The term '**key knowledge holders**' has been used in applied ethnobotany for individuals within communities who are recognized locally as being particularly knowledgeable about plants. Often, such people have received little formal education, but instead have gained their expertise through intellectual curiosity, powers of observation and long practical engagement. It has been recommended that one or more key knowledge holders should be incorporated within project teams from the outset. Their guidance will be useful in determining the directions taken by projects and in grounding them in local realities (Tuxill and Nabhan, 2001).

Research in applied ethnobotany will often involve collaboration between researchers and **resource users,** with **resource managers** occasionally as a third element. In this context, the term 'resource managers' has sometimes been defined (narrowly) as referring specifically to those responsible for protected areas (Cunningham, 2001a). '**Researchers**' are those members of project teams who are familiar with the approaches and methods of applied ethnobotany and who will generally be the instigators of studies. Unlike resource users and resource managers, researchers are commonly from outside the locality and thus will be less familiar about many aspects of local plants than at least some local people. What researchers should be able to offer to studies in applied ethnobotany are:

- knowledge of research approaches and methods that are useful in exploring relationships between people and plants;
- access to external sources of information, and key outside specialists and organizations, which might be useful for the project;
- some ability to place local conservation and development issues in wider contexts (this might include knowledge of the wider conservation status of species found in the project area, and of their uses and methods of management in other places, as well as knowledge of laws relating to plants and land).

Applied ethnobotany is an interdisciplinary subject that grades into many other disciplines which bear on conservation and sustainable development. Its application can help to reveal the conservation and development concerns of local people relating to plants. Relationships between people and plants cover so many matters that it can be daunting to know how to begin fieldwork. To help overcome this problem, Janice Alcorn (Alcorn, 1995) has suggested that asking the simple question 'What good is this plant?' can be a key to opening up dialogue and identifying key local concerns (see Figure 12.1).

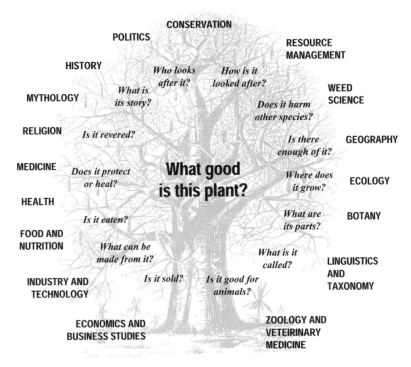

CONSERVATION

POLITICS

RESOURCE
MANAGEMENT

HISTORY

*Who looks
after it?*

*How is it
looked after?*

MYTHOLOGY

*What is
its story?*

WEED
SCIENCE

*Does it harm
other species?*

RELIGION

Is it revered?

*Is there
enough of it?*

GEOGRAPHY

MEDICINE

*Does it protect
or heal?*

**What good
is this plant?**

*Where does
it grow?*

ECOLOGY

HEALTH

Is it eaten?

*What are
its parts?*

BOTANY

FOOD AND
NUTRITION

*What can be
made from it?*

*What is it
called?*

LINGUISTICS
AND
TAXONOMY

INDUSTRY AND
TECHNOLOGY

Is it sold?

*Is it good for
animals?*

ECONOMICS AND
BUSINESS STUDIES

ZOOLOGY AND
VETEIRINARY
MEDICINE

Figure 12.1 *What good is this plant?*

Note: A simple question can lead to the identification of a wide variety of issues of conservation or development concern. The outer ring shows selected subject areas of modern knowledge and professions.
Source: adapted from Hamilton et al, 2003, based on an idea of Alcorn, 1995

PREPARATIONS FOR COMMUNITY-BASED PROJECTS

Community-based projects, including conservation projects, can take time to yield results – maybe five or ten years – and it is helpful if the individuals or organizations who wish to embark on conservation projects think in terms of **long-term commitment**, even if this cannot be initially guaranteed. Progress with community-based conservation projects can be uncertain, not least because there will be many external influences beyond the control of projects. Project teams should approach their work with modesty and not make rash promises that they cannot keep.

Community-based plant conservation projects may be full projects in their own right, or sub-projects of larger initiatives in conservation or development. An association with a larger project can have some advantages – for example, through benefits received from being able to draw upon the infrastructure of the larger project, its links to government, its research findings (for instance, the results of socio-economic surveys) or the foundations it has made with

Box 12.1 Initial steps in a community-based plant conservation project

1   Gain initial motivation to start a project based upon identification of a conservation issue(s).
2   Establish relationships with institutions and individuals as necessary or useful for project development and running. Discuss project with relevant government agencies and, if required, obtain their agreement for the project. This is an ongoing activity.
3   Select project site, using prioritization criteria if possible/applicable (see Chapter 9).
4   Through off-site research and informal visits, achieve some understanding of the site, conservation issue(s) and local people, and of how progress might be achieved.
5   Decide on initial boundaries of project – in terms of geographical extent, conservation issue(s) to be addressed, and communities and other social groups to involve.
6   Introduce the project to the community and obtain agreement for its development.
7   Form a project team, considering the range of core skills needed.

Source: Alan Hamilton and Patrick Hamilton

mobilizing communities. Conversely, a botanical component can sometimes help to provide a larger conservation or development project with direction. Experience has shown that large projects can become engrossed in internal processes and can lose sight of local realities. Community-based plant conservation approaches may uncover some key conservation and development concerns from local perspectives, and thus help to ground larger projects.

Some possible initial steps in a community-based plant conservation project are shown in Box 12.1. There will always be an **initial motivation** to start a plant conservation project, perhaps stemming from the concerns of individuals, non-governmental organizations (NGOs), communities, government agencies or industry. Whatever their origins, it is best if projects start small and are initially quite flexible. There are likely to be many unknowns when projects start – indeed, it may not even be certain that the problems initially identified really exist.

Projects can be considered as having hierarchies of **objectives**, ranging from ultimate purposes to the immediate aims of detailed practical activities (see Table 12.1). Hierarchies of objectives should be constructed at the start of projects and reviewed periodically thereafter in order to check that they remain relevant and realistic. Objective hierarchies are useful for helping to ensure that specific activities work towards a higher purpose, which can be lost in the hurly-burly of projects.

Community-based plant conservation projects should try to build **community capacity** in conservation. They should avoid creating long-term project

Table 12.1 *Example of an objective hierarchy for a conservation project*

Levels in objective hierarchy	Categories of objectives	Examples
1	Ultimate purpose	Achieve conservation of plant diversity
2	General goal	Achieve sustainable use of wild plant resources at a particular locality
3	Aim of a work programme	Provide information on the supply-and-demand sides of wild plant resources at the locality
4	Aim of a work sub-programme	Obtain information on the use of wild plant resources for a sample of households
5	Aim of a particular study	Gather information on the use of wild plant resources through semi-structured interviews

Note: Many funding agencies require presentation of logical frameworks on submission of proposals to ensure that projects are well conceived and realistic. The higher levels in the hierarchy are closer to ideals, giving philosophic orientation to the project. The lower levels are more practical and materially achievable.
Source: Alan Hamilton and Patrick Hamilton

dependency. Capacity-building is partly a matter of *individual* members of communities acquiring new knowledge, skills and attitudes, and efforts should be made to assist with education and training. In addition, the development of *community institutions* concerned with the management or use of natural resources can be one of the most valuable contributions that projects can make (McShane, 1999). **Institution-building** is crucial because of the roles that institutions play in human societies: they define the social roles and norms that regulate the behaviour of people associated with them. The solving of conservation problems with any degree of permanency requires the institutionalization of new procedures on the part of stakeholders. There are many types of community institutions that can be relevant to conservation, including families, forest user groups, groups of neighbouring households, local and traditional governments, self-help and church groups, and schools. Well-founded community institutions are needed for the negotiation and maintenance of agreements with agencies responsible for protected areas, as in joint forest management (see Chapter 11).

Capacity-building facilitated by projects need not be restricted to communities. If possible, projects should assist with the building of **capacity in plant conservation at the national level**. This can be achieved, for example, by providing opportunities for the training of postgraduate students, with the topics of research chosen so as to be useful both to the project and for educational purposes.

Projects will need to **establish relationships** with a range of institutions and key individuals, inside and outside communities, as necessary or useful for project development. Apart from existing community institutions, it may be appropriate for projects to encourage the formation of new community groups – for example, project advisory groups or groups of individuals who are interested in a particular type of plant resource. At the time when a project is being formulated, the ideas surrounding the project should normally be discussed with members of relevant government agencies, who may be able to offer helpful advice about how to proceed. They may be able to indicate particular groups or individuals within communities who would be particularly relevant or interested in becoming involved. Official permission is needed for projects in some countries. External to the project area, contacts should be established with relevant academic and scientific institutions – for example, with herbaria if it is likely that assistance will be needed in identifying plants.

Depending upon how it originated, the location and theme of a project may be predetermined or there may be choices, in which case prioritization criteria should be applied (see Chapter 9). Whatever the case, there is a need to place some initial **boundaries** on a project in terms of the geographical area to be covered, the issues to be addressed, and the communities and other social groups to engage. All these dimensions may need adjusting as the project develops. Different aspects of a project can have different geographies. For example, a project concerned with the sustainable use of a plant resource may be interested in both the area of supply of a resource (for example, a forest) and its area of end use (which may be much larger). Tackling the unsustainable harvesting of wood for the manufacture of charcoal may require intervention with government ministries far away from where the resource is harvested.

When dealing with subsistence economies, all major aspects of livelihood systems may need to be considered; in relation to professional disciplines, this can mean transgressing the traditional boundaries of specialities such as agriculture, forestry and medicine. Considerable efforts at lateral thinking may be required of members of project teams. The fodder system for stall-feeding livestock in Ayubia, Pakistan, uses plant resources found in protected areas (tree and herbaceous fodder), grass fodder grown in special 'wild' areas on farms, and plants grown or occurring incidentally in cultivated fields (crop residues and weeds) (see case study at end of Chapter 6).

Background research is needed before starting fieldwork in order to find out what is known about the site, the communities present and local conservation issues. Ideas about the approaches and methods useful for the project should be formulated – which will be greatly aided if lessons are available from the experiences of other projects. Sources of information to consult include scientific papers, official reports and unpublished reports by researchers and agencies ('grey literature') – backed up with discussions with experts. Informal visits to the project site are useful, especially if accompanied by people familiar with the place and the community. Such visits will provide opportunities for starting to get to know local people and to discuss project ideas. Fields in which some level of background knowledge is useful include physical and settlement geography (obtain maps); geology and soils; vegetation types, plant species and plant resources represented, and specific conservation issues relating to these; the fauna; types of livestock kept and their influence on the vegetation; social groups present and their cultures, economies and relationships with one another; the status and roles of men and women; types of land use and systems of agriculture; statutory and customary rules governing land tenure and access to resources; government structures and local representation of government agencies; NGOs active locally; and relevant current or recent projects.

Some knowledge of the **history** of the site, in all its aspects, is invaluable for understanding modern conditions. An historical perspective allows a better appreciation of why the flora, vegetation, patterns of land use, and systems of land tenure and resource rights are as they are, and may provide indications of what may be realistically achievable through project interventions. Sources of evidence about the past include written records, old maps, archaeological findings and fossils (for example, pollen diagrams) and traces left on the landscape (such as boundary earthworks, internal woodbanks and other surface features). Experienced ecologists will be able to read something of the past in the modern landscape. Once fieldwork starts, interviews with local people, especially elders, can be invaluable sources of information about how and why the environment has changed.

A plant conservation project is normally best implemented by a **project team**, which is a group of people dedicated to the project, involved intimately in its implementation, and which meets regularly for its review and coordination. Teams should be multidisciplinary, covering the key areas of expertise necessary to achieve project goals. In a project based on applied ethnobotany, the key fields to cover are botany/ecology and sociology/anthropology, with experts in other subjects added if the project has (or develops) specialist interests. For example, a medical doctor might be incorporated if work on health-related aspects of conservation is intended. Team members should have some understanding across disciplines, which is why knowledge of ethnobotany, straddling the botanical and social sciences, is so useful. Progress will be substantially impeded if botanists on project teams know nothing about sociology and sociologists know nothing about natural resources. Gender representation on a project team should be given attention, especially when communities are divided in their everyday lives along gender lines. Women are often strongly engaged with plants (for example, doing much of the farming or gathering of fuelwood); but men tend to dominate public decision-making. In such cases, it is essential to have **professional women** on project teams because otherwise it may be difficult or impossible to engage local women properly in project activities. Project teams should normally include members of local communities, especially key knowledge holders, the inclusion of whom has been regarded as vital for project success (Tuxill and Nabhan, 2001).

The **attitudes** of team members towards their work and local communities are crucial. Members of teams will certainly vary in their characters and specialist interests, which are not bad things in themselves; but they must have genuine respect for local knowledge and be genuinely interested in plants, otherwise these deficiencies will soon become apparent to local people and their cooperation may only be superficial. Projects should aim to build community self-confidence, which is one reason why project members must respect local culture. Gaining the support of communities requires courtesy, consideration and time (Cunningham, 2001a). Members of project teams should be sensitive to local customs and timetables, and be aware when opportunities are most likely to be favourable for the involvement of local people in project activities. Patience and perseverance are needed.

Project members will come with their own beliefs and motivations, which some of them may well find themselves reviewing as projects develop. Working with people whose lives are hard, as is so often the case in developing countries, will necessarily result among normal people in a measure of sympathy for their plights (Lagos-Witte, 1994). Plant conservationists may become drawn away from seeing local people (and the plants around them) in generalized objective ways to engaging with local people as individuals, each with his or her unique qualities and individual life (Peterson, 2001). The challenge is not to lose perspective and objectivity in analyses.

FIELDWORK

The **way that a project starts** sets the tone of its relationship with local people. Setting a good tone will assist in gaining local recognition and approval, and therefore will influence the willingness of local people to participate. Members of project teams should decide how they will represent themselves on first visits to villages or households. There may be suspicion if **introductions** are made by agencies with which there have been histories of conflict with communities (as can sometimes be the case with park or forestry departments).

The **goals** of the project and its intended approaches and methods should be discussed at formal and informal meetings (Tuxill and Nabhan, 2001). What is needed at this stage are common understandings of what the project is trying to achieve and the contributions and benefits expected of each party. The **form of agreements** with individuals and communities may evolve as projects proceed – for example, being relatively informal and unwritten at first, and perhaps becoming more formal and written later.

An initial step in a project is stakeholder analysis: the identification of the major social groups relevant to the issues at hand and the relationships between them. A **stakeholder** is defined as an individual, group or institution that is affected by, or can affect, the resource, issue or process of interest. Broadly, stakeholders can be divided into two categories – those who are resident locally and those who are not. Resident stakeholders might include the owners and managers of land and plants, the collectors of wild plant resources and the local representatives of government agencies. Non-residential stakeholders might include absentee landowners, the senior officials of forestry departments and other government agencies (responsible for setting policies), and the owners of commercial enterprises concerned with manufacturing products made from plant resources gathered locally.

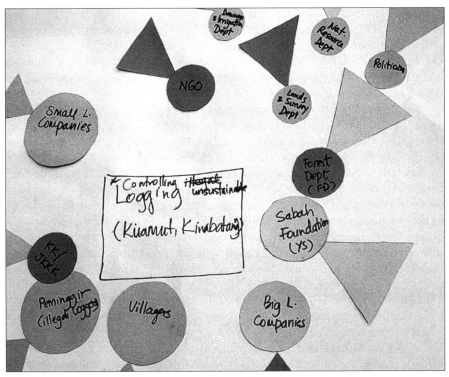

Figure 12.2 *A stakeholder analysis undertaken by conservation project workers for Kinabatang in Sabah, Malaysia*

Source: Alan Hamilton

Stakeholder analysis is best undertaken with single social groups, at least initially, because groups can vary in their perceptions of who has power over natural resources – an informative finding in its own right. Stakeholder analysis should be undertaken around key issues and to a level of detail that gives insights into use, management and power – for example, in the case of communities, perhaps down to the level of social groups that vary in their economic status, gender, age, specialized occupation and ethnicity (see Figure 12.2). Stakeholder analysis can be repeated at intervals during the lives of projects to serve different purposes. For example, at the beginning of a project it could be used to explore perceptions about who has control over forest resources, while, at a later stage, a key question might be to ascertain perceptions about who should be involved in a tree-planting initiative.

A standard way of undertaking a stakeholder analysis is through production of a Venn diagram. The procedure involves holding a workshop with a group (for example, a community), often best subdivided into sub-groups (for instance, different types of resource users or along gender lines). The participants are told to construct diagrams to show the influence of various bodies (institutions, groups, departments, programmes, etc.) over an issue or question – for example, their influence over the management of forest resources. The conventions used for the diagrams are that each body is represented by a circle, the size of which represents its importance for the issue at hand and the distances between which indicate how close bodies are linked, one to another. Strong relationships between bodies can be represented by thick lines, and weak relationships by thin ones.

Some **forms of interaction between projects and communities** are listed in Table 12.2. They vary in degrees of formality. Experience has indicated that, in the case of resource users, it is often best to work with homogeneous groups selected on the basis of an interest in a particular resource. Group size is a consideration. For example, a field exercise to assess the quality of individual fronds of a palm, *Hyphaene petersiana*, used for basket-making gave the most useful results when based on consensus reached by groups of four to five basket-makers (Cunningham, 2001a). Various types of liaison and advisory groups can be useful, with varying degrees of permanency. A People and Plants Initiative project at Ayubia, Pakistan, has had a national advisory group, liaison groups for individual villages and sets of villages, women's groups and temporary groups formed for particular purposes – for example, to trial fuel-efficient stoves. The best social structure for different purposes may take time to recognize. For instance, again in Ayubia, work to encourage community-based tree nurseries was originally based upon agreements between the project and *individual* households; but experience showed that, in this case, agreements between the project and *groups* of neighbouring households was more productive (since more was then achieved, with tasks shared between households). **Exchange visits** are extremely useful for learning from others – for instance, one community may be able to learn from another how to cultivate a particular plant.

A project can be conceived and administered as a cycle of processes, involving the setting of project objectives and activities, carrying out the activities, monitoring the results of the activities, reflecting on these results, and then setting new objectives and actions for the next round (see Figure 12.3) (CMP, 2004). A major purpose in giving a degree of formality to project management is to ensure that a project remains focused on priority issues and tackles these in the most effective way. This reduces the risk of 'project fatigue' – for example, with local people complaining about endless surveys that seem to lead nowhere. There can be links between the **project management cycle** and funding from external sources, which generally require work to be carried out in a transparent and credible way to stand a chance of continued support. Of course, the project cycle does not stand in isolation. It is preceded by a project preparation phase (already described) and should be succeeded by a process of deliberate project closure. Projects teams should try to ensure that some benefits persist once projects end, which is why a major purpose of projects should be capacity-building, individually and institutionally. Periodic independent external **evaluations** are a good idea in order to cast fresh eyes on projects, the issues that they are addressing, how far the projects have been successful and to provide advice for the future. Monitoring and evaluation form an

Table 12.2 *Some possible forms of engagement of projects with communities*

Purpose of involvement	Who?	How?
Agree to involvement in project and continuing liaison	Community representatives; liaison groups	Agreements with varying degrees of formality (may change as project evolves)
Advise project	National to local advisory groups; may include community sub-groups (for example, women's groups)	Periodic meetings to review project progress and advise on future directions and activities
Obtain overview of people–plant relationships and identify priority issues	Community as a whole or sub-groups; key knowledge holders	Participatory appraisal techniques; discussions and interviews
Detailed research on specific topics	Key knowledge holders and specialist resource users	Participatory action research
Ensure interests of those with strong ties to wild plants and natural habitats are addressed	Relevant social groups	Take steps to involve these groups (which may include more reticent people)
Feed back research findings to community	Community as a whole or representatives	Communication techniques appropriate to the audience
Learn from (and benefit) other communities	People concerned with similar conservation issues	Exchange visits
Create management plan for plant resources	Representatives of communities, including local resource users	Planning workshops

Source: Alan Hamilton and Patrick Hamilton

integral part of project management. This is one reason why projects should try to establish **baseline indicators** early in their lives, providing benchmarks against which progress can be measured.

The project management cycle has similarities to the applied conservation research cycle (see Figure 4.2) and the resource management cycle (see Figure 11.3); indeed, the three cycles should be closely linked in projects (see Figure 12.4). A major purpose of the project management cycle should be to allow progress to be made on participatory learning, and a major purpose of participatory learning should be to make progress on resource management. Thought should be given regularly to the usefulness of research in relation to its practical applications. Setting the right research questions is fundamentally important for project progress.

Many projects are likely to draw on several of the **approaches to *in situ* conservation** described in Chapter 11, as well as others. Both scientific and

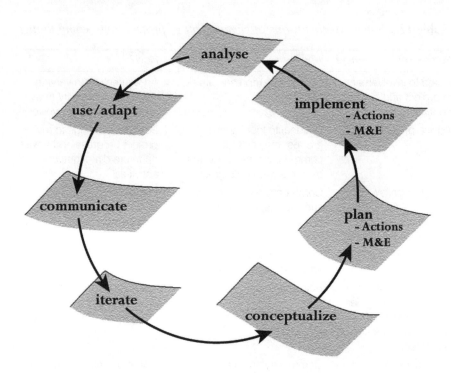

Figure 12.3 *Generalized project management cycle; the cycle consists of seven steps, each with sets of principles and tasks*

Note: Although depicted as a linear series of steps, the entire process is rarely applied in a linear fashion, as depicted. Projects normally go through complex series of back-and-forth movements. M&E = monitoring and evaluation

Source: adapted from a diagram in CMP, 2004

local perspectives on priorities should influence how projects develop, with regular reviews of priorities to ensure that projects are continuing along worthwhile lines. Experience is helpful in judging how far a project should 'stray' from dealing directly with the matters for which it was established. A willingness to respond to local concerns, provided that they have some relevance to the project, can build local confidence. The concept of an **entry point** can be useful: this is an activity carried out by a project at the request of a community that helps it gain local recognition and respect. A measure of self-doubt is almost inevitable at some stage in community-based conservation projects because, after all, such work can have many uncertainties as to what will actually be achieved. A degree of persistence in approaches has some merit, even if doubts arise as to whether they are ideal. Decisions to abandon one approach in favour of another should not be taken lightly since chopping and changing can be discouraging to project members and communities alike, and may create a feeling of unease about a project's competence.

PURPOSES

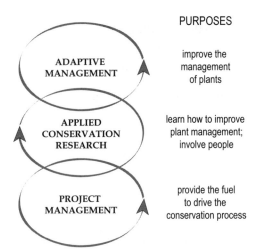

ADAPTIVE MANAGEMENT — improve the management of plants

APPLIED CONSERVATION RESEARCH — learn how to improve plant management; involve people

PROJECT MANAGEMENT — provide the fuel to drive the conservation process

Figure 12.4 *Three cycles of learning and action; the interlinked rings symbolize how projects (see Figure 12.3) can influence resource management (see Figure 11.3) through applied conservation research (see Figure 4.2)*

Source: Alan Hamilton and Patrick Hamilton

Projects should normally commence with **overview (scoping) surveys**. Scoping surveys should:

- provide general pictures of local patterns of resource use and ecological knowledge, as well as some ideas of how plants fit into local livelihood strategies;
- identify key conservation and development issues relating to plants from local perspectives;
- test preconceived ideas about conservation issues;
- provide some assurance that critical areas have not been overlooked; and
- give some ideas about the types of project approaches and activities that are most likely to be practical and effective.

Some of the data gathered will be useful as baseline information for assessing progress later in projects. Overview surveys can be valuable not only at the start of projects, but at intervals thereafter in order to determine project impacts and to re-evaluate priorities. Box 12.2 contains some suggestions about questions that might be asked in an initial overall survey, based upon a focus on wild plant resources.

Research is generally needed during projects to provide key information. Two broad classes of research methods can be recognized: participatory appraisal (PA) and participatory action research. **Participatory appraisal** techniques, such as those associated with participatory rural appraisal (PRA), can provide general pictures of community and household economies, land

---

Box 12.2 Some possible questions for an initial overview survey in a community-based plant conservation project

The emphasis in this example is on wild plant resources:

- Who are the resource users?
- What is their socio-economic and formal educational status?
- Is harvesting of wild plant resources for commercial or subsistence purposes, or both?
- Which resource categories and species are most in demand and valued (culturally and economically)? (Make an inventory of at least the major species used, with their local names.)
- What is the geography of resource use?
- What conditions of tenure relate to plant resources, who controls access to the resources (now and during the past) and what are the relevant village or other institutions?
- What are the forms of relationship between resource harvesters and landowners (and traders if resources are sold)? If there is trade, are the traders operating in competition or do they collaborate?
- What are the effects of harvesting on plant populations and which species are most vulnerable to overexploitation?
- Is overexploitation and increased resource scarcity of concern within local communities (rich *versus* poor), nationally or internationally?
- Does the harvesting of one resource adversely influence the availability of others (for example, species with multiple uses, or when harvesting adversely affects a habitat)?

Source: based largely on Cunningham, 1997a, 2001a

---

management patterns, local ecological knowledge and values placed on plants. PA can be used with sub-groups of communities – for instance, along the lines of gender, occupation or use of a particular resource. Some of the most useful PA techniques for resource management are resource listing; ranking and scoring resources; resource mapping; stakeholder analyses for particular resources; perceptions of landownership and resource rights; and construction of timelines, chronologies and seasonal calendars (see 'Field methods and tips'). PA is not just about data gathering. It can be a community development process that provides opportunities to open up dialogues about issues of development between projects and communities, as well as within communities. PA can deepen understanding of particular issues and be used for planning and evaluation. It can be used to rank problems facing a community and for preparing community plans to tackle them. PA techniques can help to build a consensus on a particular issue or to identify agreed points of disagreement. PA exercises can engage substantial numbers of people in projects, raising the chances that the results of research exercises will actually be used.

PA techniques are good for establishing general contexts and engaging communities; but, on their own, they may not provide data that are sufficiently reliable and subtle to identify precisely how resource management can be improved (Cunningham, 2001a; Tuxill and Nabhan, 2001; Campbell and Luckert, 2002). More detailed research is often needed, drawing upon the techniques of anthropology, ecology, forestry and other disciplines, though modified to ensure full 'use' is made of the knowledge, skills and wisdom of local people. This type of research (**participatory action research**) should be aimed at producing results that will be of value fairly directly for solving real-world problems. The most appropriate forms of participation of local people in the research will vary according to context.

The results of research should be presented periodically to communities for their interest, cross-checking and discussion of their implications. Methods of presentation should be chosen that are suitable for the intended audience. Publications, posters, videos, DVDs, workshops, talks and drama all have their uses. If publications are contemplated, then they should be written in styles and languages that are locally meaningful. Schools, churches and community gatherings all provide useful venues to convey and discuss results. Face-to-face meetings with officials are essential in many countries, with the mere handing-in of reports with recommendations for action unlikely to elicit much response.

Many conservation problems relating to plants are not, or only partly, solvable at the local level. Solutions may be constrained by **unfavourable government policies** or by commercial systems operating in adverse ways over which local people have little or no influence (McShane, 1999). Project teams may be able to catalyse discussion on policy reform through preparation of 'policy briefs' – that is, short, punchy reports written for policy-makers with clearly presented arguments and recommendations for policy change. Project experiences can be valuable as case studies in policy briefs to illustrate the limitations of current policies. In the case of commercial systems, it may be possible to make some headway towards more environmentally responsible business practices through creating awareness among senior figures in companies or industry associations.

Occasionally, *de facto* policy change is possible locally in a semi-legal way through introducing changes that are countenanced by local authorities, but do not conform 100 per cent to national policies. There are examples of such expediency in relation to community roles in protected areas, with local arrangements made allowing communities some rights in reserves against the strict letter of the law. This situation can arise because government policies are typically slow to change and can evolve too slowly to cope with realities on the ground. Actually, such local examples can sometimes be welcomed by governments, showing the way forward in policy and helping to accelerate policy change.

FIELD METHODS AND TIPS

Introduction

Techniques of participatory appraisal are described in several general manuals concerned with community-based conservation and development (for example, Pretty et al, 1995; Worah et al, 1999). More specific botanical aspects of participatory appraisal are covered in some People and Plants publications and videos (see Martin, 1995; Tuxill and Nabhan, 2001). Some People and Plants publications concentrate more on techniques useful in participatory action research (for example, Campbell and Luckert, 2002; Cunningham, 2001a). See also Chapter 14 for methods used in market surveys.

Thought should be used in applying methods since these will often need modification to suit local circumstances (Cunningham, 2001a). Scientific sophistication should not be the only criterion used in choosing methods. Attention should also be given to their cultural and financial appropriateness.

An individual research exercise can be considered to have **three phases**: preparation, implementation and follow-up. *Preparation* includes reviewing why the exercise is to be undertaken, obtaining the necessary tools and equipment, and making arrangements with the participants about where and when to meet. *Follow-up* includes making sure that the results have been analysed and that the whole exercise has been adequately written up. It has been suggested that initial analyses should be undertaken by all members of project teams while they are still in the field. This allows an early assessment of whether the right questions have been asked and a cross-checking of the findings. Reports from research exercises should be written up as soon as possible, bearing in mind the target audiences and what the reports are designed to achieve (Tuxill and Nabhan, 2001).

Some problems can be approached in more than one way. **Cross-checking data** by using different approaches (triangulation) is valuable for increasing the reliability of estimates (Cunningham, 2001a). For example, it might be possible to explore the amount of a plant resource being used in a village either through obtaining data through household interviews, measuring the amounts collected as they are transported from field to home, and examining fees paid for collecting licences, as found in forestry records.

People can have different views on the same subject. The solving of conservation problems can benefit greatly from having access to such **multiple perspectives**. This is why it is useful to have a range of professional disciplines represented on project teams, for project teams to include local people, for projects to interact with stakeholders of a range of different backgrounds, and for project teams to deploy a variety of information-gathering tools to explore issues (Aumeeruddy-Thomas et al, 1999; Tuxill and Nabhan, 2001).

The level of **precision** or **detail** should be sufficiently accurate for the purposes at hand. The time available may mean that trade-offs may be needed between quantity, accuracy and timeliness when collecting data (Campbell and Luckert, 2002). The fundamental questions being addressed should be kept firmly in mind, with efforts made to study *all* major aspects of the questions, without becoming 'bogged down' in collecting huge amounts of detail on some aspects only. Of course, detailed research is also valuable – for example, in allowing the detection of seasonal or long-term trends in resource harvesting or marketing, which will not be apparent from snapshot surveys (Cunningham, 2001a). On the other hand, it is not uncommon for statisticians offering advice to projects, especially when they lack direct experience themselves of field-work, to recommend laborious sampling strategies, which actually represent time poorly spent from the practical, problem-solving point of view. It is probably inevitable that there will often be continuing tensions between practical researchers and statisticians.

Equipment and recording

A wide range of **equipment** can potentially be used in community-based research (see Box 12.3). The most essential items are a notebook and pencil. *How* a notebook is used is important since this can influence how participants react to researchers – for instance, perhaps inhibiting the free flow of conversation if the process of transcription is too obtrusive or formal. Officious note taking (or questioning) may prevent the honest presentation of views, with responses tending towards those expected 'officially', especially if there are doubts about the motivations of researchers. For this reason, a small palm-sized notepad can be useful for noting down key points in conversations, with a fuller write-up later while experiences are still fresh in the mind. In any case, it is good practice to transcribe field notes into a separate notebook or computer soon after a day's work, partly as an insurance against loss of the field notebooks.

Researchers should be able to **record data systematically**, ready for storage, sorting and analysis. Computer databases are an excellent way of handling data these days and should be used if computers are available (Berjak and Grimsdell, 1999). **Standardized fields** can be valuable for recording information in systematic surveys – for example, concerning the characteristics of plant specimens, vegetation samples or households. Standardized fields can be incorporated into **field data sheets** or questionnaire forms. The choice of standardized fields is an important one in community-based research and should be discussed with local collaborators, especially key knowledge holders. Very often, the choice of fields will vary from the 'standards' found in related types of scientific investigation – for example, regarding the choice of fields to include on the labels of plant specimens, with more emphasis than normal placed on matters of ethnobotanical concern (Martin, 1995; Tuxill and Nabhan, 2001).

The **names** of plants, plant parts, categories of plants, vegetation types, places, units of weight, volume and storage should be **recorded in the local language**. They should be transcribed correctly according to convention. The **scientific equivalents** of these names and terms should be determined in due course.

BOX 12.3 SOME SUGGESTIONS FOR EQUIPMENT FOR FIELDWORK IN APPLIED ETHNOBOTANY

- *fundamental*: notebook, pencil;
- *next most fundamental*: maps, knife, machete/panga, field guides for plant identification, press, drying paper, measuring tapes (3m and 30m);
- *routinely used*: field data sheets, questionnaire forms, trowel, polythene bags (different sizes for collecting plants and other samples), string, paint and brush and/or coloured tapes and labels (for marking plots or plants), scales of different sizes (from 100g upwards, depending upon what is to be weighed), cards, flip charts and markers (for workshops), calculator;
- *routinely used, but may often be borrowed at site*: spade or other digging tool, posts (for marking plots etc.);
- *more specialist or expensive*: specimen labels, camera, compass, altimeter, global positioning system (GPS), clinometer, forestry calipers (for measuring diameters of smaller stems), diameter tape, increment borer (for studying tree rings and estimating tree age), pH kit, battery-operated electronic balance (for weighing lighter materials); soil auger.

Sources: based largely on Cunningham, 2001a; Martin, 1995, and Tuxill and Nabhan, 2001

Qualitative and quantitative approaches

Both qualitative and quantitative approaches are used in research in commnity-based plant conservation. **Qualitative methods**, such as in-depth interviews, participatory observation, informal discussions and careful observation, can reveal much about the values and uses of plants in a community when used by open-minded, experienced researchers. Such approaches are especially revealing about the underlying worldviews and mindsets that lie behind everyday thoughts and activities. Dealing with people as individuals, rather than as units of statistical populations, has the advantage that information is personalized. Variations in knowledge and attitudes can be explored in detail, including about any conflicts relating to plant resources that may exist within communities. Qualitative observations are very useful for identifying themes and hypotheses to follow up using quantitative techniques.

Some advantages of **quantitative techniques** are that they encourage the clear statement and testing of hypotheses, they provide figures (for example, on the availability or use of a resource) and they allow replication. Occasionally, quantitative analyses are based on *whole* populations (in a statistical sense), such as *all* households or *all* home gardens in a village. However, quantitative techniques normally rely upon the use of **samples**. The ways in which samples are selected will influence the methods of statistical analysis eligible for use and the types of conclusions that can be reached from the data. It is advisable to seek advice from a statistician before deciding on a sampling strategy (further comments on sampling strategies are given in Chapters 8 and 11).

Botany

An **inventory** of plants known to local people will generally be a component of a community-based plant conservation project. Inventories can be variously detailed. A list of the most important plants used by people should be compiled during the scoping survey carried out at the start of a project and further information should be added as the project continues. The degree of emphasis placed on ethnobotanical inventory will depend upon the purposes of the project. It should be noted that anything approaching a complete inventory of a community's botanical knowledge will take many years to complete and is of more academic than practical interest. Normally, after an initial scoping survey, a project will concentrate on particular aspects of the relationships between people and local plants, which will then constitute the main contexts in which the inventory is extended. Records of *who* provided information should be noted during inventories. This will be useful for analysing the **distribution of knowledge** within communities (see Chapter 4).

At least some members of project teams should be familiar with the **scientific ways of identifying and classifying plants**. It is a great advantage if a team can identify many plants in the field. This will increase the efficiency of the work and reduce the number of plant specimens that need to be collected for identification. Ethnobotanical studies require researchers to be able to **identify sterile plants** (those without flowers or fruits), as well as goods and products collected or made from plants. The sort of skill that may need to be developed is the ability to identify the types of wood in bundles of firewood or roots sold in medicinal plant markets. Key knowledge holders may be able to point out features useful in identification.

The **scientific identification** of plants encountered in the field is important because this facilitates comparisons with other sites and opens up access to scientific information about the plants, such as on their distributions, conservation status, methods of management, properties and uses. The scientific authentication of plants may require the collection of **voucher specimens**, which are samples of plants that have been flattened and dried in plant presses (for methods, see Martin, 1995). Voucher specimens are taken to herbaria and compared with specimens in the collections and descriptions and illustrations of species in books. It is very useful if there are experienced and helpful taxonomists associated with the herbaria, willing to help with identification. In any case, some members of project teams should know how to use herbaria to identify plants.

It can be helpful to retain specimens of identified plants at the project site for quick reference and for educational purposes. Research aimed at community-based conservation will generally benefit from the establishment of **ethnobotanical collections**, consisting of plant materials, goods sold (for example, samples of medicinal plants from markets) and finished products (such as baskets, with the species used identified scientifically).

Community knowledge and values of plants

Projects will often wish to explore the values that local people place on plants, including in relation to categories of use. Once lists of species placed in different **resource-use categories** have been compiled, then the ranking or scoring of these categories and the species within them can follow.

The ways in which local people identify and classify plants should be noted carefully. These may reveal information about the types and properties of plants that have escaped the attention of scientific botanists. One way to investigate local systems of plant classification is through **pile sorting**, which involves participants putting plants into groups according to how they judge their similarities with one another (Martin, 1995).

Free listing is a useful technique for exploring the categories of use of plants recognized by people, the types of plants used for different purposes and the values placed on plants. The technique can be used with people individually or in groups. As an example, participants in an exercise may be asked to list use categories for plants (for example, house construction, fuelwood and medicine) and then give the names of plants used for each of these purposes. It has been found that free listing can lead to a relatively complete itemization of local resource-use categories and also reveal the more important species used for each purpose (because these will tend to be mentioned earlier during listing).

Post-it cards (stickies) are useful devices for investigating local knowledge and values of plants within the context of workshops. Indeed, such cards are used very widely in participatory assessments in order to develop and organize ideas on many topics. An example is given by Tuxill and Nabhan (2001), involving the selection of priorities for local resource management. The authors suggest that the first step should be an oral introduction by a knowledgeable person, summarizing the topic in an unbiased way and asking anyone present for information that has been missed. If something has been missed, then this also needs to be presented orally to the participants. A facilitator then distributes blank post-it cards and asks each participant to write down plant species, sites or habitats (one per card) that they consider priorities for management. After about ten minutes, the cards are gathered together, mixed up and stuck on a wall, with the words on each card read out by the facilitator while doing so. The facilitator then asks if there are any more ideas and, if there are, they are written down on fresh cards and added. The cards are then grouped together physically into categories, chosen so that the subsequent discussion can be organized. To discuss plant resources, for instance, it might be appropriate to group the cards into categories such as 'income source for local residents', 'home subsistence products' and 'rare/difficult to find'. After this, the facilitator works through the cards in each category, asking which of the items is the most important to achieve, fostering discussion and attempting to reach a consensus. The cards are then marked with their priority numbers and placed in order – and thus a priority list is established.

There are various ways of **scoring** the values of plants – for example, in terms of their frequency of use, number of different types of use, amounts used per person or per household, or prices paid in commercial transactions. Toledo (1995) calculated scores for the values of plants based upon the number of their use types (edible, remedy, construction, etc.) and whether they were used in major or minor ways in each case (scoring 1 or 0.5, respectively). A total use value was then obtained by addition (for example, $1 + 1 + 0.5 = 2.5$ for a species with two major types of use and one minor one). A similar method had been used earlier by Phillips and Gentry (1993) based on interviews carried out within standard 1ha sample plots.

Ranking involves placing a number of items in order of preference or according to some other measure of value. For example, people could be asked to rank a list of five potential new cash crops in order of preference (1 maize; 2 coffee; 3 oranges; etc.) or problems that they think might be associated with new types of cash crop (1 difficult to cultivate; 2 susceptible to pests; 3 high labour costs; etc.). An added dimension can be introduced if ranking is carried out simultaneously along two axes (**matrix ranking**). Using the above examples, this would give a ranked list of five crops each scored according to a ranked list of problems. Matrix ranking provides added insight into local knowledge and preferences. The technique of **paired comparison** involves asking participants to place items in order of preference, based upon judgements of preference between pairs, taking each pair in turn.

Interviews, conversations, walks and participant observation

Interviews and conversations are ways of exchanging information between project teams and local residents, and building collaboration. Interviews may be conducted individually or with groups, with interviewees selected along various lines. They may be purely oral or involve the use of props – for example, pieces of wood of different species to illicit information about their merits as fuelwood or photographs of landscapes to trigger memories of environmental change. The dynamics of interviews depend upon who is involved and can be affected by such factors as age, gender and ethnicity. To take account of sensitivities and to increase the sample size, it may be best to train suitable residents in interview techniques so that they can act as interviewers.

Interviews can be held with individuals or groups. **Individual interviews** provide more personal and potentially more detailed information. Some types of information, such as about wealth or livelihoods, are sensitive and interviews concerning such matters are best carried out in private and the information received not used in ways that allow the identification of individual interviewees. Group interviews provide more general perspectives and, through the ways in which people interact with one another during discussions, some insight into community relationships. **Group interviews** can be dominated by certain participants – for example, men over women or prominent villagers over others – so that more reticent people may be inhibited from expressing their true views.

Interviews and conversations can be structured variously. **Open-ended interviews** and conversations are free-ranging discussions, with researchers taking care not to impose their ideas. A **semi-structured interview** or conversation is based on a checklist of questions held in the head of the interviewer. It is normally undertaken with one person at a time. The intention is that answers to the preformed questions will be forthcoming at some point during the discussion, but without forcing the issue or inhibiting the free flow of conversation. Various degrees of 'control' or precision can be attempted; for example, in an investigation of wild plants used for food, researchers might try to establish the names of all wild plants that are used, or perhaps a standardized list of all those eaten during the preceding 24 hours (**24-hour recall**). An advantage of a semi-structured interview is that it allows the gathering of survey information in a fairly systematic fashion. A disadvantage is that the choice of questions in the head of the interviewer can impose certain meanings on the data. For this reason, it is best not to use this approach until a reasonable understanding of people–plant relations is already available – for instance, through interaction with key knowledge holders. In this way, questions can be chosen and posed such that the greatest amount of useful information is solicited.

Walks in forests and fields with members of communities allow researchers and local residents to get to know one another and can elicit much useful information about people, plants and the land. Walks can merge into participant observation, with resource users demonstrating how they select and collect the plants that they harvest. As with interviews, walks can be undertaken with individuals or groups, selected in various ways. Furthermore, as with interviews, researchers can approach walks with completely open minds and discussions can be totally free ranging, or else the researchers can impose mental structures on the walks, which may be planned to pass certain places or plants of interest, be routed along transects or take in panoramic views of the landscape. Participants may be asked to prepare sketches of the landscape – for example, in the form of **transect diagrams**.

Participant observation is an invaluable technique for gaining an appreciation of the roles of plants in people's lives, including their symbolic meanings. It involves living with, and sharing, the experiences of people as they go about their lives. Thus, researchers may find themselves engaging in agricultural activities, helping to build a house, learning how to make handicrafts or taking part in ceremonies. Taking house construction as an example, it has been suggested that participant observation is the best way to investigate the materials used for its construction, and how these are selected, harvested and prepared for building (Alcorn, 1995). It allows the documentation of alternatives that might be used and the places where resources are harvested.

Questionnaires and household surveys

Questionnaires are especially useful for eliciting information in the case of better-educated, more literate people. They are best administered in face-to-face interactions with researchers. A wide range of questions can potentially be posed in questionnaires, including questions that are useful for interpreting the data, such as the personal circumstances of respondents in terms of age, family status, social position, occupation and institutional affiliation. Questions in the core part of the questionnaire can be either in the form of multiple choices (for example, which of the following five types of fuelwood is your most preferred?) or be open ended (for example, what are your five favourite types of firewood?). Multiple-choice questions have the advantage of ease of quantification, but pose limits to the information gathered. There are many pitfalls in devising the questions to be used in questionnaires, such as having questions that are leading, vague or make implicit assumptions. Questions should be chosen after a reasonable understanding of people–plant relationships has been achieved (for example, through in-depth interviews with key knowledge holders) and after testing pilots.

Household surveys are widely used in applied ethnobotany in order to give information on livelihoods and on the uses of plants. Both observations and interviews can be useful. One approach is to request data on the types and quantities of plants used for various purposes, the localities where they are obtained, and whether they are bought or sold. Another approach is artefact centred, taking a particular item (for instance, a house, basket or medicine) and finding out which types of plants have been used in its manufacture, where and how they were obtained and who obtained them.

Mapping and timelines

Mapping has many uses in projects (see Figure 12.5). The **stakeholder diagram** shown in Figure 12.2 is a form of social mapping. More conventional maps, drawn free hand by community members (**vernacular maps**), can reveal a great deal of information about the local environment, including aspects of local cultural significance. An alternative way to construct community maps is to ask community members to record information on published maps (**annotated maps**). There are many elements of landscapes that can be mapped, including physical features, past and present settlements, vegetation and soil types, the distribution of species and plant resources, the sites of present and past harvesting of plant resources, and the locations of landholdings. Mapping opens up many opportunities for discussion on conservation issues, including questions of management. A mapping exercise can be an excellent first step in participatory appraisal. Information from mapping exercises, combined with data from sample plots, can also be used to draw maps of the possible distribution of plant species or resources, based on extrapolation using geographic information system (GIS) techniques.

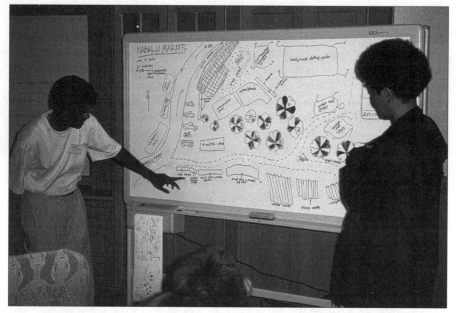

Figure 12.5 *A sketch map of a market layout, near Mount Kinabalu,
Malaysia, made by students on a course of the People and Plants Initiative*

Source: Alan Hamilton

Timelines and trend lines are methods of placing modern conditions in historical context. Many facets of ecosystems can be studied, including how landscapes have evolved, how the human population has changed, how social structures and settlements have developed, and how people's use of plant resources has varied over time. Elders and long-term residents are often particularly interested and knowledgeable about the history of their localities. Their knowledge can be recorded in various ways, such as through diagrams showing key events or trends, drawings of the landscape at intervals in the past and lists of species used at different times. Cross-checking with historical records will help to establish a reliable chronology.

Seasonal diagrams and calendars are useful for revealing local knowledge of the climate and the seasons, and annual cycles of community activities. The phases of longer-term cycles – for instance, related to swidden systems of agriculture – can also be explored.

Long-established permanent plots are invaluable for providing evidence of how the vegetation and flora have changed. Unfortunately, the locations and records about plots established in the past are often forgotten or lost. Even short-term permanent plots can provide useful information in some cases, such as plots established to determine the impacts of fires, grazing or the gathering of fodder – involving exclusion of these factors from the plots.

13

Ex Situ Conservation

Plant species and varieties can be preserved under artificial conditions away from the places where they naturally grow. *Ex situ* plant collections have a number of uses for conservation and development, including for the revitalization of plant populations and associated economies and cultures.

EX SITU AND *CIRCA SITUM* CONSERVATION

Ex situ conservation is the conservation of components of biological diversity outside their natural habitats (CBD, 1992). Some formal methods of *ex situ* conservation are listed in Table 13.1. Among these methods, the storage of seeds in **seedbanks** has some advantages for preserving species, but can only be used for species with seeds capable of remaining viable after long-term storage (known as 'orthodox' seeds). The typical technique used for seed storage is to lower the moisture content of the seeds to 2–6 per cent or less, and reduce the temperature to around 0° Celsius or lower. Collections in seedbanks should occasionally be tested for their viability. Periodically, the seeds in seedbanks should be germinated and the plants allowed to grow and produce more seeds, which are then stored as replacements for the originals. In comparison to some other common methods of *ex situ* conservation, the advantages of seed storage can include low cost, less risk of disease and more efficient use of space or land. Seed storage can be 50 to 500 times cheaper per collection than field genebanks or *in vitro* storage (Epperson et al, 1997). However, there are numerous species, including many tropical forest and temperate trees, whose seeds cannot be stored in seedbanks because they lose viability if their moisture contents are reduced to the required level ('recalcitrant' seeds).

Field genebanks can be used for conserving varieties of plants for which seedbanks are unsuitable. They are mainly used for major crop plants, such as banana *Musa*, mango *Mangifera* and yam *Dioscorea*. **Botanical gardens** and arboreta differ from seedbanks and field genebanks in that their collections usually consist of small numbers of many species, rather than many specimens of a few species (see Figure 13.1). ***In vitro* storage** refers to the maintenance of cells or tissues in sterile growth media in dishes or flasks.

Table 13.1 *Techniques of* ex situ *conservation*

Techniques	Definition
Seed storage	The collection of seed samples at one location and their transfer to a genebank for storage. The samples are usually dried to a suitable low moisture content and then kept at sub-zero temperatures.
Field genebank	The collecting of seed or living material from one location and its transfer and planting at a second site. Large numbers of accessions of a few species are usually conserved.
Botanic garden/ arboretum	The collecting of seed or living material from one location and its transfer and maintenance at a second location as living plant collections of species in a garden or for tree species in an arboretum. Small numbers of accessions of a large number of species are usually conserved.
In vitro storage	The collection and maintenance of explants (tissue samples) in a sterile, pathogen-free environment.
DNA/pollen storage	The collecting of DNA or pollen and storage in appropriate, usually refrigerated, conditions.

Source: adapted from Hawkes et al, 2001

Ex situ plant collections serve various roles in conservation and development, including (Brown and Briggs, 1991; IUCN, 2001a):

- refuges of last resort against extinction;
- sources of plants to reinstate or reinforce wild populations;
- habitat restoration;
- sources of germplasm for the selection or breeding of useful plant varieties;
- production of plant products;
- convenient collections for research; and
- public edification – for example, through displays in botanic gardens.

The *ex situ* preservation of plant species will become increasingly urgent as more and more species become extinct or threatened in the wild, as is predicted. Apart from the many other threats that species will face (see Chapter 2), that of climatic change – considered alone – poses an excellent reason for improving *ex situ* conservation facilities. Climatic change is expected to place many species in danger. *Ex situ* facilities can help by ensuring that the species do not disappear from the Earth completely and, through providing planting materials, by playing a role in the introduction of threatened species to new places where conditions for their establishment and growth are suitable. Because botanic gardens already grow so many threatened species (see Box 13.1) and have the potential to grow so many more, they certainly have a vital role to play in saving the world's plants.

There are moves towards more systematic approaches to *ex situ* plant conservation based upon the more careful targeting of species to incorporate

Figure 13.1 *The palm collection at Xishuangbanna Tropical Botanical Garden, China*

Source: Alan Hamilton

in collections, the better sampling of plant populations in order to capture more of their total genetic diversity and the integration of *ex situ* and *in situ* conservation. An early example of a more systematic approach to sampling was that of the Consultative Group on International Agricultural Research (CGIAR)/ International Plant Genetic Resources Institute (IPGRI) system, which, from the 1970s, has aimed at conserving the genetic diversity of the major food crops, principally by *ex situ* means (see 'Plant genetic resource systems'). As for wild plants, a progressive initiative was the creation of the Center for Plant Conservation in the USA, which is dedicated solely to preventing the extinction of the imperilled American flora. It is hosted by the Missouri Botanical Garden at St Louis, Missouri, and has 33 participating institutions located around the country. It maintains a National Collection of Endangered Plants, which is an *ex situ* collection of more than 600 of the country's most endangered species. This collection is intended as a back-up in case a species becomes extinct or can no longer reproduce in the wild, and for use in restoration and research.

BOX 13.1 EXAMPLES OF REPRESENTATION IN
***EX SITU* PLANT COLLECTIONS**

- The Royal Botanic Gardens, Edinburgh, grows 69 per cent of global conifer species in its four gardens in Scotland.
- About 80,000 species are grown in botanic gardens, including 10,000 threatened species.
- The Millennium Seed Bank at Wakehurst Place, part of the Royal Botanic Gardens, Kew, in the UK, aims to preserve 10 per cent of the world's seed-bearing flora by 2010, as well as all native British plants.
- There are more than 6 million accessions of plant genetic resources for food and agriculture stored in 1500 germplasm collections around the world; about 80 per cent are of major crops or their close wild relatives.

Sources: IPGRI, 2002; Young, 2002; BGCI, 2004

Britain and Ireland are countries interested in gardening (in a decorative sense) and here the National Council for the Conservation of Plants and Gardens has been established to conserve those traditional garden plants that are fast disappearing from gardens. It consists of a network of 630 National Plant Collections, each dedicated to a particular group of plants – for example, peonies or maples. Almost half of the collections are in private ownership. There should be potential for similar approaches to *ex situ* conservation in other parts of the world, based on those aspects of plant diversity that are of particular interest to local people.

In situ and *ex situ* approaches and methods to plant conservation should be seen as complementary, with methods chosen for use that are appropriate to particular cases. Unfortunately, they can sometimes be seen as *alternatives*, with potentially damaging consequences. There are some people who claim to see nothing wrong in transforming virtually the entire terrestrial surface of the planet into cities, farms or other 'productive' forms of land use, with conservation of plant genetic diversity consigned to tiny nature reserves or to seed stores within the Antarctic ice cap. This approach is radically ill conceived (for reasons outlined in Chapter 1). On a more prosaic front, one of the assessors of projects submitted for funding to the Department of Science and Technology in India informed the author Alan Hamilton in 1999 about his worries about the large number of proposals seeking funds for *ex situ* rather than *in situ* plant conservation. The proposals were often to bring a handful of specimens of a few threatened species into botanic gardens. Hardly any proposals were being received to conserve plants *in situ*. The problems here may have been partly due to the feeling that *in situ* and *ex situ* conservation are *alternatives* and the institutional positions of many concerned botanists, making it easier for them to see how they can help through *ex situ* means. This raises the question as to whether it has been wise to label the *ex situ* preservation of plants as a form of conservation, as has been the practice in plant conservation circles.

The main comparative *advantages* of *ex situ* conservation, compared with *in situ* conservation, can include less use of space, sometimes better assurance of long-term conservation of germplasm (inherited genetic material) and occasionally easier access to germplasm for use in research or development. The main comparative *disadvantages* of *ex situ* conservation are that only a part of the genetic variability of a species or wild population is conserved, the plants may be subject to loss of genetic integrity (see later), access to the material may be difficult for some users (depending upon the management system of institutions) and there can be unpredictable accidents (Hernández Bermejo, 1995). Additionally, *ex situ* conservation does not contribute to the conservation of habitats, ecosystem services or the many other forms of life associated with particular species of plants in nature. The World Conservation Union (IUCN) has advised that *ex situ* plant conservation should avoid competing directly for resources that otherwise might be channelled to the conservation of wild plant populations and habitats (IUCN, 2001a).

The distinction between *ex situ* and *in situ* conservation is not always as great as might be supposed. For one thing, there are different degrees of rigour with which these terms are applied. In its purest sense, *ex situ* conservation could possibly be restricted only to collections of species that have been properly sampled from the genetic standpoint, fully documented and actively managed in formal *ex situ* collections, such as in botanic gardens or seedbanks. Actually, many plant collections that are commonly given as examples of *ex situ* collections consist largely of essentially randomly collected specimens, quite often without proper documentation or even incorrectly labelled, and not actively managed: their claim to the *ex situ* label rests largely on their associations with botanic gardens or seedbanks. From our experience, there are many botanists in some parts of the world who find it difficult to see why the growing of *any* species that is threatened in the wild in cultivation should not be labelled as *ex situ* conservation – for example, the commercial growing by villagers of threatened medicinal plants to replace the unsustainable harvesting of wild plants from nearby forests or pastures. Similar questions arise about *in situ* conservation. In its purest form, *in situ* conservation could be taken to be the safeguarding of self-perpetuating wild populations of species in completely natural habitats. However, it is a fact that there are no 'pure natural habitats' in most parts of the world and it is likely that many 'wild' populations of species have been influenced genetically by people. Furthermore, conservationists often refer to species as 'safeguarded' if they are found within protected areas; but, in fact, this affords little guarantee that they really are safe.

The term *circa situm* conservation has been proposed for a range of practices that are intermediate between 'traditional' *in situ* and *ex situ* conservation. They are associated especially with more traditional (and biodiversity-rich) agricultural systems (Hawkes et al, 2001). They include the retention of 'wild' plants when land is cleared for agriculture or when crops are weeded, the growing of 'wild' plants in home gardens, and the storage of crop seed in granaries for later replanting. All are common practices (see Van den Eynden et al, 2003). There are many reasons why particular species (or specimens of species) may be chosen for *circa situm* conservation, including their practical uses as sources of food, medicine, fodder or other products. The selection of better varieties in *circa situm* conservation is a very common practice. Some forms of *circa situm* conservation represent insipient stages in the domestication of wild species.

Ex situ conservation faces major challenges (see, for example, Ishakia et al, 1998; Clement, 1999). A common problem is **poor genetic representation**, which can sometimes be rectified, once identified. DNA analysis of specimens of the conifer *Fitzroya cupressoides* growing in the British Isles showed that all the trees belonged to a single clone, having been propagated from cuttings. This led to a study of the genetic diversity of the species in the wild (in Chile and Argentina) and then the collection of seed from 48 'mother' trees from a wide area. The result was an expansion in the representation of genetic diversity in cultivation in the British Isles (Allnutt et al, 1999; Young, 2002). **Inadequate documentation** is another problem. For example, many of the 6 million accessions in germplasm collections worldwide lack adequate passport data (documentation associated with germplasm collections) (Hawkes et al, 2001). Another issue is **loss of genetic integrity** of the collections. This can happen either through inbreeding or out-crossing, and both in the case of free-living plants and during the regeneration of seeds in seedbanks. The genetic decline in the quality of *ex situ* collections of endangered species can make it impossible for them to survive under natural conditions and thus render them useless for reintroduction purposes (Barrett and Kohn, 1991).

The problem of loss of genetic integrity in seedbanks is illustrated by the example of varieties of oats *Avena sativa* stored in seedbanks in Germany and Austria. Some of these samples date back to 1831. All the samples have been stored in seedbanks since 1956, involving several cycles of regeneration (a normal procedure with seedbanks). Protein markers showed that 12 out of the 28 strains examined had been contaminated with genes from other strains of oats, introduced since the time of their initial storage. This was attributed to carelessness during harvesting or handling. Similar contamination has been found in samples of spelt wheat *Triticum spelta* in seedbanks. Spelt was once the most common grain in Europe, although it almost disappeared during the 20th century. In the case of these examples, contamination was attributed to a lack of sufficient funds to undertake the regeneration work with adequate care – regeneration being the most expensive work of seedbanks (Steiner et al, 1997; MacKenzie, 1998). Inadequate funding has similarly been reported as a major cause of the many problems that have been encountered with germplasm collections in Brazil and Kenya (Ishakia et al, 1998; Clement, 1999).

Plant Genetic Resource Systems

Plant genetic resources are defined by the International Plant Genetic Resources Institute (IPGRI) as the 'genetic material of plants which is of value as a resource for the present and future generations of people' (Hawkes et al, 2001). Since we cannot know which genetic material will be of value to future generations, in a broad sense the conservation of genetic resources embraces all efforts to conserve the genetic diversity of plants, including saving plant species and making utilization of plant resources more sustainable. All the various *in situ* and *ex situ* plant conservation activities described in the present book are therefore contributions to the conservation of plant genetic resources.

Over the last decades, considerable effort has been devoted to building up systems of conservation for certain types of plant genetic resources, largely through *ex situ* means (see Figure 13.2). These systematic collections are known as germplasm collections (germplasm is the material in plants that is inherited). At present, only about 2 per cent of collections in *ex situ* germplasm collections are of wild plants (Hawkes et al, 2001). The most comprehensive international system involved in systematic germplasm conservation is associated with the Consultative Group on International Agricultural Research (CGIAR), which supports several centres, each an autonomous institution and most concerned with research into food crops (see Box 13.2). The purpose of establishing this system was to meet the demands of plant breeders for fresh genetic material in the face of a mounting loss of crop landraces and surging demands for food on the part of an expanding world population. Concern about the loss of crop landraces and wild crop relatives (also sources of fresh genetic material) had been growing from the 1960s. The International Board for Plant Genetic Resources (IBPGR, now IPGRI) was established by CGIAR in 1974 with a secretariat supplied by the Food and Agriculture Organization of the United Nations (FAO). IPGRI serves as the convening centre of the CGIAR network and hosts its coordinating secretariat. At first, the focus of IPGRI was on *ex situ* conservation and so it has largely remained. *In situ* conservation (largely on-farm conservation in this case) has received little support, partly because of predictions by some agricultural experts during the 1960s and 1970s that crop landraces would soon disappear with the spread of modern high-yielding varieties. However, as it happens, many farmers in at least some parts of the world have continued to use local landraces, the loss of which has not been everywhere so massive as was once predicted. Belatedly, IPGRI started an *in situ* conservation programme in 1992/1993 and is continuing a search for suitable approaches and methodologies.

In 1996, a Global Plan of Action was adopted at a conference in Leipzig, Germany, organized by the FAO. The objective was to achieve better conservation of plant genetic resources for food and agriculture. This led to a widening in the scope of work by IPGRI, which was extended to cover relatively minor crops, including so-called 'neglected and underutilized' species. Furthermore, IPGRI began considering whether to start a programme on medicinal plants. An Economic Crop Protection/Genetic Resources (ECP/GR) Group for medicinal and aromatic plants, supported by IPGRI, convened for the first time in Slovenia in 2002, although it showed more interest in the genebanking of common culinary herbs with complicated taxonomy,

Figure 13.2 *Anji Bamboo Botanic Garden: A germplasm collection in China, specializing in bamboo*

Note: More than 300 species of bamboo are cultivated in this garden.
Source: Alan Hamilton

such as mint *Mentha*, oregano *Origanum* and thyme *Thymus*, than in conservation of threatened species of medicinal plants.

Sampling is a major consideration in building up a systematic germplasm collection. Practical advice on many aspects of collecting trips is provided by Hawkes and colleagues (2001), especially with reference to the landraces of crops. Suggestions are offered on how to decide on the number of sites to sample for a particular crop type, how these sites should be distributed and, within each site, how many plants should be sampled and how they should be selected. There are trade-offs between the various quantitative variables. As a thumbnail guide, they recommend the sampling of 50 sites per region, 50 plants at each site and 50 seeds per plant. However, readers should be warned that this level of intensity of sampling is quite inappropriate for threatened or rare wild plants. Indeed, if followed, it might well result in the demise of populations of species or even whole species. The collection of threatened or rare plants for *ex situ* collections should only be undertaken by those who have been adequately trained. Various guidelines for sampling threatened plants have been published – for example, by the Australian Network for Plant Conservation (ANPC, 1997).

Plant genetic resource systems exist at many social levels. Apart from the international CGIAR centres associated with IPGRI, formal germplasm collections are also found at national and sometimes regional levels. Efforts made

---

BOX 13.2 CONSULTATIVE GROUP ON INTERNATIONAL AGRICULTURAL RESEARCH (CGIAR) CENTRES AND THEIR RESPONSIBILITIES

CGIAR centres include the following:

- CIAT: Centro Internacional de Agricultura Tropical (Colombia);
- CIFOR: Centre for International Forestry Research (Indonesia);
- CIMMYT: Centro Internacional de Mejoramiento de Maíz y Trigo (Mexico);
- CIP: Centro Internacional de la Papa (Peru);
- ICARDA: International Centre for Agricultural Research in the Dry Areas (Syria);
- ICRAF: International Centre for Research in Agroforestry (Kenya);
- ICRISAT: International Crops Research Institute for the Semi-Arid Tropics (India);
- IFPRI: International Food Policy Research Institute (US);
- IITA: International Institute of Tropical Agriculture (Nigeria);
- ILRI: International Livestock Research Institute (Ethiopia);
- IPGRI: International Plant Genetic Resources Institute (Italy);
- IRRI: International Rice Research Institute (the Philippines);
- IWMI: International Water Management Institute (Sri Lanka);
- WARDA: West Africa Rice Development Association (Ivory Coast);
- WorldFish Centre (Malaysia).

Crops given special attention include the major cereals and food legumes, root crops (for example, cassava, potato, sweet potato and yam), banana and plantain, and oil crops (such as coconut and groundnut).

Source: Alan Hamilton and Patrick Hamilton

---

by India to conserve its agricultural biodiversity include a network consisting of a national genebank (established in 1997), 10 regional genebanks and 20 collection centres (Jayaraman, 1998). India has also established four germplasm collections specifically for medicinal and aromatic plants. Ethiopia is exceptional in that, through its Institute of Biodiversity Conservation and Research (IBCR), it established a programme in 1994 for the *on-farm* (*in situ*) conservation of landraces of its most important food crops, including teff, barley, chickpea, sorghum and faba bean. Its approach links farming communities and their agricultural varieties with its own formal genetic resource conservation efforts through the intermediary of community genebanks. This system, aimed at conservation of crop landraces at farm level (rather than in remote *ex situ* collections), allows for the continuing evolution of landraces through ongoing selection of varieties by farmers, continuing interaction of the landraces with the environment and continuing opportunities for gene exchange with wild species. Six on-farm conservation sites and community genebanks have been established for farmers' varieties in six agro-ecological

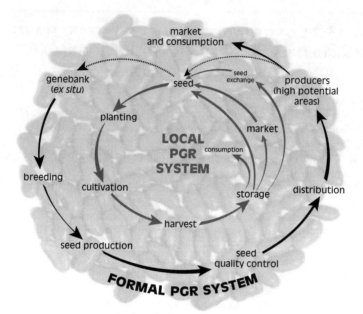

Figure 13.3 *Local and formal plant genetic resource (PGR) systems*

Source: adapted from a diagram in de Boef and Almekinders, 2000

zones. Farmer Conservator Associations have been formed for each site. IBCR documents the knowledge of farmers about their crops (such as about their methods of selection, cultivation and use of different crops and cultivars), the particular knowledge and roles in agricultural systems of women, and information on seed exchange and movement.

This Ethiopian example draws attention to the fact that, apart from formal plant genetic resource systems, there are also *local* plant genetic resource systems, such as those long used by farmers in traditional agricultural systems all over the world. In the case of crops that reproduce from seed, these systems involve cycles of planting, crop growth, seed harvest and selection of seed for planting (see Figure 13.3). Song Yiching and Jiggins (2003) have described how two parallel seed systems operate in China, one supported by the government plant-breeding establishment and the other by poor farmers. Taking the example of maize *Zea mays*, they describe how the government sector has focused on the breeding of high-yielding hybrid varieties. The objective has been to raise yields per hectare. However, this objective of maize breeding is not shared in its entirety by female farmers, who are also interested in cooking quality, taste, security of yield and other properties. Women prefer their traditional varieties because they are open-pollinating: the seed produced on their crops is viable and can be sown. In contrast, hybrid seed, associated with government maize breeding, has to be repurchased with every new planting. Open-pollination also means that new varieties can be produced on-farm through crossing.

How formal and local plant genetic resources systems interact is a matter of great conservation importance, but one that has been largely ignored. Until recently, the formal sector tended to have a rather exploitative approach to local systems, which were used as sources of germplasm for breeding new varieties of crops, but generally received nothing in return. Indeed, the system considered as a whole has tended to work *against* conservation (except in a specialist short-term sense) since the spread of modern crop varieties has caused the extinction of many landraces. Modern high-yielding crop varieties have also been a mixed blessing from the wider environmental and social points of view. On the positive side, they have contributed to raising the amount of food available on the global scale. On the negative side, their adoption has resulted in much loss of biological diversity and pollution on an industrial scale. Mechanization and other processes associated with modern high-yielding varieties have led to consolidation of landholdings and less use of labour. In turn, this has led to the displacement of many small farmers and farm labourers, who have moved to rural areas that are marginal for agriculture or else migrated into city slums.

REINTRODUCTIONS, RESTORATION AND REVITALIZATION

Plant material from *ex situ* collections can be used to replace wild populations that have become extinct, or to reinforce threatened populations. Such reintroductions can benefit from the building up of the stock of a plant through tissue culture and the growing of seedlings or cuttings for planting in nurseries. Where possible, efforts should be made to retain the original geographical patterns of genetic variation shown by species by using material for planting that has originated from the places of reintroduction. Undoubtedly, a combination of *in situ* and *ex situ* approaches to plant conservation, linked through reintroductions, will become increasingly important to conserve threatened floras. It is already considered indispensable for the survival of many species of oceanic islands, such as those of Hawaii (Meilleur, 1999).

Reintroductions are unlikely to be successful unless the original causes of threat to the species have been addressed. A major cause for the decline of many species is reduction in habitat. If this is the case, then steps are needed to restore the habitat prior to (or at the time of) the reintroduction (Figure 13.4). For example, on Mauritius, the reintroduction of many species requires that invasive species are first cleared and destructive introduced animals (such as pigs and rats) excluded from the planting sites.

Some species can *appear* to have become extinct at particular localities, but actually survive in hidden forms. It may be possible to restore them, without reintroductions, through removal of inhibiting factors. For example, the restoration of woody vegetation in degraded tropical savanna is sometimes possible through stopping excessive burning or grazing. Trees can sometimes reappear of their own accord, provided that their rootstocks are still present,

Figure 13.4 *Staff of Plantlife International planting seeds of the nationally endangered starfruit* Damasonium alisma *in a pond especially excavated for its reintroduction; Buckinghamshire, UK. Inset:* Damasonium *flowers*

Sources: Simon Williams (main photo), Bob Gibbons (inset); both images copyright © Plantlife International

concealed cryptically among the grass and shrubs, or else mother plants exist close by to spread their seeds.

Restoration of the starfruit *Damasonium alisma* in the UK shows how plants can sometimes be restored without reintroduction. By the 1980s, starfruit was known from only two ponds in Britain and was considered highly endangered. However, it was discovered at one of the sites where clearance of dense covering vegetation and exposure of bare substrate resulted in the prompt appearance of the plant in greater numbers. Subsequently, studies of its ecology contributed to the realization that its basic habitat requirements were ponds with fluctuating water tables and freedom from competing plants. Its seeds were capable of survival for long periods in pond mud, awaiting opportunities to germinate. Another British species that has been found sometimes capable of reappearing without the need for reintroduction is the fen violet *Viola persicifolia*. This species has been recorded as returning at one wetland site following clearance of scrub after an absence of almost 70 years.

The seeds of some herbs that grow in more open places in woodlands in Britain are also capable of long dormancy, with the plants able to reappear after a long time once the tree canopy has been opened up (Marren, 1999).

Times of crisis can highlight the need for some *ex situ* conservation facilities to reconsider their missions. Take the case of rice germplasm in West Africa, a region where many countries have established *ex situ* rice facilities. Unfortunately, a number of these facilities have been lost in the repeated conflicts that have wracked this region over recent decades – the *ex situ* rice germplasm collection at Rokupr, Sierra Leone, was lost in 1980 when cold storage equipment was looted, facilities for rice research in Liberia were extensively damaged by fighting and became no longer functional, and one national agricultural research institute was once a guerrilla training camp. These conflicts have also been responsible for the uprooting of rural populations, which have been forced to abandon their farms along with their local varieties of rice. Local plant genetic resource systems have been undermined even further in several ways. Sometimes, farmers, in desperation, have sold or eaten their planting seed, inappropriate varieties of rice have been provided by relief agencies (often more interested in volume than variety) and trust among rural populations has been undermined by the conflicts. Trust forms the traditional platform upon which local systems of seed exchange are built (Richards and Ruivenkamp, 1997).

A problem with these *ex situ* rice facilities (apart from their looting!) is that they were established essentially just to serve the needs of plant breeders. According to Richards and Ruibenkamp (1997), they should be reorientated to also serve the needs of local livelihoods. Links between these *ex situ* facilities and local *in situ* and *circa situm* systems of plant conservation should be explored. To achieve this, it will be necessary to establish good links with local communities and to build up the *ex situ* collections taking more interest in the local socio-economic, cultural and agronomic contexts of plant conservation. Richards and Ruibenkamp (1997) write about a vision of *ex situ* facilities forming a 'socio-technical ensemble, linking seed reserves, seed systems, information networks, technical facilities, relief agencies and farmer groups in a transnational web of mutually reinforcing interactions (the regional seed safety web)'.

An example of how *ex situ* collections can connect with *circa situm* conservation is shown by a People and Plants project, involving reintroduction of the pepper-bark tree *Warburgia salutaris* into Zimbabwe. The pepper-bark tree is the most highly prized medicinal plant in Southern Africa. Over-harvesting of this plant over the years eventually resulted in its national extinction in Zimbabwe – a matter of great concern for traditional doctors and their patients alike. However, Tony Cunningham and the Southern Alliance for Indigenous Resources (SAFIRE), a Zimbabwean non-governmental organization (NGO), managed to reintroduce this species successfully from nurseries in South Africa through the distribution of rooted cuttings to the home gardens of traditional doctors and local farmers, who knew and valued the species. The rationale was that these particular individuals would be sufficiently motivated to guard the plants with the attention that would certainly be required. It was thought that reintroduction of the species back into its natural forest habitat would not work because it would most likely soon disappear again, given that the causes of threat had not been removed (Cunningham, 2001b).

There is much to learn about how best to conserve crop diversity in agricultural landscapes. One necessity is to move away from the pervading premise underlying much agricultural research that the purpose of developing farming systems is overwhelmingly to increase agricultural production. There should be a philosophical reorientation among the agricultural research community towards the notion that maintaining, restoring or creating sustainable livelihoods is a focus of at least equal importance (Clement, 1999). Another concept that requires some adjustment is the idea that plant conservation is necessarily about maintaining the same species or varieties in the places where they are currently present. Traditional farming systems are dynamic, with the frequent gaining and losing of landraces. A dynamic notion of plant conservation should be developed for landraces.

Techniques suggested to encourage the *in situ* conservation of agricultural crop diversity include (Tuxill and Nabhan, 2001):

- drawing attention to the unique qualities of local landraces – for example, their uses in ceremonies or as definers of cultural or personal identity;
- linking crop diversity to social prestige – for example, through farm competitions for the greatest crop diversity;
- stressing the economic values of traditional varieties (they will typically be less valuable than modern high-yielding varieties in the market; but there are growing consumer interests in diet and organic foods);
- increasing farmers' access to landrace diversity through encouraging community seedbanks, and formal and informal exchanges of germplasm;
- tailoring formal crop-selection programmes to maintaining landrace diversity, as through **participatory varietal selection** (farmers select pre-existing varieties) or **participatory plant breeding** (farmers extensively involved in crop-breeding programmes, which are fundamentally orientated by farmers' perspectives of what is desirable). Of these, the second is more likely to result in retention of on-farm crop diversity since participatory varietal selection can result in the uptake of a single variety replacing several traditional landraces.

Conservation attention needs to be given to the woody elements of agricultural systems, as well as to herbaceous crops. The conservation and development of useful trees in agroforestry systems is a largely neglected subject, but one of great economic importance in developing countries. Trees in farming land often provide many products valued for subsistence use or sale. Leakey and colleagues (Leakey, 2003; Leakey et al, 2003; Simons and Leakey, 2004) report that, very often, varieties of trees found in villages are suboptimal from the livelihoods perspective and substantial benefits are possible from programmes of tree crop improvement. Apart from attention being given to *individual* types of trees, there is also a need to look at the overall contribution of trees in the *landscape* since, very often, rather few species are grown and fewer products produced than potentially possible. These authors have suggested that programmes of domestication be developed, based on the principle of scientists who specialize in tree selection and breeding working directly with farmers. The idea is to carry out such work village by village, basing selection on farmers' own preferences and, to an extent, using germplasm that is local to each place. Leakey believes that this approach could help to revitalize the farming landscape, while conserving something of the genetic diversity of the species.

Restoration is a growing issue in conservation. It is estimated that 17 per cent of the Earth's vegetated land has been degraded by human activities since 1945 (Robinson, 2003). Desertification, salination and soil erosion are rife in many places, with ecosystems much less usefully productive than they could be. Restoration of vegetation will become an increasingly important aspect of conservation. Normally, it should involve a diversity of plants, which is one of the reasons why it is so vital to ensure the survival now of a wide variety of species and of genetic diversity within species. Successful restoration will be a socio-economic and cultural, as well as a biological, achievement – hence, the usefulness of the concept of revitalization: the return again of biological and human vibrancy to places from which it has been eroded.

14

Plant Trade

If products from wild plants become traded, then pressure on populations of wild plants can dramatically increase. Interventions in favour of conservation are possible at various points along the chains of production and trade. Actions taken at 'higher' points in trade systems – away from the sites where the plants grow – must feed back to the production level if they are to have conservation benefits.

WILD PLANTS IN TRADE

Today, there is a vast trade in plants and plant products, with many impacts on ecosystems. Monetary economies now penetrate nearly everywhere on Earth and there are few people who do not purchase any products with botanical ingredients. Here, attention is directed especially to trade based on *wild* plant resources because of the particularly close connections that can exist with the conservation of plant diversity. The term 'wild' is used here for plants that are uncultivated, unmanaged or not owned, though recognizing that none is necessarily an absolute condition. Most tropical timber is, in effect, taken from wild plants because, although governments commonly grant rights of logging to concessionaires with rules attached, in practice these arrangements are often weak. There are many wild plants that are collected for local subsistence use and, generally, this does not result in major conservation concerns. It is the pressure of trade that creates major problems.

Many of the commodities being considered here are non-timber forest products (NTFPs). An idea of some of the items involved can be gained from Box 1.3 in Chapter 1. The major trade sectors include timber, construction, furniture, crafts, energy, food, herbal medicine and horticulture. The term medicinal and aromatic plants (MAPs) is widely used in trade circles to describe plants used for the manufacture of medicines, spices, essential oils and similar products. The term botanicals is used, especially in the USA, for certain types of products with specific botanical ingredients (but excluding mainstream foods), including herbal medicines, perfumes, colouring agents, detergents, health foods, food supplements, herbal teas and other items related to health and personal care, especially those marketed as 'green' or 'healthy'. Box 14.1 gives some examples of valuable products based on wild fungi.

BOX 14.1 SOME EXAMPLES OF HIGH-PRICED MUSHROOMS
AND OTHER PRODUCTS BASED ON FUNGI

- **Truffles** are underground fruiting bodies produced by certain fungi (notably *Tuber melanosporum*) that form mycorrhizal associations with the roots of some temperate forest trees, particularly species of oak *Quercus*. Especially known from France and Italy, truffles are highly prized gastronomically and can fetch very high prices.

- Various species of **morels** Morchella are harvested for export to Europe from the temperate forest zone on mountains from Turkey to Nepal. This is one of the highest priced non-timber forest products (NTFPs) in the region and there are many collectors. In Pakistan, morels are sold for 5000 to 6000 rupees (US$60 to $75) per kilogram dry weight by collectors, compared with approximately 6 rupees (US$0.08) per kilogram for many medicinal and aromatic plants (MAPs) (1999 figures) (Zahid Waheed, personal communication, 2000). In Pakistan, which exports 99 per cent of its production, the morel is not regarded as a food, though there are some medicinal uses (analgesic, aphrodisiac and for the treatment of rheumatoid arthritis).

- **Matsutake** *Tricholoma matsutake* is a mushroom found in East Asia. It forms mycorrhizal associations with various temperate forest trees, such as species of oak *Quercus* and pine *Pinus*. It is very highly regarded in Japan, where the quantity available from internal sources has fallen from 10,000 to 12,000 tonnes per annum (mid 1800s) to 3000 tonnes per annum (today). Imports to Japan from other parts of East Asia have soared since 1986. It is estimated that 80 per cent of the population of Shangri-la County in Yunnan, China, is engaged in some way in the business, which accounts for 50 to 80 per cent of income at community level (He Jun, 2004).

- **Yarsagumba** (Tibetan; *Cordyceps sinensis*), otherwise known as *dong chong xia cao* in Chinese, is a very high-priced product very popular in China, where 1 gram (two to three pieces) can cost US$4 (2002 prices). Consisting of a fungus attached to a parasitized caterpillar, its collection forms a major source of income for many in the Himalayas. In 2003, 10,000 to 15,000 people 'invaded' Shey Phoksundo National Park in Nepal, coming from all over the country, in response to an announcement by Maoist insurgents that the species was available for collection on payment of an (illegal) licence fee. The number of permanent inhabitants in this park is only about 3000.

- **Gaharu** (agarwood or eaglewood) is a fragrant substance with aromatic, medicinal and religious uses that is found in the wood of some specimens of at least 15 species of *Aquilaria* (and *Gyrinops*, which is taxonomically very close), a genus of Southeast Asian forest trees. Gaharu is produced after wounding and/or fungal infection of the wood. This is a very high-priced commodity that can sell for over US$10,000 per kilogram in end-use markets. Historically, supplies for the global trade came mainly from India; but today India is an importer and several countries in Southeast Asia have also been depleted. Papua New Guinea is said to carry the world's last substantial reserves (Zich and Compton, 2001).

Some of the stresses placed on plants by trade can be illustrated with reference to the orchids, the largest family of flowering plants (approximately 20,000 species). The collection of showy species of orchids to sell to the horticultural trade or specialist collectors is a major conservation problem in many parts of the world. In addition, there are many orchids collected for medicine, for example 299 (24 per cent) out of the 1240 species of orchids known from China (more or less all of these medicinal orchids are now endangered in China) (Yan Zhi-Jian, 2004). In some countries, serious problems have developed because of the collection of orchids for food or beverages. In Turkey, the tubers of nearly 40 species of wild orchids are collected on a large scale to make *salep*, a type of flour used in the manufacture of food and beverages, including a popular type of ice cream. Around 1000 orchids are needed to make 1 kilogram of the flour (Özhatay et al, 1997). Another case of harvesting wild orchid tubers for food is known from the Southern Highlands in Tanzania, including the Kitulo Plateau, where up to 85 species of wild orchids (especially species of *Disa*, *Habenaria* and *Satyrium*) are dug up for export to Zambia, where they are used in the preparation of *chikanda* (*kinaka*), a much prized type of sausage. The trade has increased greatly since 1997, reaching 2 million orchids per year (2002). It now constitutes a major threat to the survival of some species (WCS, 2002).

The medicinal plant sector is noteworthy for the large number of species traded. Trade exists at all levels from the local to the international. Four hundred species of indigenous medicinal plants are sold in markets in Natal, 346 taxa of MAPs are collected for sale in Turkey and at least 2000 species are traded in Europe (Cunningham, 1997a; Özhatay et al, 1997; Lange, 1998a, b). In the European case, species traded originate from 120 countries, the majority of species (1200 to 1300) from Europe itself, where at least 90 per cent of the species are wild collected. In terms of volume, 30 to 40 per cent of MAPs traded in Europe come from wild sources. Germany is by far the leading country for imports and exports in Europe, with south-east European countries being the main source of European supply (Lange, 1998a, b). It has been estimated that the number of MAPs in *international* trade may be about 2500 species (Schippmann et al, 2002). In terms of species, the great bulk of supplies produced in many countries comes from wild plants, approaching 100 per cent in Albania, Pakistan, Turkey and virtually all of Africa (Lange, 1998a, b; Hamilton, 2004).

Some conservation concerns relating to trade in wild plants refer to the *particular species* that are selected as resources. There can be fears of genetic erosion and extinction, on scales varying from the local to the global. There are likely to be associated developmental concerns. Other problems relating to wild plant trade refer to the *collateral* damage that can be sustained by natural habitats or human communities. Damage to *habitats* can be extensive if the plants harvested are keystone species or if damaging methods of harvesting are employed. One of the concerns about the harvesting of *muhuhu* (*Brachylaena huillensis*) for the woodcarving industry in coastal East Africa is the damage that this can cause to a highly threatened type of lowland forest (see case study at end of this chapter). The harvesting of the endangered orchid *Ansellia* in Kenya is doubly damaging because not only is it disappearing itself, but its host trees are often felled for its collection (Khayota, 1999). On the *social* side, the introduction of trade in wild plant resources can disrupt village institutions traditionally responsible for resource management (see Chapter 2) (Sheldon et al, 1997).

There has been a major growth in trade of some NTFPs in recent years – for instance, in the case of rattan (Rao, 1997). Taking MAPs as an example, it is estimated that this market sector grew at a rate of 10 to 25 per cent annually in Europe and North America during the 1990s (Lange, 1998a, b; ten Kate and Laird, 1999; Srivastava, 2000). The recent rate of expansion of the market for Ayurvedic medicines in India is said to have been around 20 per cent annually, while the quantity of medicinal plants obtained from just one province of China (Yunnan) has grown tenfold in the last ten years (Pei Shengji, 2002; Subrat, 2002). This increased consumption must have transmitted pressures down trade chains back to the places where the plants have been gathered; but the effects of increased pressure of harvesting have been little studied *directly* in the field. One response of collectors – recorded for the Gori valley in the Indian Himalayas – has been to increase the annual period devoted to harvesting MAPs from two to five months (Uniyal et al, 2002).

One reason why trade in NTFPs has grown is because the world now has more people. Both rich and poor buy products made from NTFPs, the prices of which can vary enormously (see Table 14.1). Today, there are more wealthy people in the world than ever before and some NTFPs are so highly prized that some people will pay virtually anything to obtain them. Meanwhile, the number of very poor people living in cities has expanded greatly, fuelled by in-migration from the countryside. Many people who formerly collected wild plant resources for their own use now have to buy them. Despite the theoretical availability of some alternative products or services in cities, such as pharmaceutical drugs for medicine or electricity for energy, in practice many poor city dwellers continue to use wild-collected resources, largely because they are cheaper. This can lead to very high levels of demand in cities for fuelwood, charcoal, herbal medicines and other products. As for the collectors of NTFPs, the business of their collection continues to be an attractive option in some rural economies in which there can be few other opportunities to earn some money. Therefore, there are a number of reasons why we can expect the NTFP trade sector to continue to expand in coming years.

Plant trade obviously has a financial aspect, which must be considered in conservation efforts. However, there is more to trade than money. For one thing, many of those working in trade systems have substantial other interests in their occupations, apart from the mercenary. It is to the benefit of conservation that some resource managers take pride in adopting more sustainable systems of production and that some manufacturers try to obtain their raw materials from well-managed sources. There are consumers who will pay premium prices for ecologically friendly products, thus opening up the possibility of using consumer pressure to influence manufacturers and, through

Table 14.1 *Estimated quantities in trade and market prices of some medicinal plants in Nyingchi County, Tibet, China, during 2003–2004*

Species	Unit price (yuan per kilogram)	Unit price (US$ equivalent per kilogram)	Quantity traded (kilograms per annum)	Total value (US$ thousands per annum)
Cordyceps sinensis	30,000	3700	1500–2000	5500–7500
Gastrodia elata	500	60	700–800	42–48
Gymnadenia conopsea	220	27	4000–5000	108–135
Gentiana tibetica	12	1.5	10,000–15,000	15–23
Rhodiola fastigiata	15	1.9	7000–8000	13–15

Source: Liu Yujin et al, 2004

them, systems of production. Trading relationships can be complex. Collectors sometimes prefer to sell to traders whom they know, even if prices are relatively low, because then they might be able to receive cash in advance or ask for financial or political assistance at times of trouble. Traders can be influential people in their societies.

Official statistics and other information on NTFPs are typically of poor quality (Peters, 1994). Every aspect is under-researched: the species involved and why they are chosen; how they are collected and in what quantities; the impacts of collection on the species themselves and on their habitats and local social systems; the structures of trade chains and nature of trading relationships; and financial values at different points in trading systems. Paucity of reliable information is a major reason for the frequent low level of awareness among decision-makers of the significance of the trade in terms of conservation impacts and human livelihoods.

From the time that it started, trading has been one of the major routes through which individuals and social groups have been able to gain influence and wealth. Trade systems based on wild plant resources are typically marked by gross inequities in the distribution of financial benefits (see, for example, Clarke et al, 1996). Where licences are granted for exclusive rights to harvest resources over extensive areas – as is often the case with timber – then there can be questions about how the licences are issued. A transparent public bidding system is often much needed. Generally, the amounts received by the collectors of wild plant resources are very low compared with prices further along trade chains. In Pakistan, the collectors of the medicinal plant *Valeriana wallichiana* receive 20 rupees per kilogram dry weight, compared with 70 rupees by local traders and 399 rupees by wholesalers (Hassan Sher, personal communication, 2000). Harvesters of the woody vine cat's claw *Uncaria tomentosa* in Peru receive US$0.30–$0.65 per kilogram, yet bulk unprocessed cat's claw fetches US$11 per kilogram in the US (King et al, 1999).

Local people sometimes work as labourers for companies which happen to be temporarily harvesting produce in their areas. This provides them with a transient source of income that can, of course, be very welcome in rural communities in developing countries. However, in the long term, the benefits can be short lived. Communities can lose out fundamentally if the companies leave behind an environment that is heavily degraded, as is often the case with logging operations in tropical forest. Although perspicacious members of communities can resist the entry of companies into their areas to avoid this calamity, not everyone can afford to take the longer view and, anyway, there are sometimes influential members of communities prepared to assist companies (for example, through granting them permission to extract resources) in exchange for private benefits.

Inequity in benefits received from trade in wild plants extends to the international level. Generally, the pattern is for developing countries to supply raw materials and for richer countries to produce end products (although there are a few countries classified as 'developing' which are also major manufacturers). Unless producer countries have monopolies, importers of raw materials can often hold prices down by bargaining off producing countries one against another. The history of trade in agricultural commodities, such as coffee, shows how difficult it is for poor countries to remain solidly united to force richer countries to pay higher (and fairer) prices.

There are millions of people globally with large measures of financial dependency upon the collection of wild plant resources for sale (see Figure 14.1). It has been estimated that 50 to 100 per cent of households in the northern part of central Nepal and 25 to 50 per cent in the middle part of the same region are involved in collecting medicinal plants for sale, the materials being mostly traded on to wholesale markets in Delhi, India. The money received represents 15 to 30 per cent of the annual income of poorer households (Olsen, 1997). Although information is scarce, these figures are probably typical for poor remote areas. Wild plant harvesting is especially the preserve of the most economically marginalized people in rural societies, often including landless people, minorities and women. Migrant pastoralists and those practising transhumance are other groups who frequently pick MAPs to sell for supplementary income, as they guard their flocks or herds.

The collection and post-harvest treatment of NTFPs can be monopolized by particular genders, age groups or ethnicities. Charcoal burners and transporters in Africa seem to be exclusively male. Woodcarvers in Kenya are virtually all men and, furthermore, nearly all members of one tribe, the Wakamba. It is estimated that about 50 per cent of collectors of MAPs in Pakistan are women, 40 per cent children and 10 per cent men (Z. K. Shinwari, personal communication, 2000). In Sengezi settlement, Zimbabwe, substantial differences in gender and age have been noted according to types of product. Women are almost exclusively responsible for selling thatching grass and medicinal plants, men for selling poles and fibres, men and boys for selling firewood, children for selling wild or exotic fruits (to one another), and men and women equally for selling fish and wild vegetables (Goebel, 2003). In eastern Amazonia, it is men who collect medicinal plants from the forest, but women who process them into medications (Shanley and Luz, 2003).

Figure 14.1 *Commercial collector of medicinal plants drying plants in Shaktikjor Village Development Community, Nepal*

Source: Alan Hamilton

HISTORICAL DYNAMICS OF TRADE IN WILD PLANTS

Projects aimed at tackling conservation problems posed by trade in wild plants are faced with choices about what they should do (see 'Actions in favour of conservation'). These decisions will benefit from a good understanding of how the trade systems of interest are structured and function, and predictions about how they might develop 'naturally' – that is, assuming no interventions aimed at conservation. Studies of the history of trade are useful for developing models to predict future trends and plan interventions.

Conservationists need to be alert to the appearance of new trade pressures on wild plants. These can be expected to arise frequently as new uses for plant resources emerge or entrepreneurs spot market opportunities.

The market is selective in terms of the number of species traded, especially on the larger scale. Many more species are used for local subsistence than are traded, and many more species are traded on smaller than larger scales. It is estimated that, out of perhaps 3000 plant species that have been domesticated for food, only 150 have entered world commerce (Hawkes et al, 2001). From his considerable experience, Cunningham (2001a) has concluded that it tends to be the plants yielding goods or products of higher quality that tend to become commercialized – for instance, the fruits with the best taste or species providing the most powerful medicines.

However, other considerations emerge when trade systems become dominated by only a few companies, as is increasingly the case nationally and internationally. New features of plant products become significant, such as tolerance to long-distance transport or long periods of storage, ease of grading and market appeal. Market appeal is not left to chance, being readily manipulable by advertising and subliminal messaging.

Once a product enters the market, then pressure on the plants supplying resources used in its production will mount. The impact upon populations of a species will vary according to its abundance, resilience to harvest and adequacy of management. If pressure is excessive, then the resource will develop a widening area of depletion as it becomes stripped out of more accessible locations. Such has been the fate of timber in Southeast Asia over recent decades, as loggers have 'finished off' one country after another. In 1950, the Philippines was a major exporter of timber, supplying 93 per cent of imports into Japan from Southeast Asia; but today the Philippines has little forest cover remaining and is a net timber importer. So, too, is Thailand, where forest cover is reported to have declined from 93 to 18 per cent between 1961 and 1988. Logging pressure then moved to Indonesia and other countries (Dudley and Stolton, 1994). Logging companies from Southeast Asia are now active in Africa and South America. It was calculated in 1998 that foreign companies from Southeast Asia and elsewhere owned or controlled about 4.5 million hectares of the Brazilian Amazon (Laurance, 1998).

Similar patterns have been experienced with other resources. Malaysia and Indonesia were once the major international suppliers of rattan; but later new sources emerged in Myanmar, Papua New Guinea and Vietnam (Rao, 1997). The collection of wood for the manufacture of charcoal has caused expanding halos of deforestation around Kinshasa, Yaoundé and other cities in Africa. In the case of Dar es Salaam, the distance from the city at which charcoal has been manufactured increased from 50km in 1970 to 200km in 1996 (Misana et al, 1996; Dounias et al, 2000; Wilkie and Laporte, 2001). Depletion of some medicinal plants in Pakistan has led to a large increase in imports from Afghanistan during recent years (Williams, personal communication, 1998). The botanically closely related and much prized medicinal plants *Litsea monopetala* and *Neolitsea chinensis*, which used to be common in Pakistan, are now extremely rare in that country and today are imported from India.

Commercial harvesting can either weaken or strengthen resource tenure (Cunningham, 2001a). Weakening is possible if traditional systems responsible for the management of natural resources are undermined. Conversely, strengthening may follow if particular individuals or social groups are able to gain stronger control over the resources in demand. Stricter privatization of land is a major trend today in many developing countries (see Chapter 10). Protected areas, being under government control, may fail to benefit from increased individualized interest in the long-term fate of natural resources, and accordingly can become vulnerable to destructive harvesting. Plant resources in protected areas can represent islands of still available commodities set among oceans of privatized land.

As a wild plant resource in demand becomes rare or costly, the supply might become transferred to cultivated plants *of the same species*. Another possibility is for substitution *with other species* – wild collected or cultivated. If wild collected, then they, too, may be subject to over-harvesting. Successive substitution of one wild-harvested species after another has been recorded for the woodcarving industry in Kenya (see case study at end of this chapter). If demand is transferred to cultivation (of whatever species), then this may provide a significant source of income for farmers. Fifty per cent of charcoal supplied to Nairobi is today produced from commercially grown trees, such as *Eucalyptus*, planted on high-value farmland (Cunningham, 1997a).

Two examples are given here of the histories of discovery of new products and subsequent developments. The rosy periwinkle *Catharanthus roseus* is a native of Madagascar, but has long been grown throughout the tropics as a garden ornamental and for medicinal properties. It is used in folk medicine to treat a wide range of maladies. During the 1950s, researchers at the University of Western Ontario in Canada became interested in the use of the rosy periwinkle to treat diabetes in Jamaica. Attempts in the laboratory to verify the efficacy of this use failed; but, instead, these researchers and others at the pharmaceutical company Eli Lilly found that the species contains numerous alkaloids, two of which – vinblastine and vincristine – proved to be very useful for the treatment of childhood leukaemia. It is reported that Eli Lilly initially tried to obtain supplies of the plant for drug manufacture from wild and cultivated plants in Madagascar; but this proved difficult due to an uncertain political climate and problems with standardization. Instead, the company switched to farm production in Texas. Today, supplies are available from farms in several countries (Noble, 1990; Sheldon et al, 1997).

The yew *Taxus* is a genus of trees native to north temperate forests in Eurasia and America. There are several rather similar-looking species. The discovery of the anti-cancer properties of yew was the result of random screening by the National Cancer Institute in the USA during the early 1960s. It was found that the bark of the Pacific yew *Taxus brevifolia*, found in the northwest temperate forests of the USA, contains a compound (taxol) with a previously unknown mechanism for curtailing the growth of cancer. Taxol is found in all parts of the plant except the aril (the red fleshy cup holding the seed), but is present in its least perishable and thus most useful form in the bark. Heavy pressure on the Pacific yew to collect the bark quickly followed, with illegal collection proving difficult to prevent. Extensive bark collection is fatal to the slow-growing trees of this genus. Several Asian yews *Taxus chinensis*, *T. wallichiana* and *T. yunnanensis* came under collection pressure during the 1990s, resulting in a ban on the export of *T. wallichiana* by India in 1994 and a ban on the harvesting of wild trees of all types of yew by China in the same year. *Taxus wallichiana* was placed on Appendix II of the Convention on the International Trade in Endangered Species of Wild Fauna and Flora (CITES) in 1995, with proposals later made to list other species of yew that look similar. Synthetic production of taxol has been achieved; but it is not commercially attractive due to the very large number of chemical steps involved. Today, commercial production comes largely from the synthesis of taxol from a precursor compound found in the leaves. Concentrations of this chemical in the leaves are low; but the overwhelming advantage is that leaves can be collected non-destructively. Today, there is a market for clippings from yew hedges in France and substantial plantations are being developed in China (Sheldon et al, 1997).

ACTIONS IN FAVOUR OF CONSERVATION

When trade results in the over-harvesting of a wild resource, then interventions in favour of conservation are possible at several points along trade chains (see Figure 14.2). Such actions become increasingly desirable as the scale of trade increases because resource depletion then becomes a national, regional or even global conservation and development concern, not just a local problem. The chance of total elimination of a species increases. If collection causes degradation of a habitat, then this will be experienced on an expanded scale. If a resource is *still* required after stocks have been run down, then it will be necessary to restore populations of the species, which will almost certainly prove more expensive and otherwise more demanding than managing the species properly in the first place.

It should be borne in mind that any actions taken at 'higher' levels of trade chains – that is, away from the actual or potential sites of production – *must* feed back to the production level if they are to produce any material benefits for conservation. For example, if efforts are made to heighten the awareness of consumers about the consequences of their purchasing decisions, then benefits *must* accrue at the local level (for instance, in terms of strengthened systems of resource management) if any real conservation advantage is to be gained.

There are various ways in which projects can assist those involved in trade systems to adopt practices more favourable to conservation. The encouragement of **peer-level exchanges** is useful since some people are more progressive and innovative than others, or have valuable techniques to share. For example, one community may have good expertise in how to harvest resources sustainably and be prepared to help others raise their standards. Projects can also encourage **participatory action research** as a way of investigating and finding answers to problems (see Chapter 4). Participatory action research might

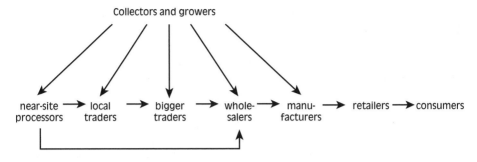

Figure 14.2 *Some categories of people who may be involved in trade chains for plant resources; the arrows denote movements of raw materials or products*

Note: More than one country may be involved.
Source: Alan Hamilton

reveal common concerns among various parties, such as producers, traders and manufacturers – for example, an interest in the sustainability of resource supply. Projects can then follow up by providing **opportunities for different stakeholders to come together** to see if they can define common, conservation-related, goals and agree to complementary activities.

An **exploration of where power resides in trade systems** will help in the identification of key points of intervention. Zerner (1999) has asked: 'At what point in these complex systems of extraction is value attached, how, by whom, and to whose benefit?' He points out that there has sometimes been a tendency among conservationists and resource managers to blame or lecture local people for degrading the environment through collecting wild plant resources to sell. However, in actuality, local people are often desperately poor, politically weak and have little influence on the functioning of the trade systems of which they are part. With the best of intentions, they may see little option than to harvest wild plant resources to sustain their lives, even if they know this to be unsustainable or illegal. Zerner has further remarked that, in some places, communities can be the objects of intimidation and extortion by ramifying bureaucracies of government or military officials, or powerful businessmen. He concludes from this that, at the *local* level, conservationists may sometimes be able to assist with local improvements for conservation and livelihoods (for example, with the establishment of cultivation to take the pressure off nearby natural areas); but, at the same time, they should try and identify key points for intervention at *higher* levels in political, bureaucratic or commercial systems, where changes have the possibility to be more fundamental and wide ranging in their effects.

Various technical steps may be possible at the **production level** to improve how resources are managed and harvested (see Chapter 11). They include steps to ensure that the right species are harvested (adulteration is a common problem with medicinal plants), that they are harvested at the right time to ensure good quality, and that the methods of harvesting are such as to minimize damage and encourage regeneration. A project of the World Wide Fund for Nature-Pakistan (WWF-Pakistan) and the People and Plants Initiative at Miandam, Swat, in Pakistan has helped to improve the quality of collection of maidenhair fern *Adiantum capillus-veneris*, sold for medicinal purposes. Collectors were instructed not to rip up whole plants, as had previously been the practice, but rather to harvest the fronds after the sori had opened and released their spores. In this way, the plants will not be killed when the harvest is made and there is a chance for the fern population to expand through the establishment of new plants from the spores. The collectors followed this advice and the clean material that resulted fetched a better than normal price with Qarshi Industries (a local botanicals company which has stated an interest in the sustainable production of medicinal plants).

Improving **access to information is invaluable for communities**. There are many aspects, some relating to technical matters (such as those referred to above) and others concerned with financial affairs – for example, information on the current prices paid for materials by traders and manufacturers, and about which potential buyers might be prepared to pay better prices for better-quality material. There may be a need to **develop community organizations** with respect to the management and post-harvest treatment of resources, and also for the purposes of bargaining more effectively with traders and negotiating with government officials. One of the problems that communities can face is that the quantities of products that they produce may be small compared with the demands of even quite small companies. There are, therefore, questions about how they can best unite into associations or co-operatives to augment their bargaining power (Peters, 1994; Martin, 1995).

Value can often be added to commodities at the community level if there is proper **post-harvest treatment**. With MAPs, post-harvest treatment might include removing adulterating material and ensuring good cleaning, drying, storage and packaging. The therapeutic properties of medicinal plants may be reduced, or may even be totally lost, if storage of plant material is inadequate. An experiment on 19 species of medicinal plants in Nepal found that, six years after storage, 3 had lost all therapeutic activity, 10 were only partially active and only 6 retained their full strength (Griggs et al, 2001).

There may be ways of moving some processing down trade chains to maximize economic benefits closer to sources of production. In some cases, it may be possible to undertake some **initial product processing** at community level – for example, primary distillation of aromatic plants to extract their essential oils. If such processes are contemplated, then checks should be made on market demands (including required standards of quality) before making investments. With commodities traded internationally, the same principle of trying to maximize economic benefits closer to production applies at *national* level. For example, with respect to MAPs, national-level companies in developing countries should be encouraged to undertake primary processing and not just export raw materials.

Encouragement of **cultivation** is often advocated as a way of taking the pressure off wild plants (see discussion on issues in Schippmann et al, 2002). On the part of *producers*, advantages of cultivation over wild harvest may include easier management (in terms of having stricter control over production systems), the possibility of growing varieties preferred by the market and higher production per unit area. The selection of the right varieties for the commercial market is important (Leakey, 2003). Advantages to *manufacturers* can include less risk of adulteration, greater uniformity in product quality and greater assuredness of supply. On the other hand, cultivation may *not* actually reduce or remove pressure on wild resources. Frequently, those communities or members of communities who are well placed to take up cultivation are *not* the people who have been collecting wild plants to sell. For instance, they may be relatively prosperous farmers with surplus land who have never been engaged in wild plant collection. For this reason, projects promoting cultivation should pay special attention to the needs of those who currently harvest resources from wild plants.

There can be **advantages in retaining wild harvest** if this can be made sustainable. One major advantage is that this maintains links between local people and details of local natural habitats – foundations of interest upon which conservation involvement can potentially be built. In the context of certification, one advantage of persevering with trying to make wild harvest more sustainable is that some relevant categories of resources (foods, medicinal plants) are more likely to meet organic standards than plants that have been cultivated in terms of lacking or having reduced quantities of pesticides, heavy metals and other poisons.

There are various actions that **traders and manufacturers** can take that can strengthen systems of resource production for the benefits of conservation and local livelihoods. The benefits for the traders and manufacturers themselves can include enhanced long-term commercial security and better-respected businesses. Memoranda of understanding (MOUs) or more formal agreements may be useful to spell out understandings between traders or manufacturers and producers. Ways in which traders or manufacturers can support more sustainable production of resources and local livelihoods can include (Laird et al, 2005):

- guaranteed payment of minimum (or premium) prices for commodities produced in sustainable ways;
- provision of technical assistance to communities to ensure proper post-harvest treatment;
- financial (and, if appropriate, technical) assistance for the experimental development of systems of sustainable production, for subsequent adoption by communities;
- in the case of manufacturers, incorporation of a requirement for sustainable sourcing on their specification sheets (used to specify how their products are made).

The ideological positioning of **governments** will influence how, and to what extent, they become involved in trade systems based upon plant resources. Most governments today accept that private enterprise has a major role to play in the economy. In this ideological context, some of the principal tools available to governments to influence trade systems include economic instruments, rules and regulations (and their enforcement), and public procurement policies. Governments commonly also provide some technical services, run public-awareness campaigns and authorize educational curricula. The **economic and legal instruments** at the disposal of governments cover such matters as the setting of conditions for wild harvest (including quotas and fees) and the establishment of rules and fees relating to the transport, sale and export of commodities. There are many types of **technical services** offered by governments that are potentially useful, including assisting with the development of improved systems for the management of wild plant resources (through forestry departments) and the development and promotion of cultivation (through agricultural departments).

There is much room for improvement in the **documentation required for export or import** of plants and plant materials. At present, the documentation required can be minimal. In the case of MAPs, it has been suggested that international trade controls might be much improved if the precise species being traded had to be stated, as well as their places of origin (Lange, 1997). Customs officials have many duties and can be lax in their enforcements of trade regulations relating to plants and plant products (except for certain high-profile illegal products, such as cocaine!). Efforts have been made in some countries to assist customs officers through providing guides to identifying listed plants and plant products, such as on how to distinguish between wild and cultivated orchids.

Governments sometimes **impose bans** to suppress destructive harvesting. The Dominican Republic closed all sawmills in the country during 1970 in order to reduce pressure on its few remaining forests. China banned logging in natural forests in the headwaters of the Yangtze and Mekong rivers in 1998 after scientists blamed logging for major flash floods along the Yangtze. Various countries, such as Bhutan and Tanzania, have forbidden the export of round logs, a measure intended additionally to encourage local processing. Similar steps have been taken by Indonesia and Malaysia with respect to rattan, with prohibitions on its export in raw or semi-processed forms (Rao, 1997). Thailand banned the export of wild orchids in 1998, stopping a legal trade in wild-collected specimens of such highly prized genera as *Dendrobium*, *Paphiopedilum* and *Vanda*, previously amounting to 200,000 to 550,000 plants per year. Of course, national bans can result in the dispatching of problems to other countries, as has happened in the case of the Chinese logging ban, with markets (and conservation concerns) now switched to sources in Russia, Southeast Asia and elsewhere (Kirk, 2004). China imported 1 million cubic metres of logs in 1997 and 16 million cubic metres in 2002. By 2010, only half of its demand for industrial wood will be met from domestic production. Jepson et al (2001) have given thought to the influence of bans, with special reference to the case of timber in Indonesia, where destructive logging is rife. In this case, they argue that a ban would gain little or no support within Indonesia and would probably prove unenforceable. They add that foreign pressure for such a ban might engender a nationalist backlash and kill the few initiatives in Indonesia that have the potential to foster sustainable forest management.

Some species that are traded as living plants or goods derived from plants have been included in the CITES Appendices due to concerns over the impact of international trade on their conservation status (see Box 3.2). There are traders and manufacturers who are aware of species covered by CITES and avoid their trade or use. The types of plant species in CITES are very selective in terms of commodity categories, a matter that would repay some attention. Many of the plant species listed seem to have been included because of concerns posed by the *horticultural* trade. As of now, there is a heavy emphasis on plants traded as ornamentals, with, for instance, all species of cacti and orchids being listed on either Appendix I or II, representing approximately 70 per cent of the roughly 28,370 species on these Appendices. In contrast, it has been estimated that only 233 plant species included in these Appendices are *medicinal*, though only 16 of these were listed specifically because of concerns about their trade as medicinals, rather than trade for other purposes (Schippmann, 2001). There are only 19 *timber* species (Oldfield, 2002). Some concerns have been expressed about CITES with respect to orchids, such as about the difficulties that *bona fide* nurseries can face in becoming registered by governments for the commercial sale of artificially propagated material of species listed on Appendix I (allowed under CITES). This is said to have discouraged nurseries from establishing trade in artificially propagated specimens of these species, as might prove useful for taking pressure off wild populations (IUCN and SSC, 1996). However, even if the bureaucracy can be reduced – as would appear to be desirable – there will *still* remain a need for independent verification in order to ensure that licences are not abused.

STANDARDS AND CERTIFICATION

One way in which sustainable production of plant resources can be promoted is through the setting of standards, certification and labelling. Currently, it can be very difficult for environmentally minded consumers to know which products to buy, confronted as they are with a multitude of claims about products which they cannot verify. Many products are labelled as environmentally friendly, and consumers are left to make their own judgements about whether to believe them.

There are three types of **certification**, varying from self-claims to independently audited assessments (see Box 14.2). Whether or not first-party claims are regarded as credible depends greatly upon the trust that consumers attach to particular companies or trademarks. Third-party claims are intrinsically more trustworthy. To gain credibility, individuals or groups of companies (in

> **BOX 14.2 TYPES OF CERTIFICATION FOR THE SUSTAINABLE PRODUCTION OF PLANT RESOURCES**
>
> Types of certification include the following:
>
> - **First-party certification**: claims are made by individuals or organizations on their own behalf.
> - **Second-party certification**: standards are set by groups, such as trade or manufacturer associations. Members are encouraged or required to adhere, with varying levels of verification or checks.
> - **Third-party certification**: standards are set by independent entities, such as governments, non-governmental organizations (NGOs) or private certification agencies. Products must pass inspection by independent individuals or organizations (often accredited by an umbrella organization) for retention of certification.
>
> *Source:* Alan Hamilton and Patrick Hamilton

associations) may decide to draw up **ethical codes or guidelines** for their operations. These are best published so that the public can assess whether they have been followed. Some international organizations have also published **guidelines** to encourage environmentally friendly production, such as *Guidelines on the Conservation of Medicinal Plants* (currently being revised) and *Guidelines on Good Agricultural and Collection Practices for Medicinal Plants* (WHO et al, 1993; WHO, 2003).

The **setting of standards** is useful for first- and second-party certification, and essential for third-party certification. Sustainability standards can apply either to producers (gatherers of wild plants or farmers) or to production areas (see Chapter 11). While it is the standards of management of wild resources or of cultivation that are of particular interest to conservationists, there are many manufacturers and members of the general public who are more interested in *other* types of standards, which can also have their own systems of certification and labelling. For instance, in the case of herbal medicines, consumers will almost certainly have a strong interest in **product quality** because they will want to know whether the products they purchase are effective for treating their medical conditions or, at least, are not toxic. **Social justice** is also a matter of concern to many people. Today, conservationists face the challenge of achieving recognition for their own standards in the face of a proliferation of standard-setting by parties with fundamentally different interests (Pierce and Laird, 2003).

The proliferation of standard-setting by various parties has resulted in a multiplication of labels and consumer confusion. From the perspective of promoting ecological sustainability, Pierce and Laird (2003) have suggested that it would be useful if there was a simplification of certification and labelling systems. This might prove welcome, not only to consumers, but also to producers and manufacturers who otherwise can be faced with the costs of registration and administration associated with joining and adhering to several certification schemes related to diverse uncoordinated standards. They have advocated the development of **comprehensive standards** that combine the three basic categories of interest – product quality, social justice and ecological sustainability. With respect to MAPs, Pierce and Laird (2003) point out that ethical manufacturers are *not* going to abandon their fundamental interest in the medical quality of their products. However, they also note that there are commonalities between ensuring product quality (from the medical point of view) and ecological sustainability. These include correct species identification and acceptably low levels of contamination by pesticides, heavy metals and other poisons. They note that efforts to improve product quality and ecological sustainability both depend upon knowing how the plant materials used in manufacture have originated – that is they *both depend upon traceability*. Pierce and Laird suggest that it will not be possible to develop a single comprehensive standard that will work for all types of product in the herbal sector because of the complexities of standard-setting. However, they suggest that it may be possible to evolve comprehensive standards for particular MAP product sub-sectors.

The **Forest Stewardship Council** (FSC) is one of several certifiers concerned with promoting good standards for forest management. FSC standards refer to environmental responsibility, social responsibility and economic viability. FSC is an international organization that accredits certification bodies, which, in turn, are charged with guaranteeing the authenticity of claims (that is, FSC is a third-party certification system). FSC has established a set of ten principles for good forest management, each with a set of criteria (see www.fsc.org). Products that have been produced according to FSC standards may carry the FSC label, which is becoming well recognized in some markets. In order to carry the FSC label, all stages of production, distribution and sale of the product must be independently evaluated. It must be demonstrable that the wood in a product (or other forest material used in its manufacture) has been tracked all the way from the forest of its origin to the retail shelf. The forest where the wood or other material originated must have been awarded a certificate for forest stewardship by an FSC-accredited certification body. The requirement for tracking means that all those handling the wood along a trade system (traders, transporters, wholesalers, manufacturers, distributors and retailers) must be formally involved in the certification system and conform to the standards required. The process whereby proof is provided that a product really has been derived from an FSC-certified forest is known as **chain-of-custody verification**.

MARKET SYSTEMS AND THEIR STUDY

Studies of markets in conservation projects can reveal much about trade in NTFPs (Cunningham, 2001a) An overview survey of local markets should normally be undertaken as part of the initial scoping survey recommended for community-based plant conservation projects (see Chapter 12). The results of market surveys can help in the identification of priority species for project attention. Quantitative data collected in markets can contribute to mathematical evaluations of sustainability (see Chapter 11), though Cunningham (2001a) advises against spending much time in collecting information in markets on amounts sold, since (he states) other ways of collecting data about sustainability can be more cost effective.

A marketing or **trade system** for charcoal in Zambia is depicted in Figure 14.3. Similar types of diagrams can be constructed by projects for the trading systems of interest. There are many variants of trade systems, depending upon the types of commodity, scales of trade and sophistication of manufacturing processes. Some of the principal categories of people involved are producers, traders (various types, including local traders and wholesalers), transporters, manufacturers and retailers. **Bulking-up centres** are of particular interest in conservation projects because they bring together produce from more or less extensive areas and therefore allow an impression to be gained of the overall scale of trade. Bulking-up centres will normally lie close to the sources of plant produce for bulky items (such as fuelwood), but may draw upon supplies from extensive areas for scarce commodities (such as some medicinal plants).

Much trade in NTFPs falls into the informal sector of the economy and, although it can be substantial, is generally poorly recorded in government statistics. Even so, **government records** and any other relevant documents should be studied at the initiation of trade surveys. Categories of information that can be available include the names of commodities collected, names of licensees, the manufacturers involved and the financial values of the commodities at different points (for example, in terms of payments for collecting licences, the values of exports and estimates of the economic value of sectors as a whole).

Trading systems generally involve **markets**, meaning (in this context) physical sites where goods are bought and sold. Markets for NTFPs are not always easy to classify and several types can sometimes be found in close proximity at the same locality. The main types, listed from the most extensive to the most local, are:

- **regional markets**, which cover the largest areas;
- **central markets**, usually found at strategic points in transportation networks, characterized by *wholesale* exchanges of goods;
- **marketplaces (standard markets)**, which are the starting points for flows of rural produce upwards and end points for the sale of items imported from urban centres; in addition, local exchanges occur in marketplaces;
- **minor markets**, characterized by *local* exchanges of goods between local people.

A full study of markets involved in trade in NTFPs in a region should start conceptually with research to characterize the **whole market system**. Local knowledge will be very valuable for finding out where the markets are, what they are called, how they should be classified and their relationships with one

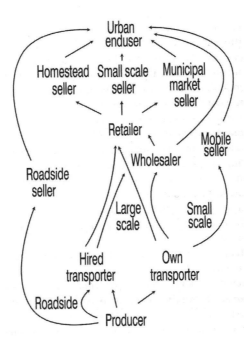

Figure 14.3 *The marketing system for charcoal in Zambia*

Sources: Brigham et al, 1996; figure reproduced with permission from Campbell, 1996

another. It is a good idea to start by surveying regional markets because this should provide an overview of the major items that are traded over wide areas. After this, researchers should work down market systems, eventually resulting in identification of areas of production – which can then be visited in order to study how the resources are being collected and to verify the identification of species.

Some important variables to determine for an **individual market** (considered as part of a larger market system) are its size, periodicity, seller composition and whether products are sold or bartered. There are three basic types of market schedule: open every day, periodic (for example, weekly) and special (the least frequent and most irregular). Traders in different markets may vary by gender and ethnicity, and according to the permanency of their attendance. Some traders are permanently stationed in particular markets, others move between small markets following market schedules, while others, yet again, travel to different markets but are based at their homes. Repeat surveys of markets are invaluable since trade in many NTFPs is seasonal, so single visits may not give a reasonable overall picture of what is traded.

A survey of an individual **marketplace** is best made after finding out who controls the market and after obtaining permission for the survey. A market controller will be able to reveal much information about how the market is organized and why this is the case, and about the types, numbers and distribution of sellers. Sketch maps should be prepared. Unless the market is very large, it should be possible to count or estimate the total number of vendors and, with local assistance, to get an idea of the cultural and socio-economic status of people visiting the market to buy various categories of products. If it is intended to follow up with detailed interviews with vendors, then initial visits to markets will help to clarify the types of information that will be useful to collect, given the objectives of the research and the time available. Survey forms can then be designed.

Information collected in discussions and observations with **individual vendors** might concern the types of species sold (collect voucher specimens), their names, parts of plants used, whether sold fresh or dry, their prices and places of origin (currently and in previous years), the people involved in their trade, and whether there has been substitution of one species by others. Vendors can be asked about the products that they sell in other seasons.

CASE STUDY: CERTIFICATION OF WOODCARVING IN KENYA

There has been a substantial growth in commercial woodcarving in Kenya over recent years (see Figure 14.4). The entire industry was started by one person, Mutisya Munge, who learned his carving skills from the Makonde people while serving with the army in former German East Africa (now Tanzania) during World War I. Today, there are an estimated 60,000 woodcarvers in Kenya, with co-operatives at Mombassa, Nairobi, Wamunyu and elsewhere. During the late 1990s, exports made up 30 per cent of all sales through co-operatives (60 per cent to the USA), with retail sales to tourists accounting for a further 40 per cent and 30 per cent going to local wholesalers. Since then, a decline in tourism in Kenya has led to a fall in sales to tourists, by 75 per cent according to some carvers.

Altogether, some 50 species of tree have been used for carving over recent years. The first carving centre, at Wamunyu, initially used *mpingo* (*Dalbergia melanoxylon*), which was then available nearby; but, as local supplies became exhausted, *mpingo* became harvested at ever greater distances. Other species, such as *muhuhu* (*Brachylaena huillensis*), started to be used. Today, logs for Wamunyu are obtained from sites well over 100km distant. Over recent years, much *mpingo* and *muhuhu* has been imported into Kenya from northern Tanzania (at least partly illegally).

Apart from loss of the carving species themselves, the collection of wood for the carving industry in Kenya has caused local degradation of forests close to the coast. These forests are small in size and belong to a vegetation type that is endemic rich, highly threatened and recognized as a global conservation priority. Until recently, the Forest Department has found it difficult to enforce regulations and decrees, such as a 1986 presidential ban on the logging of indigenous timber, due to widespread governmental corruption. Accordingly, there has been little economic incentive on the part of carvers not to carve the

Figure 14.4 *Logs of* muhuhu *at Lungalunga, Kenya, awaiting carving;
this carving centre is within a few hundred metres of the border with
Tanzania, the source of the logs*

Source: Alan Hamilton

slow-growing indigenous species to which they have become accustomed.
However, more recently, the government has proved more effective in clamping down on the transport of illegal logs.

Since 1995, a project of the People and Plants Initiative has been promoting a shift towards fast-growing exotic trees, such as neem *Azadirachta indica* and mango *Mangifera indica*. The awareness of carvers, tourists and importers about the conservation issues has been raised through drama, videos and other means. In 1997, stakeholders agreed on a 20-year vision to put the woodcarving industry on a sustainable footing, accepting a market-led approach as one avenue to explore. Subsequently, in 1999, it was agreed to develop a certification system. The preconditions for certification seemed to be good: large stocks of 'good woods' (such as neem) were available, some carvers had been carving 'good woods' for some years and the co-operatives were keeping very detailed records, which appeared to need little modification to trace certified logs through the carving process. With regard to buyers, special attention was directed at importers in Europe and the USA because it was thought that a few large orders for 'good woods' from such importers would send powerful signals to the co-operatives and carvers. It was thought easier to convince a few buyers in Europe and the USA to buy certified carvings than large numbers of individual tourists buying at numerous retail outlets. It was decided to work towards certification through the Forest Stewardship Council (FSC), one reason being its relative familiarity to potential buyers (adopting a lesser-known scheme might be confusing).

Working towards FSC certification has drawn attention to some constraints of the FSC for cases of this type, where wood (neem and mango, in this case) is produced on a multitude of small farms, each yielding very few trees per year. The FSC system was designed primarily to ensure sustainable production from large natural forests (later including large plantations). The application of FSC certification to small operations is complex, particularly in developing countries, where levels of awareness of certification issues are low, ownership rights can be tenuous and financial resources to pay for certification are scarce. Only 0.1 per cent of all forests certified under FSC are composed of areas under 100ha. More FSC certificates are issued and more forest hectares certified in 'Northern' countries than in the 'South'. FSC clearly needs to evolve unless it is to cause further economic disadvantage to small producers and developing countries.

FSC is an area-based scheme. Considering the large number of small producers in the Kenyan example, the model adopted (with the help of the Woodmark Soil Association) followed the principles of **group certification**, which allow several sites to be evaluated and certified under a single certificate, resulting in significant cost savings per site. Membership of a group can change without changing the FSC status of the group itself. The key to group certification is that a **group manager** is responsible for ensuring that all the sites within the group are managed to meet the certification standards and for demonstrating this to a certification body. Tree cutters and carving co-operatives have to operate under **chain-of-custody requirements** before FSC-labelled carvings can be sold. This requires complete traceability of logs from farm to shelf, with a resource or group manager responsible for managing the relationships between the farmers, cutters and carvers.

One issue for FSC in relation to woodcarving in Kenya is that the environmental benefits derived from certification are expected to be mostly off-site (that is, less pressure on the coastal forests), rather than where the 'good wood' trees are grown (on farms). The coastal forests are under several other threats apart from woodcarving. Therefore, it will be impossible to assess the precise contribution of introducing 'good woods' on relieving pressures on the forests. Problems of this type have led the FSC to modify some of its criteria for small and low-intensity managed forests (SLIMFs). SLIMFs tackles a number of constraints that apply to the Kenya case, including the high cost of ordinary certification (especially burdensome for developing countries) and the need for a system of evaluation. Despite the many complexities, FSC certification was achieved in March 2005 for a group of 576 farmers, a 3000-member carving co-operative and a marketing organization.

Source for case study: Schmitt and Maingi, 2005

Acronyms and abbreviations

ANPC	Australian Network for Plant Conservation
ARC	Alliance of Religions and Conservation
ASEAN	Association of Southeast Asian Nations
ASSI	Area of Special Scientific Interest
AU	African Union
BGCI	Botanic Gardens Conservation International
BINP	Bwindi Impenetrable National Park
BMA	British Medical Association
CALM	Department of Conservation and Land Management (Western Australia)
CAMP	Conservation Assessment and Management Plan
CAN	Comunidad Andina
CAP	Common Agricultural Policy
CBD	Convention on Biological Diversity
CBR	community biodiversity register
CFC	chlorofluorocarbon
CFR	Cape Floristic Region (South Africa)
CGIAR	Consultative Group on International Agricultural Research
CI	Conservation International
CIAT	Centro Internacional de Agricultura Tropical (Colombia)
CIFOR	Centre for International Forestry Research (Indonesia)
CIMMYT	Centro Internacional de Mejoramiento de Maiz y Trigo (Mexico)
CIP	Centro Internacional de la Papa (Peru)
CITES	Convention on the International Trade in Endangered Species of Wild Fauna and Flora
CJD	Creutzfeldt-Jakob disease
cm	centimetre
CO_2	carbon dioxide
COP	Conference of the Parties
DFID	Department for International Development (UK)
DNA	deoxyribonucleic acid
ECP/GR	Economic Crop Protection/Genetic Resources Group
ESON	Ethnobotanical Society of Nepal
EU	European Union
EWNHS	Ethiopian Wildlife and Natural History Society
FAO	United Nations Food and Agriculture Organization

FFI	Fauna and Flora International
FRLHT	Foundation for Revitalization of Local Health Traditions (India)
FSC	Forest Stewardship Council
g	gram
GELA	Grupo Etnobotánico Latinoamericano
GIS	geographic information system
GM	genetically modified
GMO	genetically modified organism
GPS	global positioning system
GSPC	Global Strategy for Plant Conservation
ha	hectare
HAA	Himalayan Amchi Association
HFC	Healing Forest Conservancy
IBCR	Institute of Biodiversity Conservation and Research (Ethiopia)
IBPGR	International Board for Plant Genetic Resources (now IPGRI)
ICARDA	International Centre for Agricultural Research in the Dry Areas (Syria)
ICD	integrated conservation and development
ICDP	integrated conservation and development project
ICRAF	International Centre for Research in Agroforestry (Kenya)
ICRISAT	International Crops Research Institute for the Semi-Arid Tropics (India)
IFPRI	International Food Policy Research Institute (US)
IIED	International Institute for Environment and Development
IITA	International Institute of Tropical Agriculture (Nigeria)
ILRI	International Livestock Research Institute (Ethiopia)
IP	intellectual property
IPA	Important Plant Area
IPGRI	International Plant Genetic Resources Institute (Italy)
IPR	intellectual property right
IRRI	International Rice Research Institute (the Philippines)
ISA	Important Species Area
IUCN	World Conservation Union (formerly International Union for the Conservation of Nature and Natural Resources)
IWMI	International Water Management Institute (Sri Lanka)
KBA	Key Biodiversity Area
kg	kilogram
km	kilometre
m	metre
MAB	Man and the Biosphere Programme
MAP	medicinal and aromatic plant
MDG	Millennium Development Goal
M&E	monitoring and evaluation
MOU	memorandum of understanding

MPCA	Medicinal Plant Conservation Area
NAFTA	North American Free Trade Agreement
NGO	non-governmental organization
NMCP	Northern Mountains Conservation Project (Nepal)
NTFP	non-timber forest product
NWFP	non-wood forest product
PA	participatory appraisal
PEEN	Pan-European Ecological Network
PEK	Projek Etnobotani Kinabalu (Malaysia)
PGR	plant genetic resource
PPI	People and Plants Initiative
ppm	parts per million
PRA	participatory rural appraisal
RBG	Royal Botanic Gardens, Kew
SAARC	South Asian Association for Regional Cooperation
SABONET	Southern Africa Botanical Diversity Network
SAC	Special Area for Conservation
SADC	Southern African Development Community
SAFIRE	Southern Alliance for Indigenous Resources (Zimbabwe)
SBSTTA	Subsidiary Body on Scientific, Technical and Technological Advice (UN)
SLIMF	small and low-intensity managed forest
SPNP	Shey Phoksundo National Park (Nepal)
SSC	Species Survival Commission (IUCN)
SSSI	Site of Special Scientific Interest
TCM	Traditional Chinese Medicine
TMP	traditional medical practitioner
TPO	Tree Preservation Order
TRAMIL	Traditional Medicine for the Islands (Caribbean and Central America
TSE	Tanzanian Society on Ethnoscience
TRIPS	Trade Related Aspects of Intellectual Property Rights
TWINSPAN	Two-Way Indicator Species Analysis
UK	United Kingdom
UMIYAC	Union of Indigenous Doctors of the Colombian Amazon
UN	United Nations
UNEP	United Nations Environment Programme
UNESCO	United Nations Educational, Social and Cultural Organization
UPOV	International Union for the Protection of New Varieties of Plants
US	United States
USAID	US Agency for International Development
UWB	University of Wales at Bangor
VIDCO	Village Development Committee (Zimbabwe)
WARDA	West Africa Rice Development Association (Ivory Coast)

WCPA	World Commission on Protected Areas
WHO	World Health Organization
WSSD	World Summit on Sustainable Development
WTO	World Trade Organization
WWF	World Wide Fund for Nature (formerly World Wildlife Fund)

References

Adams, W. M. and Hulme, D. (2001) 'If community conservation is the answer in Africa, what is the question?', *Oryx*, vol 35, pp193–200

Agee, J. K., Bahro, B., Flinney, M. A., Omi, P. N. and Sapsis, C. N. (2000) 'The use of shaded fuelbreaks in landscape fire management', *Forest Ecology and Management*, vol 127, pp55–66

Ahlback, A. J. (1986) *Industrial Plantation Forestry in Tanzania*, Ministry of Natural Resources and Tourism, Dar es Salaam, Tanzania

Akeroyd, J. (1999) 'Conserving the Mediterranean flora: A way forward', *PlantTalk*, vol 18, 23–28

Alcorn, J. B. (1995) 'The scope and aims of ethnobotany in a developing world', in R. E. Schultes and S. von Reis (eds) *Ethnobotany: Evolution of a Discipline*, Chapman and Hall, London, UK, pp23–39

Allnutt, T. R., Newton, A. C., Lara, A., Premoli, A., Armesto, J. J., Vergara, R. and Gardner, M. (1999) 'Genetic variation in *Fitzroya cupressoides* (*alerce*), a threatened South American conifer', *Molecular Ecology*, vol 8, p975

Almquist, A. (2001) 'Horticulture and hunting in the Congo basin', in W. Weber, L. J. T. White, A. Vedder and L. Naughton-Treves (eds) *African Rain Forest Ecology and Conservation*, Yale University Press, New Haven, US, pp334–343

Anderson, S. (2002) *Identifying Important Plant Areas – A Site Selection Manual for Europe, and a Basis for Developing Guidelines for Other Regions of the World*, Plantlife International, Salisbury, UK

ANPC (Australian Network for Plant Conservation) (1997) *An Introduction to the Principles and Practices for Seed and Germplasm Banking of Australian Species*, ANPC, Canberra, Australia

Appleton, M. (1997) 'Conservation in a conflict area', *Oryx*, vol 31, pp153–154

Arnold, M. L. (2004) 'Natural hybridization and the evolution of domesticated, pest and disease organisms', *Molecular Ecology*, vol 13, pp997–1007

Asfaw, Z. and Tadesse, M. (2001) 'Prospects for sustainable use and development of wild food plants in Ethiopia', *Economic Botany*, vol 55, pp47–62

Ashford, R. W., Reid, G. D. F. and Butynski, T. M. (1990) 'The intestinal faunas of man and mountain gorilla in a shared habitat', *Annals of Tropical Medicine and Parasitology*, vol 84, pp337–340

Aumeeruddy-Thomas, Y. (1994) *Local Representations and Management of Agroforests on the Periphery of Kerinci Seblat National Park, Sumatra, Indonesia*, People and Plants Initiative, People and Plants Working Paper no 3, Division of Ecological Sciences, UNESCO, Paris, France

Aumeeruddy-Thomas, Y., Saigal, S., Kapoor, N. and Cunningham, A. B. (1999) *Joint Management in the Making – Reflections and Experiences*, People and Plants Working Paper no 7, Division of Ecological Sciences, UNESCO, Paris, France

Aumeeruddy-Thomas, Y., Shinwari, Z. K., Ayaz, A. and Khan, A. A. (2004) *Ethnobotany and the Management of Fodder and Fuelwood at Ayubia National Park, North West Frontier Province, Pakistan*, People and Plants Working Paper no 13, WWF–UK, Godalming, UK

Baillie, J. E. M., Hilton-Taylor, C. and Stuart, S. N. (eds) (2004) *2004 IUCN Red List of Threatened Species: A Global Species Assessment*, IUCN, Gland, Switzerland, and Cambridge, UK, p191

Baldauf, S. L., Roger, A. J., Wenk-Sierfert, I. and Doolittle, W. F. (2000) 'A kingdom-level phylogeny of eukaryotes based on combined protein data', *Science*, vol 290, pp972–977

Balick, M. J. and Cox, P. A. (1996) *Plants, People and Culture: The Science of Ethnobotany*, The Scientific American Library, New York, US

Balick, M. J. and Mendelsohn, R. (1992) 'Assessing the economic value of traditional medicines from tropical rainforests', *Conservation Biology*, vol 45, pp725–729

Balkwill, K. and McCallum, D. A. (1999) 'A review of Africa forest Acanthaceae', in J. Timberlake and S. Kativu (eds) *African Plants: Biodiversity, Taxonomy and Uses*, Royal Botanic Gardens, Kew, UK, pp11–21

Bandeira, S. O., Hatton, J. C., Munisse, P. and Izidine, S. (1994) 'The ecology and conservation status of plant resources in Mozambique', in B. J. Huntley (ed) *Botanical Diversity in Southern Africa*, National Botanical Institute, Pretoria, South Africa

Baranga, J. (1999) 'Natural change and human impact in Madagascar', *African Journal of Ecology*, vol 37, pp372–373

Barnosky, A. D. (1989) 'The late Pleistocene event as a paradigm for widespread mammalian extinction', in S. J. Donovan (ed) *Mass Extinctions: Processes and Evidence*, Belhaven Press, London, UK, pp235–254

Barrett, S. C. H. and Kohn, J. R. (1991) 'Genetic and evolutionary consequences of small population size in plants: Implications for conservation', in D. A. Falk and K. E. Holsinger (eds) *Genetics and Conservation of Rare Plants*, Oxford University Press, New York, US, and Oxford, UK, pp3–30

Barsh, L. L. R. and Bastien, K. (1997) *Effective Negotiation by Indigenous People: An Action Guide with Special Reference to North America International Labour Organization*, Ottawa, Canada

Barthlott, W., Lauer, W. and Placke, A. (1996) 'Global distribution of species diversity in vascular plants: Towards a world map of phytodiversity', *Erdkunde*, vol 50, pp317–327

Basset, Y., Novotny, V., Miller, S. E., Weiglen, G. D., Missa, O. and Stewart, A. J. A. (2004) 'Conservation and biological monitoring of tropical forests: The role of parataxonomists', *Journal of Applied Ecology*, vol 41, pp163–174

Bawa, K. S. and Dayanandan, S. (1998) 'Global climate change and tropical forest genetic resources', *Climatic Change*, vol 39, pp473–485

Bekele, T., Berhan, G., Zerfu, S. and Yeshitela, K. (2001) 'Perspectives of the Ethiopian Orthodox Tewahido Church in forest biodiversity conservation', in *Biodiversity Conservation in Ancient Churches and Monastery Yards in Ethiopia*, Ethiopian Wildlife and Natural History Society, Ethiopia, and University of Bangor, UK, pp5–9

Berjak, M. and Grimsdell, J. (1999) *Botanical Databases for Conservation and Development*, WWF, Godalming, UK

Berlin, B. (1978) 'Ethnobiological classification', in E. Roschy, B. B. Lloyd and N. J. Hillsdale (eds) *Cognition and Categorization*, Lawrence Erlbaum Associates, distributed by John Wiley and Sons, pp11–16

Berlin, B. (1992) *Ethnobiological Classification*, Princeton University Press, Princeton, New Jersey, US

Berlin, B. and Berlin, E. A. (2000) 'Improving health care by coupling indigenous and modern medical knowledge: The scientific bases of Highland Maya herbal medicine in Chiapas, Mexico', in A. M Cetto (ed) *Science for the Twenty-First Century: A New Commitment*, UNESCO, Paris, France, pp338–349

BGCI (Botanic Gardens Conservation International) (2004) www.bgci.org.uk

Bingham, M. G. (1998) 'The conservation status of Zambian vegetation', in R. P. Adams and J. E. Adams (eds) *Conservation and Utilization of African Plants*, Missouri Botanical Gardens Press, St Louis, Missouri, US

Bird, M. I. and Cali, J. A. (1998) 'A million-year record of fire in sub-Saharan Africa', *Nature*, vol 394, pp767–769

Birks, H. J. B. (1996) 'Contributions of Quaternary palaeoecology to nature conservation', *Journal of Vegetation Science*, vol 7, pp89–98

Bissonnette, L. D. (2003) 'The basket makers of the central California interior', in P. L. Howard (ed) *Women and Plants*, Zed Books, London, UK, pp197–210

Bobbink, R., Hornung, M. and Roelofs, J. G. M. (1998) 'The effects of air-borne nitrogen pollutants on species diversity in natural and semi-natural European vegetation', *Journal of Ecology*, vol 86, pp717–738

Bond, W. J. (1994) 'Do mutualisms matter? Assessing the impact of pollinator and disperser disruption on plant extinction', *Philosophical Transactions of the Royal Society of London*, B, vol 344, pp83–90

Bonell, M. (1998) 'Possible impacts of climate variability and change on tropical forest hydrology', *Climatic Change*, vol 39, pp215–272

Botanical Society of America (1995) *Botany for the Next Millennium*, Botanical Society of America, Columbus, Ohio, US

Boutin, C. and Harper, J. L. (1991) 'A comparative study of the population dynamics of five species of *Veronica* in natural habitats', *Journal of Ecology*, vol 79, pp199–221

Bowe, L. M., Coat, G. and de Pamphilis, C. W. (2000) 'Phylogeny of seed plants based on all three genomic compartments: Extant gymnosperms are monophyletic and Gnetales' closest relatives are conifers', *Proceedings of the National Academy of Sciences*, vol 97, pp4092–4097

Bramwell, D. (2002) 'How many plant species are there?', *PlantTalk*, vol 28, pp32–34

Braun, K. P. and Dlamini, G. M. (1994) 'Swaziland's plant diversity and its conservation', in B. J. Huntley (ed) *Botanical Diversity in Southern Africa*, National Botanical Institute, Pretoria, pp117–124

Brickell, C. (ed) (1999) *A–Z Encyclopedia of Garden Plants*, Dorling Kindersley Ltd, London, UK, p1080

Brigham, T., Chihongo, A. and Chidumayo, E. (1996) 'Trade in woodland products from the miombo region', in B. Campbell (ed) *The Miombo in Transition: Woodlands and Welfare in Africa*, Centre for International Forestry Research, Bogor, Indonesia, pp137–174

Brokaw, N. (1998) 'Fragments past, present and future', *Trends in Ecology and Evolution*, vol 13, 382–383

Brown, A. H. D. and Briggs, J. D. (1991) 'Sampling strategies for genetic variation in *ex situ* collections of endangered plant species', in D. A. Falk and K. E. Holsinger (ed) *Genetics and Conservation of Rare Plants*, Oxford University Press, New York, US, and Oxford, UK, pp99–119

Bruna, E. M. (1999) 'Seed germination in rainforest fragments', *Nature*, vol 402, p139

Bruner, A. G., Gullison, R. E., Rice, R. E. and da Fonseca, G. A. B. (2001) 'Effectiveness of parks in protecting tropical biodiversity', *Science*, vol 291, pp125–128

Budowski, G. (1988) 'Building bridges between botanists and developers in tropical countries', *Symbolae Botanicae Upsalienses*, vol 28, pp281–286

Bukenya-Ziraba, R. (1999) 'Some non-cultivated edible plants of Uganda', in J. Timberlake and S. Kativu (eds) *African Plants: Biodiversity, Taxonomy and Uses*, Royal Botanic Gardens, Kew, UK,

Burgess, N. D., Clarke, G. P. and Rodgers, W. A. (1998) 'Coastal forests of eastern Africa: Status, endemism patterns and their potential causes', *Biological Journal of the Linnean Society*, vol 64, pp337–367

Burgess, N. D., D'Amico Hales, J., Dinerstein, E., Olson, D., Itoua, I., Schipper, J., Ricketts, T. and Newman, K. (in prep) *Terrestrial Ecoregions of Africa and Madagascar: A Continental Assessment*, Island Press, Washington, DC

Bush, M. B. (1994) 'Amazonian speciation: A necessarily complex model', *Journal of Biogeography*, vol 21, pp5–17

Butynski, T. M. (1984) *Ecological Survey of the Impenetrable (Bwindi) Forest, Uganda, and Recommendations for Its Conservation and Management*, Unpublished report for the Government of Uganda, Uganda

Byfield, A. (1995a) 'Forest's disappearing heathland flora', in J. Newton (ed) *Proceedings of the First European Conference on the Conservation of Wild Plants*, Plantlife, Hyères, France, p157

Byfield, A. (1995b) 'Important Plant Areas in Turkey: A model for Europe', in J. Newton (ed) *Proceedings of the First European Conference on the Conservation of Wild Plants*, Plantlife, Hyères, France, pp229–232

Byfield, A. (2001) 'Applying the IPA guidelines in Turkey', Paper presented at Third European Conference on the Conservation of Wild Plants, Czech Republic

Byfield, A. and Özhatay, N. (1997) *A Future for Turkey's Peatlands: A Conservation Strategy for Turkey's Peatland Heritage*, DHKD, Istanbul, Turkey, and Fauna and Flora International, Cambridge, UK

Caldwell, M. M., Teramura, A. H. and Manfred, T. (1989) 'The changing solar ultra-violet climate and the ecological consequences for higher plants', *Trends in Ecology and Evolution*, vol 4, pp363–367

Cameron, C. M. (1994) 'Opening address', in B. J. Huntley (ed) *Botanical Diversity in Southern Africa*, National Botanical Institute, Pretoria, South Africa

Campbell, B. (ed) (1996) *The Miombo in Transition: Woodlands and Welfare in Africa*, CIFOR, Bogor, Indonesia

Campbell, B. M. and Luckert, M. K. (2002) *Uncovering the Hidden Harvest: Valuation Methods for Woodland and Forest Resources*, Earthscan, London, UK

Cano, J. H. and Volpato, G. (2004) 'Herbal mixtures in the traditional medicine of eastern Cuba', *Journal of Ethnopharmacology*, vol 90, pp293–316

Cavallli-Sforza, L. L., Menozzi, P. and Piazza, A. (1994) *The History and Geography of Human Genes*, Princeton University Press, Princeton, US

CBD (Convention on Biological Diversity) (1992) www.biodiv.org

CBD (2002) www.biodiv.org/programmes/cross-cutting/plant/targets.shtml

CBD (2004) www.biodiv.org/programmes/cross-cutting/ecosystem/default.asp

Celis, C., Scurrah, M., Cowgill, S., Chumbiauca, S., Green, J., Franco, J., Main, G., Kiezebrink, D., Visser, R. G. F. and Atkinson, H. J. (2004) 'Environmental biosafety and transgenic potato in a centre of diversity for this crop', *Nature*, vol 432, pp222–225

Chase, M. (2001) 'Consequences of polyploidy', *Kew Scientist*, vol 3, p3

Chater, A. O. (2003) 'The treatment of infraspecific taxa in local Floras', *Watsonia*, vol 24, pp281–286

Chidumayo, E. and Frost, P. (1996) 'Population ecology of miombo trees', in B. Campbell (ed) *The Miombo in Transition: Woodlands and Welfare in Africa*, Centre for International Forestry Research, Bogor, Indonesia, pp59–71

Chidumayo, E., Gambiza, J. and Grundy, I. (1996) 'Managing miombo woodlands', in B. Campbell (ed) *The Miombo in Transition: Woodlands and Welfare in Africa*, Centre for International Forestry Research, Bogor, Indonesia

Childes, S. L. (1999) 'Invasive alien plants – a serious biological threat to the grasslands and microphyllous shrubland of the Nyanga National Park, Zimbabawe', in J. Timberlake and S. Kativu (eds) *African Plants: Biodiversity, Taxonomy and Uses*, Royal Botanic Gardens, Kew, UK

Christal, A., Davies, D. H. K., Warren, J. and Macintosh, J. (2001) 'The management of long-term set-aside for conservation', *Botanical Journal of Scotland*, vol 49, pp415–424

CI (Conservation International) (2005) *Biodiversity Hotspots*, CI, www.biodiversity-hotspots.org

Clarke, J., Cavendish, W. and Coote, C. (1996) 'Rural households and miombo woodlands: Use, value and management', in B. Campbell (ed) *The Miombo in Transition: Woodlands and Welfare in Africa*, Centre for International Forestry Research, Bogor, Indonesia, pp101–135

Clement, C. R. (1999) '1492 and the loss of Amazonian crop genetic resources, 1: The relation between domestication and human population decline', *Economic Botany*, vol 53, pp188–202

CMP (Conservation Measures Partnership) (2004) *Open Standards for the Practice of Conservation*, CMP, CMPinfo@ConservationMeasures.org

Coley, P. D. (1998) 'Possible effects of climate change on plant/herbivore interactions in moist tropical forests', *Climatic Change*, vol 39, pp455–472

Colvinvaux, P. A., De Oliveira, P. E., Moreno, J. E., Miller, M. C. and Bush, M. B. (1996) 'A long pollen record from lowland Amazonia: Forest and cooling in glacial times', *Science*, vol 274, pp85–88

Connell, J. H. and Slayter, R. O. (1977) 'Mechanisms of succession in natural communities and their role in community stability and organisation', *American Naturalist*, vol 111, pp1119–1144

Constanza, R., d'Arge, R., De Groot, R., Farber, S., Hannon, B., Limburg, K., O'Neill, S. N., Paruelo, J., Raskin, R. G., Sutton, P. and van den Belt, M. (1997) 'The value of the world's ecosystem services and natural capital', *Nature*, vol 387, pp253–260

Cottrell, J. E., Munro, R. C., Tabbener, H. E., Gillies, A. C. M., Forrest, G. I., Deans, J. D. and Lowe, A. J. (2002). 'Distribution of chloroplast DNA variation in British oaks (*Quercus robur* and *Q. petraea*): The influence of postglacial colonisation and forest management' *Forest Ecology and Management*, vol 156, 181–195

Cowling, R. M. (1999) 'Planning for persistence – systematic reserve design in Southern Africa's Succulent Karoo desert', *Parks*, vol 9, pp17–30

Cowling, R. M. and Hilton-Taylor, C. (1994) 'Patterns of plant diversity and endemism in southern Africa: An overview', in B. J. Huntley (ed) *Botanical Diversity in Southern Africa*, National Botanical Institute, Pretoria, South Africa, pp31–52

Cowling, R. M. and Pressey, R. L. (2003) 'Conservation planning in the Cape Floristic Region, South Africa', *Biological Conservation*, vol 112, pp1–297

Cowling, R. M., Rundel, P. W., Lamont, B. B., Arroyo, M. K. and Arianoutsou, M. (1996) 'Plant diversity in Mediterranean-climatic regions', *Trends in Ecology and Evolution*, vol 11, pp362–366

Critchley, C. N. R. (2003) 'Conservation of lowland semi-natural grasslands in the UK: A review of botanical monitoring results from agri-environment schemes', *Biological Conservation*, vol 115, pp263–278

Cully, A.C., Cully Jr, J. F. and Hiebert, R. D. (2003) 'Invasion of exotic plant species in tallgrass prairie fragments', *Conservation Biology*, vol 17, p990

Cunningham, A. B. (1993) *Ethics, Biodiversity and New Natural Products Development*, WWF International, Gland, Switzerland

Cunningham, A. B. (1996a) 'Medicinal plants and miombo woodland: Species, symbolism and trade', in B. Campbell (ed) *The Miombo in Transition: Woodlands and Welfare in Africa*, Centre for International Forestry Research, Bogor, Indonesia, p166

Cunningham, A. B. (1996b) *People, Park and Plant Use*, People and Plants Working Paper no 4, Division of Ecological Sciences, UNESCO, Paris, France

Cunningham, A. B. (1997a) 'Review of ethnobotanical literature from eastern and southern Africa', *Bulletin of the African Ethnobotany Network*, vol 1, pp23–87

Cunningham, A. B. (1997b) 'The "Top 50" listings and the Medicinal Plants Action Plan', *Medicinal Plant Conservation*, vol 3, pp5–7

Cunningham, A. B. (2001a) *Applied Ethnobotany: People, Wild Plant Use and Conservation*, Earthscan, London, UK

Cunningham, A. B. (2001b) 'Return of the pepper-bark', *Medicinal Plant Conservation*, vol 7, pp21–22

Cunningham, A. B., Belcher, B. and Campbell, B. (2005) *Carving Out a Future: Forests, Livelihoods and the International Woodcarving Trade*, Earthscan, London

Cunningham, M., Cunningham, A. B. and Schippmann, U. (1997) *Trade in* Prunus africana *and the Implementation of CITES*, German Federal Agency for Nature Conservation, Bonn, Germany

Davis, E. W. (1995) 'Ethnobotany: An old practice, a new discipline', in R. E. Schultes and S. von Reis (eds) *Ethnobotany: Evolution of a Discipline*, Chapman and Hall, London, UK, pp40–51

Davis, M. B. (1976) 'Pleistocene biogeography of temperate deciduous forests', *Geoscience and Man*, vol 13, pp13–26

Davis, S. D., Droop, S. J. M., Gregerson, P., Henson, L., Leon, C. J., Villa-Lobos, J. L., Synge, H. and Zantovska, J. (1986) *Plants in Danger: What Do We Know?*', IUCN, Gland, Switzerland, and Cambridge, UK

Davis-Merlen, G. (1998) 'New introductions and a special law for Galapagos', *Aliens*, vol 7, pp10–11

Dawson, I. K. and Powell, W. (1999) 'Genetic variation in the Afromontane tree *Prunus africana*, an endangered medicinal species', *Molecular Ecology*, vol 8, pp151–156

de Boef, W. and Almekinders, C. (eds) (2000) *Encouraging Diversity and Development of Plant Genetic Resources*, Intermediate Technology Publications, London, UK

de Klemm, C. (1990) *Wild Plant Conservation and the Law*, World Conservation Union, Gland, Switzerland

Dessalegn, R. (1998) 'Environmentalism and conservation in Wällo before the revolution', *Journal of Ethiopian Studies*, vol 31, pp43–86

Dev, S. (1999) 'Ancient–modern concordance in Ayurvedic plants: Some examples', *Environmental Health Perspectives*, vol 107, pp783–789

Dhar, U., Manjkhola, S., Joshi, M., Bhatt, A., Bisht, A. K. and Joshi, M. (2002) 'Current status and future strategy for development of medicinal plants sector in Uttaranchal, India', *Current Science*, vol 83, pp956–964

Diáz-Betancourt, M., Ghermandi, L., Ladio, A., López-Moreno, I., Raffaele, E. and Rapoport, E. H. (1999) 'Weeds as a source for human consumption: A comparison between tropical and temperate North America', *Revista de Biologica Tropical*, vol 47, pp329–338

Dogan, Y., Baslar, S., Mert, H. H. and Ay, G. (2003) 'Plants used as natural dye sources in Turkey', *Economic Botany*, vol 57, pp442–453

Dold, T. and Cocks, M. (2001) 'The trade in medicinal plants in the Eastern Cape Province, South Africa', *TRAFFIC Bulletin*, vol 18, pp11–13

Donaldson, J. (ed) (2003) *Cycads: Status Survey and Conservation Action Plan*, IUCN and SSC Cycad Specialist Group, Gland, Switzerland, and Cambridge, UK, p86

Dounias, E., Rodrigues, W. and Petit, C. (2000) 'Review of ethnobotanical literature for Central and West Africa', *Bulletin of the African Ethnobotany Network*, vol 2, pp5–117

Dransfield, J. and Johnson, D. (1991) 'The conservation status of palms in Sabah (Malaysia)', in D. Johnson (ed) *Palms for Human Needs in Asia: Palm Utilization and Conservation in India, Indonesia, Malaysia and the Philippines*, Balkema, Rotterdam, The Netherlands, pp175–179

Duckworth, J. (2002). 'The pink waxcap *Hydrocybe calyptriformis*', *Plantlife*, Autumn 2002, p16

Dudley, N. and Stolton, S. (1994) *The East Asian Timber Trade*, WWF, Godalming, UK

Du Plessis, M. A. (1995) 'The effects of fuelwood removal on the diversity of some cavity-using birds and mammals in South Africa', *Biological Conservation*, vol 74, pp77–82

DWAF (Department of Water Affairs and Forestry) (2002) *Working for Water Programme*, DWAF, South Africa. www.dwaf.gov.za

Eggleston, H. S. and Irwin, J. G. (1995) 'Trends in sulphur and nitrogen emissions', in R. W. Battarbee (ed) *Acid Rain and Its Impact: The Critical Loads Debate*, Ensis Publishing, London, UK, pp15–16.

Eilu, G. (1999) 'Climbers from tropical rain forests at the Albertine Rift, Western Uganda', *Lidia*, vol 4, pp93–120

Eilu, G. and Bukenya-Ziraba, R. (2004) 'Local use of climbing plants of Budongo Forest Reserve, Western Uganda', *Journal of Ethnobiology*, vol 24, pp307–327

Eldin, S. and Dunford, A. (1999) *Herbal Medicine in Primary Care*, Butterworth-Heinemann, Oxford, UK

Ellen, R. (2002) 'Ethnobotany of Seram people in Malakum, East Indonesia', Paper presented at sixth Congress of the International Society for Ethnobiology, 16–20 September 2002, Addis Ababa, Ethiopia

Epperson, J. E., Pachico, D. H. and Guevara, L. (1997) 'A cost analysis of maintaining cassava plant genetic resources', *Crop Science*, vol 37, pp1641–1649

Ertug, F. (2003) 'Gendering the tradition of plant gathering in central Anatolia (Turkey)', in P. L. Howard (ed) *Women and Plants*, Zed Books, London, UK, pp183–196

Erwin, T. I. (1991) 'How many species are there? Revisited', *Conservation Biology*, vol 5, pp1–4

Eyzaguirre, P. (2002) 'Land tenure and community institutions in the management of fruit and nut tree diversity in Uzbekistan', Paper presented at the sixth Congress of the International Society of Ethnobiology, 16–20 September 2002, Addis Ababa, Ethiopia

FAO (United Nations Food and Agriculture Organization) (1992) *Forest Products Yearbook*, FAO, Rome, Italy

FAO (2004) *International Undertaking on Plant Genetic Resources for Food and Agriculture*, www.fao.org/ag/cgrfa

Farjon, A. and Page, C. N. (1999) *Conifers: Status Survey and Conservation Action Plan*, IUCN and SSC Conifer Specialist Group, IUCN, Gland, Switzerland and Cambridge, UK

Farnsworth, N. R., Akerele, O. and Bingel, A. S. (1985) 'Medicinal plants in therapy', *Bulletin of the World Health Organization*, vol 63, pp965–981

Farnsworth, N. R. and Soejarto, D. D. (1991) 'Global importance of medicinal plants', in O. Akerele, V. Heywood and H. Synge (eds) *The Conservation of Medicinal Plants*, Cambridge University Press, Cambridge, UK, pp25–51

Ferris, C., Oliver, R. P., Davy, A. J. and Hewitt, G. M. (1993) 'Native oak chloroplasts reveal an ancient divide across Europe', *Molecular Ecology*, vol 2, pp337–344

Fischer, M. and Matthies, D. (1998) 'Effects of population size on performance in the rare plant *Gentianella germanica*', *Journal of Ecology*, vol 86, pp195–204

Fjeldså, J. and Lovett, J. C. (1997a) 'Biodiversity and environmental stability', *Biodiversity and Conservation*, vol 6, pp315–323

Fjeldså, J. and Lovett, J. C. (1997b) 'Geographical patterns of old and young species in African forest biota: The significance of specific montane areas as evolutionary centres', *Biodiversity and Conservation*, vol 6, pp325–346

Frost, P. (1996) 'The ecology of miombo woodlands', in B. Campbell (ed) *The Miombo in Transition: Woodlands and Welfare in Africa*, Centre for International Forestry Research, Bogor, Indonesia, pp11–57.

Gao Lishi (1998) *On the Dais' Traditional Irrigation System and Environmental Protection in Xishuangbanna*, Yunnan Nationality Press, Kunming, China

Galloway, D. J. (1992) 'Biodiversity: A lichenological perspective', *Biodiversity and Conservation*, vol 1, pp312–323

García, R., Mejía, M. and Zanoni, T. (1994) 'Composicion floristica y principales asociaciones vegetales en la Reserve Cientificia Ebano Verde, Cordillera Central, Republica Dominicana', *Moscosoa*, vol 8, pp86–130

Gärdenfors, U. (2001) 'Classifying threatened species at national versus global levels', *Trends in Ecology and Evolution*, vol 16, pp511–516

Geddes, P. (1915) *Cities in Evolution*, Williams and Norgate, London

Ghimire, S. K. and Aumeeruddy-Thomas, Y. (2004) 'Approach to in situ conservation of threatened Himalayan plants: A case study from Shey-Phoksundo National Park, Dolpo', in *Himalayan Medicinal and Aromatic Plants: Balancing Conservation and Use*, Proceedings of the Regional Workshop on Wise Practice and Experiential Learning in the Conservation and Management of Himalayan Medicinal Plants, WWF Nepal Programme, Ministry of Forest and Soil Conservation, HMGN, IDRC, Medicinal and Aromatic Plants Program in South Asia and WWF/UNESCO People and Plants Initiative, Kathmandu, Nepal

Godfray, H. C. J. (2002) 'Challenges of taxonomy', *Nature*, vol 417, pp17–19

Goebel, A. (2003) 'Gender and entitlements in the Zimbabwean woodlands: A case study of resettlement', in P. L. Howard (ed) *Women and Plants*, Zed Books, London, UK, pp115–129

Goldammer, J. G. and Price, C. (1998) 'Potential impacts of climate change on fire regimes in the tropics based on MAGICC and a GISS GCM-derived lightning model', *Climatic Change*, vol 39, pp273–296

Gómez Pompa, A. and Jiménes-Osornio, J. (1999) 'Conservation of biodiversity in ethnic groups: A case study of the Maya', in *Proceedings of the 16th International Botanical Congress*, St Louis, US, p167

Govaerts, S. (2001) 'How many species of seed plants are there?', *Taxon*, vol 50, pp1085–1090

Graham, L. and Wilcox, L. (2000). *Algae*, Prentice Gall, Englewood Cliffs, NJ

Granger, A. (1996) 'Forest environments', in W. M. Adams, A. S. Goudie and A. R. Orme (eds) *The Physical Geography of Africa*, Oxford University Press, Oxford, UK, pp173–195

Green, M. J. B., Murray, M. G., Buntin, G. C. and Paine, J. R. (1997) *Priorities for Biodiversity Conservation in the Tropics*, World Conservation Monitoring Centre Biodiversity Bulletin, Cambridge, UK

Green, M. J. B. and Paine, J. (1997) *State of the World's Protected Areas at the End of the Twentieth Century*, World Commission on Protected Areas, IUCN, Gland, Switzerland

Grifo, F. and Rosenthal, J. (eds) (1997) *Biodiversity and Human Health*, Island Press, Washington, DC, US

Griggs, J. K., Manandhar, N. P., Towers, G. H. N. and Taylor, R. S. L. (2001) 'The effects of storage on the biological activity of medicinal plants from Nepal', *Journal of Ethnopharmacology*, vol 77, pp247–252

Guijt, I. (1998) 'Valuing wild plants with economic and participatory methods: An overview of the hidden harvest methodology', in H. D. V. Prendergast, D. R. Etkin, D. R. Harros and P. J. Houghton (eds) *Plants for Food and Medicine*, Royal Botanic Gardens, Kew, UK, pp223–235

Guijt, I., Hinchcliffe, F. and Melnyk, M. (1995) *The Hidden Harvest: The Value of Wild Resources in Agricultural Systems – A Summary*, International Institute for Environment and Development, London, UK

Hadley, M. (2002) *Biosphere Reserves: Special Places for People and Nature*, UNESCO, Paris, France

Hallingbäck, T. and Hodgetts, N. (eds) (2000) *Mosses, Liverworts and Hornworts: Status Survey and Conservation Action Plan for Bryophytes*, IUCN/SSC Bryophyte Specialist Group, IUCN, Gland, Switzerland, and Cambridge, UK

Hamilton, A. C. (1974) 'Distribution patterns of forest trees in Uganda and their historical significance', *Vegetatio*, vol 29, pp21–35

Hamilton, A. C. (1981) 'The Quaternary history of African forests: Its relevance to conservation', *African Journal of Ecology*, vol 19, pp1–6

Hamilton, A. C. (1984) *Deforestation in Uganda*, Oxford University Press, Nairobi, Kenya

Hamilton, A. C. (1997) 'Threats to plants: An analysis of Centres of Plant Diversity', in D. H. Touchell and K. W. Dixon (eds) *Conservation into the 21st Century*, Proceedings of the fourth International Botanic Gardens Conservation Congress, Kings Park and Botanic Garden, Perth, Australia, pp309–322

Hamilton, A. C. (1999) 'Vegetation, climate and soil: Altitudinal relationships on the East Usambara Mountains, Tanzania', in J. Timberlake and S. Kativu (eds) *African Plants: Biodiversity, Taxonomy and Uses* (eds), Royal Botanic Gardens, Kew, UK, pp165–169

Hamilton, A. C. (2001) *Human Nature and the Natural World: From Traditional Societies to the Global Age*, Privately published, Godalming, UK

Hamilton, A. C. (2004) 'Medicinal plants, conservation and livelihoods', *Biodiversity and Conservation*, vol 13, pp1477–1517

Hamilton, A. C. (2005). 'Resource assessment for sustainable harvesting of medicinal plants', Paper at Sustainable Supply Chain Management of Medicinal and Aromatic Plants workshop organized by D. Lange, K. Dürbeck and C. Franz as a side-event at the International Botanical Congress, Vienna, 21–22 July 2005 (see also www.plantlife.org.uk)

Hamilton, A. C., Baranga, J. and Tindigarukayo, J. (1990) *Proposed Bwindi (Impenetrable) National Park: Results of a Public Inquiry and Recommendations for Its Establishment*, Report for Uganda National Parks, Kampala, Uganda

Hamilton, A. C. and Bensted Smith, R. (1989) *Forest Conservation in the East Usambara Mountains*, Tanzania IUCN, Gland, Switzerland, and Cambridge, UK

Hamilton, A. C., Cunningham, A., Byarugaba, D. and Kayanja, F. (2000) 'Conservation in a region of political instability', *Conservation Biology*, vol 14, pp1722–1725

Hamilton, A. C., Pei Shengji, Kessy, J., Khan, A. A., Lagos-Witte, S. and Shinwari, Z. K. (2003) *The Purposes and Teaching of Applied Ethnobotany*, People and Plants Working Paper no 11, WWF, Godalming, UK

Hamilton, A. C. and Taylor, D. (1991) 'History of climate and forests in tropical Africa during the last 8 million years', *Climatic Change*, vol 19, pp65–78

Hamilton, A. C., Taylor, D. and Howard, P. (2001) 'Hotspots in African forests as Quaternary refugia', in W. Weber, L. J. T. White, A. Vedder and L. Naughton-Treves (eds) *African Rain Forest Ecology and Conservation*, Yale University Press, New Haven, US, pp57–67

Hamlin, C. C. and Salick, J. (2003) 'Yanesha agriculture in the upper Peruvian Amazon: Persistence and change fifteen years down the "road"', *Economic Botany*, vol 57, pp163–180

Hart, T. B. (2001) 'Forest dynamics in the Ituri basin (DR Congo)', W. Weber, L. J. T. White, A. Vedder and L. Naughton-Treves (eds) *African Rain Forest Ecology and Conservation*, Yale University Press, New Haven, US, and London, UK, pp154–162

Hart, T. B., Hart, J. A., Dechamps, R., Fournier, M. and Ataholo, M. (1996) 'Changes in forest composition over the last 4000 years in the Ituri basin, Zaire', in L. J. G. van der Maesen (ed) *The Biodiversity of African Plants*, Kluwer Academic Publishers, The Netherlands, pp541–563

Hasbagan, A. and Chen Shan (1996) 'The cultural importance of animals in traditional Mongolian plant nomenclature', in C. Humphrey and D. Sneath (eds) *Culture and Environment in Inner Asia*, vol 2 (*Society and Culture*), The White Horse Press, Cambridge, UK, pp25–29

Hasbagan, A. and Pei Shengji (2001) 'Ethnobotanical study of the Arhorchis Mongolian folk medicinal plants', in Chen Shan and A. Hasbagan (eds) *Ethnobotanical Studies in the Mongolian Plateau*, Inner Mongolian Education Press, China, pp164–169

Hatfield, G. (1999) *Memory, Wisdom and Healing: The History of Domestic Plant Medicine*, Sutton Publishing, Stroud, UK

Hawkes, J. G., Maxted, N. and Ford-Lloyd, B. V. (2001) *The Ex Situ Conservation of Plant Genetic Resources*, Kluwer Academic Publishers, London, UK

Hawkes, N. (2004) 'Nature's trick helps millions of children beat disease', *The Times*, 20 October 2004, London, UK

Hawksworth, D. L. (2001) 'The magnitude of fungal diversity: The 1.5 million species estimate revisited', *Mycological Research*, vol 105, pp1422–1432

Hawksworth, D. L., Kirk, P. M., Sutton, B. C. and Pegler, D. N. (eds) (1995) *Ainsworth and Bisby's Dictionary of the Fungi*, 8th edn, CAB International, Wallingford, UK

Hawthorne, W. D. and Parren, M. P. E. (2000) 'How important are forest elephants to the survival of woody plant species in Upper Guinea forests?' *Journal of Tropical Ecology*, vol 16, pp133–150

He Jun (2004) *Globalized Forest Products: Commodification of Matsutake Mushroom in Tibetan Villages*, Yunnan Province, China, www.iascp2004.org.mx

He Shan-an, Yin Gu, and Pang Zi-jie (1997) 'Resources and prospects of *Ginkgo biloba* in China', in T. Hori, R. W. Ridge, W. Tulecke, P. Del Tredici, J. Trémouillaux-Guiller and H. Lobe (ed) Ginkgo biloba: *A Global Treasure*, Springer-Verlage, Tokyo, Japan

Hedberg, I. (1993) 'Botanical methods in ethnopharmacology and the need for conservation of medicinal plants', *Journal of Ethnopharmacology*, vol 38, pp121–128

Hernández Bermejo, E. J. (1995) 'Habitat and species management', in J. Newton (ed) *First European Conference on the Conservation of Wild Plants*, Plantlife, Hyères, France, pp127–137

Hernández Cano, J. and Volpato, G. (2004) 'Herbal mixtures in the traditional medicine of Eastern Cuba', *Journal of Ethnopharmacology*, vol 90, pp293–316

Hewitt, G. M. (1993) 'Postglacial distribution and species substructure: Lessons from pollen, insects and hybrid zones', in D. R. Lees and D. Edwards (eds) *Evolutionary Patterns and Processes*, Academic Press, London, UK, pp97–123

Hewitt, G. M. (1999) 'Post-glacial re-colonization of European biota', *Biological Journal of the Linnean Society*, vol 68, pp87–112

Heywood, V. (2000) 'Management and sustainability of the resource base for medicinal plants', in S. Honnef and R. Melisch (eds) *Medicinal Utilization of Wild Species: Challenge for Man and Nature in the New Millennium*, WWF-Germany/TRAFFIC Europe-Germany, EXPO 2000, Hannover, Germany

Hilton-Taylor, C. (1994) 'Western Cape Domain (succulent karoo)', in S. D. Davis, V. Heywood and A. C. Hamilton (eds) *Centres of Plant Diversity*, vol 1, IUCN, Gland, Switzerland, and Cambridge, UK, pp204–217

Hilton-Taylor, C. (2000) *2000 IUCN Red List of Threatened Species*, IUCN, Gland, Switzerland, and Cambridge, UK

Holmstedt, B. R. and Bruhn, J. G. (1995) 'Ethnopharmacology – a challenge', in R. E. Schultes and S. von Reis (eds) *Ethnobotany: Evolution of a Discipline*, Chapman and Hall, London, UK, pp338–342

Holsinger, K. E. and Gottlieb, L. D. (1991) 'Conservation of rare and endangered plants: Principles and prospects', in D. A. Falk and K. E. Holsinger (eds) *Genetics and Conservation of Rare Plants*, Oxford University Press, New York, US, and Oxford, UK, pp195–208

Howard, P. C. (1991) *Nature Conservation in Uganda's Tropical Forest Reserves*, IUCN, Gland, Switzerland

Howard, P. C., Davenport, T. R. B., Kigenyi, F. W., Viskanic, P., Baltzer, M. C., Dickinson, C. J., Lwanga, J., Matthews, R. A. and Mupada, E. (2000) 'Protected area planning in the tropics: Uganda's national system of forest nature reserves', *Conservation Biology*, vol 14, pp1–19

Huenneke, L. F. (1991) 'Ecological implications of genetic variation in plant populations', in D. A. Falk and K. E. Holsinger (eds) *Genetics and Conservation of Rare Plants*, Oxford University Press, New York, US, and Oxford, UK, pp31–44

Hulme, M. (1996) 'Climate change within the period of meteorological records', in W. M. Adams, A. S. Goudie and A. R. Orme (eds) *The Physical Geography of Africa*, Oxford University Press, Oxford, UK, pp88–102

Huntley, B. J. (1999) 'SABONET: Into the new millennium', *SABONET News*, pp174–175

Huntley, B. J. and Matos, E. M. (1994) 'Botanical diversity and its conservation in Angola', in B. J. Huntley (ed) *Botanical Diversity in Southern Africa*, National Botanical Institute, Pretoria, South Africa, pp53–74

Idle, E. T. (1996) 'Conflicting priorities in site management in England', in N. S. Cooper and R. C. J. Carling (eds) *Ecologists and Ethical Judgements*, Chapman and Hall, London, UK, pp151–159

IPGRI (International Plant Genetic Resources Institute) (2002) *Neglected and Underutilized Plant Species: Strategic Action Plan of the International Plant Genetic Resources Institute*, IPGRI, Rome, Italy

Ishakia, M., Obunga, E., Morimoto, Y., Kamondo, B. and Aman, R. (1998) 'Conservation of plant genetic resources in Kenya', in R. P. Adams and J. E. Adams (eds) *Conservation and Utilization of African Plants*, Missouri Botanical Gardens Press, St Louis, Missouri, US

IUCN (World Conservation Union) (1994a) *IUCN Red List Categories*, IUCN, Gland, Switzerland

IUCN (1994b) *Parks for Life: Action for Protected Areas in Europe*, Commission on National Parks and Protected Areas, IUCN, Gland, Switzerland, and Cambridge, UK

IUCN (2001a) *IUCN Policy on the Management of* Ex Situ *Populations for Conservation*, IUCN, Gland, Switzerland.

IUCN (2001b) *IUCN Red List Categories and Criteria: Version 3.1*, Species Survival Commission, IUCN, Gland, Switzerland and Cambridge, UK

IUCN (2004b) IUCN website species information service, www.iucn.org/themes/ssc/programs/sisindex.htm.

IUCN and WCMC (World Conservation Monitoring Centre) (1983) *List of Rare, Threatened and Endemic Plants in Europe*, 2nd edition, Council of Europe, Strasbourg, France

IUCN and SSC (Species Survival Commission) (1996) *Orchids – Status Survey and Conservation Action Plan*, Orchid Specialist Group, IUCN, Gland, Switzerland.

Jabeen, A. (1999) *Ethnobotany of Fodder Species of Ayubia National Park, Nathia Gali, Its Conservation Status and Impacts on Environment*, MPhil thesis, Quaid-I-Azam, Islamabad, Pakistan

James, L. (1998) *The Rise and Fall of the British Empire*, Little, Brown and Co., London, UK

Jayaraman, K. S. (1998) 'India seeks tighter controls on germplasm', *Nature*, vol 392, p536

Jenkins, M. and Kapos, V. (2000) *Biodiversity Indicators for Monitoring GEF Programme Implementation and Impacts*, World Conservation Monitoring Centre, Cambridge, UK

Jennings, S. B., Brown, N. D., Boshier, D. H., Whitmore, T. C. and Lopes, J. D. C. A. (2000) 'Ecology provides a pragmatic solution to the maintenance of genetic diversity in sustainably managed tropical rain forests', *Forest Ecology and Management*, vol 5394, pp1–10

Jepson, P. (2001) *Biodiversity and Protected Area Policy: Why Is It Failing in Indonesia*, PhD thesis, Oxford University, Oxford, UK

Jepson, P. and Canney, S. (2001) 'Biodiversity hotspots: hot for what?', *Global Biogeography and Ecology*, vol 10, pp225–227

Jepson, P., Jarvie, J. K., MacKinnon, K. and Monk, K. A. (2001) 'The end for Indonesia's lowland forests?', *Science*, vol 292, pp859–861

Johansson, O. (1995) 'The integration of forestry with plant conservation', in J. Newton (ed) *First European Conference on the Conservation of Wild Plants*, Plantlife, Hyères, France, Salisbury, UK, pp114–115

Johnson, D. and the IUCN/SSC Palm Specialist Group (eds) (1996) *Palms: Their Conservation and Sustained Utilization. Status Survey and Conservation Action Plan*, IUCN, Gland, Switzerland, and Cambridge, UK, p116

Johnson, S. D. and Steiner, K. E. (2000) 'Generalization versus specialization in plant pollination systems', *Trends in Ecology and Evolution*, vol 15, pp140–143

Jolly, D., Taylor, D., Marchant, R., Hamilton, A., Bonnefille, R., Buchet, G. and Riollet, G. (1997) 'Vegetation dynamics in central Africa since 18,000 BP: Records from the interlacustrine highlands of Burundi, Rwanda and western Uganda', *Journal of Biogeography*, vol 24, pp495–512

Jones, G. E. and Garforth, C. (2002) 'The history, development and future of agricultural extension', www.fao.org.docrep/W5830E/w5830e03.htm

Jonsell, R. (2003) 'Swedish provincial Floras – A survey of their history and present status', *Watsonia*, vol 24, pp331–336

Ka Hou Chu, Ji Qu, Zu-Guo Yu and Vo Anh (2004) 'Origin and phylogeny of chloroplasts revealed by a simple correlation analysis of complete genomes', *Molecular Biology and Evolution*, vol 21, pp200–206

Kalema, H. and Bukenya-Ziraba, R. (2005) 'Patterns of plant diversity in Uganda', *Biologiske Skrifter*, vol 55, pp331–341

Karlen, A. (1995) *Plague's Progress*, Indigo, Cassell Group, London, UK

Karron, J. D. (1991) 'Patterns of genetic variation and breeding systems in rare plant species', in D. A. Falk and K. E. Holsinger (eds) *Genetics and Conservation of Rare Plants*, Oxford University Press, New York, US, and Oxford, UK, pp87–98

Kasenene, J. (2001). 'Lost logging: Problems of tree regeneration in forest gaps in Kibale Forest, Uganda', in W. Weber, L. J. T. White, A. Vedder and L. Naughton-Treves (eds) *African Rain Forest Ecology and Conservation*, Yale University Press, New Haven, US, pp480–490

Kelleher, C. T., Hodkinson, T. R., Kelly, D. L. and Douglas, G. C. (2004) 'Characterisation of chloroplast DNA haplotypes to reveal the provenance and genetic structure of oaks in Ireland', *Forest Ecology and Management*, vol 189, pp123–131

Keller, L. F. and Waller, D. M. (2002) 'Inbreeding effects in wild populations', *Trends in Ecology and Evolution*, vol 17, pp230–241

Kellman, M., Tackaberry, R., and Rigg, L. (1998) 'Structure and function in two tropical gallery forest communities: Implications for forest conservation in fragmented ecosystems', *Journal of Applied Ecology*, vol 35, pp195–206

Kenk, G. K. (1992) 'Silviculture of mixed-species stands in Germany', in M. G. R. Cannell, D. C. Malcolm and P. A. Robertson (eds) *The Ecology of Mixed-Species Stands of Trees*, Blackwell, Oxford, UK, pp53–63

Kerr, W. A., Hobbs, J. E. and Yampoin, R. (1999) 'Intellectual property protection, biotechnology and developing countries: Will the TRIPS be effective?', *AgBioForum*, vol 2, pp203–211

Kessy, J. F. (1998) *Conservation and Utilization of Natural Resources in the East Usambara Forest Reserves: Conventional Views and Local Perspectives*, Wageningen Agricultural University, Wageningen, The Netherlands

Kevles, D. J. (2001) 'Patenting life: An historical overview of laws, interests and ethics', www.yale.edu/law/ltw/papers/ltw-kevles.pdf

Khan, A. U. (1994a) *Harappa: Site for Conservation of a Threatened Botanical Interest*, WWF-Pakistan, Lahore, Pakistan

Khan, A. U. (1994b) 'History of decline and present status of natural thorn forest in Punjab', *Biological Conservation*, vol 67, pp205–210

Khayota, B. N. (1999) 'Notes on systematics, ecology and conservation of *Ansellia* (Orchidaceae)', in J. Timberlake and S. Kativu (eds) *African Plants: Biodiversity, Taxonomy and Uses*, Royal Botanic Gardens, Kew, UK

Kier, G. and Barthlott, W. (2001) 'Measuring and mapping endemism and species richness: A new methodological approach and its application on the flora of Africa', *Biodiversity and Conservation*, vol 10, pp1513–1529

Kiew, R. (1991) 'The limestone flora', in R. Kiew (ed) *The State of Nature Conservation in Malaysia*, Malayan Nature Society, Kuala Lumpur, Malaysia, pp42–50

King, S. R., Meza, E. N., Carlson, T. J. S., Chinnock, J. A., Moran, K. and Borges, J. R. (1999) 'Issues in the commercialization of medicinal plants', *HerbalGram*, vol 47, pp46–51

Kirk, C. (2004) *Green Nature*, 11 November, www.greennature.com/article287.html

Kitching, R. L., Mitchell, H., Morse, G., and Thebaud, D. (1997) 'Determinants of species richness in assemblages of canopy arthropods in rainforests', in N. E. Stork, J. Adis and R. K. Didham (eds) *Canopy Arthropods*, Chapman and Hall, London, UK, pp131–150

Knight, V. (2004) '*Rubus* breeding worldwide and the raspberry breeding programme at East Malling', Presentation at the The Linnean Society, London, UK

Körner, C. (1998) 'Tropical forests in a CO_2-rich world', *Climatic Change*, vol 39, pp297–315

Kothari, B. (2003) 'The invisible queen in the plant kingdom: Gender perspectives in medical ethnobotany', in P. L. Howard (ed) *Women and Plants*, Zed Books, London, UK, pp150–164

Kozlowski, G., Landergott, U., Holderegger, R. and Schneller, J. J. (2002) 'The importance of recent population history for understanding genetic diversity in natural populations of endangered *Dryopteris cristata*', *The Fern Gazette*, vol 16, p465

Küchi, C. (1996/1997) *Changing Forest Use and Management in the Alps and the Himalayas: A Comparison between Switzerland and Nepal*, Overseas Development Institute, London, UK

Ladio, A. H. and Lozada, M. (2000) 'Edible wild plant use in a Mapuche community of northwestern Patagonia', *Human Ecology*, vol 28, pp53–71

Lagos-Witte, S. (1994) 'La etnobotánica en un enfoque interdisciplinario: Un acercamiento a la participación de comunidades indígenas y campesinas?', in R. Fortunato and N. Bacigalupo (eds) *VI Congreso Latinoamericano de Botánico, Mar del Plata, Argentina*, vol 68, Missouri Botanical Garden Press, St Louis, Missouri, US, pp215–224

Laird, S. A. (2002) *Biodiversity and Traditional Knowledge: Equitable Partnerships in Practice*, Earthscan, London, UK

Laird, S. A. and Pierce, A. R. (2002) *Promoting Sustainable and Ethical Botanicals: Strategies to Improve Commercial Raw Material Sourcing*, Rainforest Alliance, New York, US

Laird, S. A., Pierce, A. R. and Schmitt, S. F. (2005) 'Sustainable raw materials in the botanical industry: Constraints and opportunities', *Acta Horticulturae*, vol 676, pp111–117

Laird, S. A. and ten Kate, K. (2002) 'Linking biodiversity prospecting and forest conservation', in S. Pagiola, J. Bishop and N. Landell-Mills (eds) *Selling Forest Environmental Services*, Earthscan, London, UK, pp151–172

Lama, Y. C., Ghimire, S. K. and Aumeeruddy-Thomas, Y. (2001) *Medicinal Plants of Dolpo: Amchis' Knowledge and Conservation*, WWF-Nepal Program, Kathmandu, Nepal

Langdale-Brown, I., Osmaston, H. A. and Wilson, J. G. (eds) (1964) *The Vegetation of Uganda (excluding Karamoja) and Its Bearing on Land Use*, Government Printer, Entebbe, Uganda

Lange, D. (1997) 'Trade in plant material for medicinal and other purposes', *TRAFFIC Bulletin*, vol 17, pp21–32

Lange, D. (1998a) *Europe's Medicinal and Aromatic Plants: Their Use, Trade and Conservation – An Overview*, TRAFFIC International, Cambridge, UK

Lange, D. (1998b). 'Status and trends of medicinal and aromatic plant trade in Europe: An overview', in *Medicinal Plant Trade in Europe: Conservation and Supply*, TRAFFIC Europe, Brussels, pp1–4

Langton, M. (2000) 'Indigenous knowledge and conservation policy: Aboriginal fire management of protected areas', in A. M. Cetto (ed) *Science for the Twenty-First Century: A New Commitment*, UNESCO, Paris, France, pp435–436

Laurance, W. F. (1998) 'A crisis in the making: Responses of Amazonian forests to land use and climate change', *Trends in Ecology and Evolution*, vol 13, pp411–415

Laurance, W. F. (2001) 'Future shock: Forecasting a grim fate for the Earth', *Trends in Ecology and Evolution*, vol 16, pp531–533

Laurance, W. F., Albernaz, A. K. M., Fearnside, P. M., Vasconcelos, H. L. and Ferreira, L. V. (2004) 'Deforestation in Amazonia', *Nature*, vol 304, p1109

Laurance, W. F., Powell, G. and Hansen, L. (2002) 'A precarious future for Amazonia', *Trends in Ecology and Evolution*, vol 17, pp251–252

Lawrence, A. and Hawthorne, W. (eds) (in preparation) *Plant Identification, Conservation and Management: Methods for Producing User-Friendly Field Guides*, Earthscan, London

Lawton, R. O., Nair, U. S., Pielke Sr, R. A. and Welch, R. M. (2001) 'Climatic impact of tropical lowland deforestation on nearby montane cloud forests', *Science*, vol 294, pp584–587

Leakey, R. R. B. (2003) 'The domestication of indigenous trees as the basis of sustainable land use in Africa', in J. Lemons, R. Victor and D. Schaffer (eds) *Conserving Biodiversity in Arid Regions*, Kluwer, Boston, US, pp27–40

Leakey, R. R. B., Tchoundjeu, Z., Schreckenberg, K., Simons, T., Shackleton, S., Mander, M., Wynberg, R., Shackleton, C. and Sullivan, C. (2003) 'Trees and markets for agroforestry tree products: Targeting poverty reduction and enhanced livelihoods', in *World Agroforestry and the Future*, ICRAF, Nairobi, Kenya, pp1–34

Leaman, D. J., Arnason, J. T., Yusuf, R., Sangat-Roemantyo, H., Soedjito, H., Angerhofer, C. K. and Pezzuto, J. M. (1995) 'Malaria remedies of the Kenyah of the Apo Kayan, East Kalimantan, Indonesian Borneo: A quantitative assessment of local consensus as an indicator of biological efficacy', *Journal of Ethnopharmacology*, vol 49, pp1–16

Lebbie, A. R. and Guries, R. P. (1995) 'Ethnobotanical value and conservation of sacred groves of the Kpaa Mende in Sierra Leone', *Economic Botany*, vol 49, pp297–308

Lemars, B. O. L. and Gornall, R. J. (2003) 'Identification of British species of *Callitriche* by means of isozymes', *Watsonia*, vol 24, pp389–399

Lenné, J. M. and Wood, D. (1991) 'Plant diseases and the use of wild germplasm', *Annual Review of Phytopathology*, vol 28, pp35–63

Lewis, W. H. (2003) 'Pharmaceutical discoveries based on ethnomedicinal plants: 1985 to 2000 and beyond', *Economic Botany*, vol 57, pp126–134

Liu Yujin, Yang Xiaolin, Bao Longyou, Lan Xiaozhong and Tsering, D. (2004) *A Report on Resource Investigations and Market Analyses of Medicinal Plants in Nyingchi County, China*, WWF, Godalming, UK

Livingstone, D. A. (1967) 'Postglacial vegetation of the Ruwenzori Mountains in equatorial Africa', *Ecological Monographs*, vol 37, pp25–52

Lock, R. (1996) 'The future of biology beyond the compulsory schooling age or whither post-16 biology', *Journal of Biological Education*, vol 30, pp3–6

Long Chunlin and Pei Shengji (2003) 'Cultural diversity promotes conservation and application of biological diversity', *Acta Botanica Yunnanica*, vol 14 (supplement), pp11–22

Lovejoy, T. (2002) 'Biodiversity: Dismissing scientific process', *Scientific American*, vol 286, pp69–71

Lundberg, B., Payiatas, G. and Argyrou, M. (1999) 'Notes on the diet of the Lessepsian migrant herbivorous fishes, *Siganus luridus* and *S. rivulatus*, in Cyprus', *Israel Journal of Zoology*, vol 45, pp127–134

Luoga, E. J., Witkowski, E. T. F. and Balkwill, K. (2000) 'Differential utilization and ethnobotany of trees in Kitulanghalo Forest Reserve and surrounding communal lands, eastern Tanzania', *Economic Botany*, vol 54, pp328–343

Mabey, R. (1997) *Flora Brittanica*, Chatto and Windus, London

Macilwain, C. (1998) 'When rhetoric hits reality in debate on bioprospecting', *Nature*, vol 392, pp535–540

Macilwain, C. (2004) 'Organic: Is it the future of agriculture?', *Nature*, vol 428, pp792–793

MacKenzie, D. (1998) 'Growing problems in seed banks', *New Scientist*, vol 11, 14 February

MacKinnon, J. (1997) *Protected Areas Systems Review of the Indo-Malayan Realm*, The Asian Bureau for Conservation Limited, Hong Kong

MacKinnon, J. and Artha, M. B. (1982) *National Conservation Plan for Indonesia: Introduction, Evaluation, Methods and Overview of National Nature Riches*, UNDP/FAO National Parks Development Project, Bogor, Indonesia

Maheshwari, J. K. (1995) 'Ethnobotanical resources of hot, arid zones of India', in R. E. Schultes and S. von Reis (eds) *Ethnobotany: Evolution of a Discipline*, Chapman and Hall, London, UK, pp235–249

Mahli, Y. and Grace, J. (2000) 'Tropical forests and atmospheric carbon dioxide', *Trends in Ecology and Evolution*, vol 15, pp332–337

Maini, J. S. (1993) 'Sustainable development of forests: A systematic approach to defining criteria, guidelines and indicators', Paper presented to the seminar of the Conference on Security and Cooperation in Europe (CSCE) experts on sustainable development of boreal and temperate forests, Montreal, Canada

Maldonado, S., Sala, O. and Montenegro, G. (2003) 'Latin American Plant Sciences Network: A program for the development of plant sciences and conservation of biodiversity in Latin America', in J. Lemons, R. Victor and D. Schaffer (eds) *Conserving Biodiversity in Arid Regions*, Kluwer, Boston, US, pp373–386

Maley, J. (2001) 'The impact of arid phases on the African rain forest through geological history', in W. Weber, L. J. T. White, A. Vedder and L. Naughton-Treves (eds) *African Rain Forest Ecology and Conservation*, Yale University Press, New Haven, US, and London, UK, pp68–87

Marchant, N. (2000) 'Name that plant', *Landscape*, vol 15, pp35–40

Marren, P. (1999) *Britain's Rare Flowers*, Academic Press, London, UK

Marshall, F. (2001) 'Agriculture and use of wild and weedy greens by the Piik ap Oom Okiek of Kenya', *Economic Botany*, vol 55, pp32–46

Marshall, N. T. (1998) *Searching for a Cure: Conservation of Medicinal Wildlife Resources in East and Southern Africa*, TRAFFIC-International, Cambridge, UK

Martin, G. (1995) *Ethnobotany: A Methods Manual*, Chapman and Hall, London, UK

Martin, G., Lee Agama, A., Beaman, J. H. and Nais, J. (2002) *Projek Etnobotani Kinabalu: The Making of a Dusun ethnoflora (Sabah, Malaysia)*, People and Plants Working Paper no 7, Division of Ecological Sciences, UNESCO, Paris

Martin, W., Stoebe, B., Goremykin, V., Hansmann, S., Hasegawa, M. and Kowallik, K. V. (1998) 'Gene transfer to the nucleus and the evolution of chloroplasts', *Nature*, vol 393, pp162–165

Masood, E. (1998) 'Social equity versus private property: Striking the right balance', *Nature*, vol 392, p537

Mathew, R. (2000) 'Educating today's youth in indigenous ecological knowledge: New paths for traditional ways', in A. M. Cetto (ed) *Science for the Twenty-First Century: A New Commitment*, UNESCO, Paris, France, pp439–441

Mathews, S. and Donaghue, M. J. (1999) 'The root of angiosperm phylogeny inferred from duplicate phytochrome genes', *Science*, vol 286, pp947–950

Matyssek, R. and Samdermann Jr., H. (2003) 'Impact of ozone on trees: An ecophysiological perspective', *Progress in Botany*, vol 64, pp349–404

Maundu, P., Berger, D., ole Saitabau, C., Nasieku, J., Kukutia, M., Kipelian, M., Kone, S., Mathenge, S., Morimoto, Y. and Höft, R. (2001) *Ethnobotany of the Loita Maasai: Towards A Community Management of the Forest of the Lost Child*, People and Plants Working Paper no 8, UNESCO, Paris, France

Maundu, P., Johns, T., Eyzaguirre, P. and Moromoto, Y. (2002) 'African leafy vegetables: Diversity and use in sub–Saharan Africa', Unpublished paper presented to the sixth Congress of the International Society of Ethnobiology, 16–20 September 2002, Addis Ababa, Ethiopia

Maundu, P. M., Ngugi, G. W. and Kabuye, C. H. S. (1999) *Traditional Food Plants of Kenya*, National Museums of Kenya, Nairobi, Kenya

Mavi, S. (1994) 'Medicinal plants and their uses in Zimbabwe', in B. J. Huntley (ed) *Botanical Diversity in Southern Africa*, National Botanical Institute, Pretoria, South Africa, pp169–174

May, R. M. (1998) 'Understanding diversity in the natural world and in higher education', *Bulletin of the British Ecological Society*, pp8–9

May, T. (2000) 'Respuesta de la vegetación en un "calimental" de *Dicropteris pectinata* después de un fuego, en la parte oriental de la Cordillera Central, República Dominicana', *Moscosoa*, vol 11, pp113–132

McDonald, D. J. and Boucher, C. (1999) 'Towards mapping the fynbos for the revised vegetation map of South Africa', in J. Timberlake and S. Kativu (eds) *African Plants: Biodiversity, Taxonomy and Uses*, Royal Botanic Gardens, Kew

McKenna, D. J., Luna, L. E. and Towers, G. N. (1995) 'Biodynamic constituents in ayahuasca admixture plants: An uninvestigated folk pharmacopeia', in R. E. Schultes and S. Von Reis (eds) *Ethnobotany*, Chapman and Hall, London, UK, pp349–361

McShane, T. O. (1999) *Voyages of Discovery: Four Lessons from the DGIS–WWF Tropical Forest Portfolio*, WWF, Gland, Switzerland

Meffe, G. K., Carroll, C. R. and Pulliam, H. R. (1997) *Principles of Conservation Biology*, second edition, Sinauer Associates, Inc, Sunderland, Massachusetts, US

Meikle, R. D. (1984) *Willows and Poplars of Great Britain and Ireland*, Botanical Society of the British Isles, London, UK

Meilleur, B. A. (1998) 'Clones within clones: Cosmology and esthetics and Polynesian crop selection', *Anthropologica*, vol 40, pp71–82

Meilleur, B. A. (1999) 'Hawaii: How the SSC Plants Program informs conservation and recovery of one of the world's most imperilled floras', in *Proceedings of the 16th International Botanical Congress*, St Louis, US, p207

Mendelsohn, R. and Balick, M. J. (1989) 'Valuing undiscovered pharmaceuticals in tropical forests', *Economic Botany*, vol 51, p328

Mendelsohn, R. and Balick, M. J. (1995) 'The value of undiscovered pharmaceuticals in tropical forests', *Economic Botany*, vol 49, pp223–228

Menges, E. S. (1991) 'The application of minimum viable population theory to plants', in D. A. Falk and K. E. Holsinger (eds) *Genetics and Conservation of Rare Plants*, Oxford University Press, New York, US, and Oxford, UK, pp45–61

Menges, E. S. (2000) 'Population viability analysis in plants: Challenges and opportunities', *Tree*, vol 15, pp51–56

Meyers, J.-Y. (1998) 'Polynesia's green cancer', *World Conservation*, vol 2, no 98, p7

Misana, S., Mung'ong'o, C. and Mukamuri, B. (1996) 'Miombo woodlands in the wider context: Macro-economic and inter-sectoral influences', in B. Campbell (ed) *The Miombo in Transition: Woodlands and Welfare in Africa*, Centre for International Forestry Research, Bogor, Indonesia, pp73–99

Moerman, D. E. (1998) 'Native North American food and medicinal plants: Epistemological considerations', in H. D. V. Prendergast, N. L. Etkin, D. R. Harris and P. J. Houghton (eds) *Plants for Food and Medicine*, Proceedings of the Joint Conference of the Society for Economic Botany and the International Society for Ethnopharmacology, London, 1–6 July 1996, Royal Botanic Gardens, Kew, UK, pp69–74

Moore, P. D. (1998) 'Frondless ferns lie low to survive', *Nature*, vol 392, pp661–662

Morimoto, Y. (2002) 'Safeguarding the biodiversity of the bottle-gourd in Kenya', Unpublished paper presented to the sixth Congress of the International Society for Ethnobiology, 16–20 September 2002, Addis Ababa, Ethiopia

Morimoto, Y. and Omari, P. (2002) 'Highlights of the Kenya Society of Ethnoecology Sacred Site Conservation efforts', Unpublished paper presented to the sixth Congress of the International Society for Ethnobiology, 16–20 September 2002, Addis Ababa, Ethiopia

Mutume, G. (2004) 'Accords save trade talks from collapse', *Africa Renewal*, vol 18, p3

Mwasumbi, L. B., Burgess, N. D. and Clarke, G. P. (1994) 'Vegetation of Pande and Kiono coastal forests, Tanzania', *Vegetation*, vol 113, pp71–81

Myers, N. (1996) 'The biodiversity crisis and the future of evolution', *The Environmentalist*, vol 16, pp37–47

Myers, N., Mittermeier, R. A., Mittermeier, C. G., da Fonseca, G. A. B. and Kent, J. (2000) 'Biodiversity hotspots for conservation priorities', *Nature*, vol 403, pp853–858

Nais, J. and Wilcok, C. C. (1998) 'The *Rafflesia* conservation incentive scheme in Sabah, Malaysian Borneo', *Sabah Parks Nature Journal*, vol 1, pp9–17

Nakashima, D. (1998) 'Conceptualizing nature: The cultural context of resource management', *Nature and Resources*, vol 34, pp8–22

Nakashima, D. (2000) 'What relationship between scientific and traditional systems of knowledge: Some introductory remarks', in A. M. Cetto (ed) *Science for the Twenty-First Century: A New Commitment*, UNESCO, Paris, France, p432

Naughton-Treves, L. (2001) 'Overview of Part IV', in W. Weber, L. J. T. White, A. Vedder and L. Naughton-Treves (eds) *African Rain Forest Ecology and Conservation*, Yale University Press, New Haven, US, and London, UK, pp311–313

Naughton-Treves, L. and Weber, W. (2001) 'Human dimensions of the African rain forest', in W. Weber, L. J. T. White, A. Vedder and L. Naughton-Treves (eds) *African Rain Forest Ecology and Conservation*, Yale University Press, New Haven, US, and London, UK, pp30–43

Newberry, D. M., Renshaw, E. and Brünig, E. F. (1986) 'Spatial pattern of trees in kerangas forest, Sarawak', *Vegetatio*, vol 65, pp77–89

Noble, R. L. (1990) 'The discovery of the vinca alkaloids – chemotherapeutic agents against cancer', *Biochemistry and Cell Biology*, vol 68, pp1344–1351

Northwood, P., Darter, C. and Drake, K. (2001) *Heathland Restoration at Hindhead*, Hindhead Commons Committee, National Trust, UK

Norton, T. E., Melkonian, M. and Andersen R. A. (1996) 'Algal biodiversity', *Phycologia*, vol 35, pp308–326

Noss, A. J. (2001) 'Conservation, development and "the forest people"', in W. Weber, L. J. T. White, A. Vedder and L. Naughton-Treves (eds) *African Rain Forest Ecology and Conservation*, Yale University Press, New Haven, US, and London, UK, pp313–333

Novacek, M. J. and Cleveland, E. E. (2001) 'The current biodiversity extinction event: Scenarios for mitigation and recovery', *Proceedings of the National Academy of Sciences of the United States of America*, vol 98, pp5466–5470

Ogle, B. M., Ho Thi Tuyet, Duyet, H. N. and Dung, N. N. X. (2003) 'Food, feed or medicine: The multiple functions of edible wild plants in Vietnam', *Economic Botany*, vol 57, pp103–117

Oldfield, S. (1995) 'The appropriateness of species listings', in J. Newton (ed) *First European Conference on Conservation of Wild Plants*, Plantlife, Hyères, France, pp72–74

Oldfield, S. (ed) (1997) *Cactus and Succulent Plants – Status Survey and Conservation Action Plan*, IUCN and SSC Cactus and Succulent Specialist Group, IUCN, Gland, Switzerland, and Cambridge, UK, p212

Oldfield, S. (2002) *The Trade in Wildlife: Regulation for Conservation*, Earthscan, London, UK

Oldfield, S., Lusty, C. and MacKinven, A. (1998) *The World List of Threatened Trees*, World Conservation Press, Cambridge, UK.

Olsen, C. S. (1997) *Commercial Non-timber Forestry in Central Nepal: Emerging Themes and Priorities*, PhD thesis, Royal Veterinary and Agricultural University, Copenhagen, Denmark

Olson, D. M. and Dinerstein, E. (1998) 'The Global 200: A representation approach to conserving the Earth's most biologically valuable ecoregions', *Conservation Biology*, vol 12, pp502–515

Olson, D. M. and Dinerstein, E. (2002) 'The Global 200: Priority ecoregions for global conservation', *Annals of the Missouri Botanical Garden*, vol 89, pp199–224

Oteng-Yeboah, A. A. (1998) 'Research and training needs in conservation approaches in Africa', in R. P. Adams and J. E. Adams (eds) *Conservation and Utilization of African Plants*, Missouri Botanical Gardens Press, St Louis, Missouri, US

Özhatay, N., Koyuncu, M., Atay, S. and Byfield, A. (1997) *The Wild Medicinal Plant Trade in Turkey*, Dogal Hayati Korumja Dernegi, Istanbul, Turkey

Page, S. E., Siegert, F., Rieley, J. O., Boehm, H.-D.V. and Limin, S. (2002) 'The amount of carbon released from peat and forest fires in Indonesia during 1997', *Nature*, vol 420, pp61–65

Palmer, M. (1995) 'Legal measures for plant conservation at a national level – submission from the United Kingdom', in J. Newton (ed) *First European Conference on Conservation of Wild Plant Species*, Plantlife, Hyères, France, pp65–71

Pant, R. (2002) *Customs and Conservation: Cases of Traditional and Modern Law in India and Nepal*, Kalpavriksh and International Institute of Environment and Development, Puni, India

Papageorgiou, A. C. (1997) 'Genetic structure of Mediterranean tree species as influenced by human activities', *Bulletin of the International Union of Forestry Research Organizations: Special Programme for Developing Countries*, summer, pp12–16

Paye, G. D. (2000) *Cultural Uses of Plants: A Guide to Learning about Ethnobotany*, New York Botanical Garden Press, Bronx, New York, US

PEEN (Pan-European Ecological Network) (2001) 'The Strategy Guide: In support of the Pan-European Biological and Landscape Diversity Strategy', www.strategy guide.org/at1/at1_inde.html

Peguero, B., Salazar, J. and Castillo, D. (2000) 'Usos en artesania de productos no maderables del bosqueen la artesanía, Santo Domingo, Repéblica Dominicana', *Moscosoa*, vol 11, pp189–220

Pei Shengji (2001) 'Ethnobotanical approaches of traditional medicine studies: Some experiences from Asia', *Pharmaceutical Botany*, vol 39, pp74–79

Pei Shengji (2002) 'Ethnobotany and modernisation of Traditional Chinese Medicine', Paper presented to a Workshop on Wise Practices and Experiential Learning in the Conservation and Management of Himalayan Medicinal Plants, Kathmandu, Nepal, 15–20 December 2002, supported by the Ministry of Forest and Soil Conservation, Nepal, the WWF-Nepal Program, Medicinal and Aromatic Plants Program in Asia (MAPPA), IDRC, Canada, and the WWF–UNESCO People and Plants Initiative

Pennington, R. T., Cronk, Q. C. B. and Richardson, J. A. (2004) 'Introduction and synthesis: Plant phylogeny and the origin of major biomes', *Philosophical Transactions of the Royal Society*, B, Biological Sciences, vol 359, pp1455–1464

Peterken, G. F. (1974) 'A method for assessing woodland flora for conservation using indicator species', *Biological Conservation*, vol 6, pp239–245

Peterken, G. F. (1996) *Natural Woodland: Ecology and Conservation in Northern Temperate Regions*, Cambridge University Press, Cambridge, UK

Peters, C. M. (1994) *Sustainable Harvest of Non-timber Plant Resources in Tropical Moist Forest: An Ecological Primer*, Biodiversity Support Programme, World Wide Fund for Nature, Washington, DC, US

Peters, C. M., Gentry, A. H. and Mendelsohn, R. O. (1989) 'Valuation of an Amazonian rainforest', *Nature*, vol 339, pp655–656

Peters, C. M., Purata, S. E., Chibnik, M., Brosi, B. J., López, A. M. and Ambrosio, M. (2003) 'The life and times of *Bursera glabrifolia* (H. B. K.) Engl. in Mexico: A parable of ethnobotany', *Economic Botany*, vol 57, pp431–441

Peterson, R. B. (2001) 'Conservation – for whom? A study of immigration onto DR Congo's Ituri forest frontier', in W. Weber, L. J. T. White, A. Vedder and L. Naughton-Treves (eds) *African Rain Forest Ecology and Conservation*, Yale University Press, New Haven, US, and London, UK, pp355–368

Petit, R. J., Csaikl, U., Bordacs, S., Burg, K., Coart, E., Cottrell, J., Deans, D., Dumolin-Lapegue, S., Fineschi, S., Finkeldey, R., Gillies, A., Goicoechea, P., Jensen, J., Konig, A., Lowe, A., Madsen, S. F., Matyas, G., Oledska, I., Pemonge, M., Slade, F. P. D., Tabbener, H., Taurchini, D., Van Dam, B., Ziegenhagen, B. and Kremer, A. (2002) 'Chloroplast DNA variation in European white oaks: Phylogeography and patterns of diversity based on data from over 2600 populations', *Forest Ecology and Management*, vol 156, pp5–26

Phillips, O. L. and Gentry, A. H. (1993) 'The useful plants of Tambopata, Peru: 1. Statistical hypotheses tests with a new quantitative technique', *Economic Botany*, vol 47, pp15–32

Phillips, O. L. and Gentry, A. H. (1994) 'Increasing turnover through time in tropical forests', *Science*, vol 263, pp954–958

Phillips, O. L., Gentry, A. H., Reynel, C., Wilkin, P. and B. G.-D. (1994) 'Quantitative ethnobotany and Amazonian conservation', *Conservation Biology*, vol 8, pp225–248

Phillips, O. L. and Meilleur, B. A. (1998) 'Usefulness and economic potential of the rare plants of the United States: A statistical survey', *Economic Botany*, vol 52, pp57–67

Pickersgill, B. (1998) 'Crop introductions and the development of secondary areas of diversity', in H. D. V. Prendergast, N. L. Etkin, D. R. Harris and P. J. Houghton (eds) *Plants for Food and Medicine*, Proceedings of the Joint Conference of the Society for Economic Botany and the International Society for Ethnopharmacology, London, 1–6 July 1996, Royal Botanic Gardens, Kew, UK, pp93–105

Pierce, A. R. and Laird, S. A. (2003) 'In search of comprehensive standards for non-timber forest products in the botanicals trade', *International Forestry Review*, vol 5, pp138–147

Plana, V. (2004) 'Mechanisms and tempo of evolution in the African Guineo-Congolian rainforest', *Philosophical Transactions of the Royal Society*, B, Biological Sciences, vol 359, pp1585–1594

Plant Europa (2001) *Important Plant Areas in Europe*, Plantlife, London

Plantlife (2004) *Identifying and Protecting the World's Most Important Plant Areas*, Plantlife International, Salisbury, UK

Plumptre, A. J. (2001) 'The effects of habitat change due to selective logging on the fauna of forests in Africa', in W. Weber, L. J. T. White, A. Vedder and L. Naughton-Treves (eds) *African Rain Forest Ecology and Conservation*, Yale University Press, New Haven, US, and London, UK, pp463–479

Porembski, S. and Barthlott, W. (2000) 'Biodiversity research in botany', *Progress in Botany*, vol 61, pp335–362

Prance, G. T. (1995) 'Ethnobotany today and in the future', in E. Schultes and S. von Reis (eds) *Ethnobotany: Evolution of a Discipline*, Chapman and Hall, London, pp60–68

Prance, G. T. (2003) 'Plants of northeastern Brazil: A programme in sustainable use of plant resources', in J. Lemons, R. Victor and D. Schaffer (eds) *Conserving Biodiversity in Arid Regions*, Kluwer, Boston, US, pp291–314

Prance, G. T., Beentje, H., Dransfield, J. and Johns, R. (2000) 'The tropical flora remains undercollected', *Annals of the Missouri Botanical Garden*, vol 87, pp67–71

Pressey, R. L. (1999) 'Editorial: Systematic conservation planning for the real world', *Parks*, vol 9, pp1–6

Pressey, R. L., Humphries, C. J., Margules, C. R., Vane-Wright, R. I. and Williams, P. H. (1993) 'Beyond opportunism: Key principles for systematic reserve selection', *Trends in Evolution and Ecology*, vol 8, pp124–128

Preston, C. D. (2003) 'Perceptions of change in English county Floras, 1660–1960', *Watsonia*, vol 24, pp287–304

Preston, C. D., Pearman, D. A. and Dines, T. D. (eds) (2002) *New Atlas of the British and Irish Flora*, Oxford University Press, Oxford, UK, p210

Pretty, J. N. (1998). 'Supportive policies and practice for scaling up sustainable agriculture', in N. G. Röling and M. A. E. Wagemakers (eds) *Facilitating Sustainable Agriculture: Participatory Learning and Adaptive Management in Times of Environmental Uncertainty*, Cambridge University Press, Cambridge, UK, pp23–45

Pretty, J. N., Guijt, I., Scoones, I. and Thompson, J. (1995) *A Trainer's Guide for Participatory Learning and Action*, International Institute for Environment and Development, London, UK

Pryer, K. M., Schneider, H., Smith, A. R., Cranfill, R., Wolf, P. G., Hunt, J. S. and Sipes, S. D. (2001) 'Horsetails and ferns are a monophyletic group and the closest living relatives to seed plants', *Nature*, vol 409, pp618–622

Pullin, A. S. and Knight, T. M. (2001) 'Effectiveness in conservation practice: Pointers from medicine and public health', *Conservation Biology*, vol 15, pp50–54

Rackham, O. (1976) *Trees and Woodland in the British Landscape*, J. M. Dent and Sons, Ltd, London, UK

Rackham, O. (1980) *Ancient Woodland: Its History, Vegetation and Uses in England*, Edward Arnold, London

Rackham, O. (1992) 'Mixtures, mosaics and clones: The distribution of trees within European woods and forests', in M. G. R. Cannell, D. C. Malcolm and P. A. Robertson (eds) *The Ecology of Mixed-Species Stands of Trees*, Special Publication Number 11, Blackwell Scientific Publications, Oxford, UK, pp1–20

Raeymaekers, G. and Synge, H. (1995) 'International organisations and agreements for the conservation of wild plants in Europe', in J. Newton (ed) *First European Conference on the Conservation of Wild Plants*, Plantlife, Hyères, France, pp1–22

Rajaram, S. and Braun, H.-J. (2001) 'Half a century of international wheat breeding', in P. D. S. Caligari and P. E. Brandham (eds) *Wheat Taxonomy: The Legacy of John Percival*, The Linnean Society of London, London, UK, pp137–163

Ramón-Laca, L. (2003) 'The introduction of cultivated citrus to Europe via Northern Africa and the Iberian Peninsula', *Economic Botany*, vol 57, pp502–514

Rao, A. N. (1997) 'A review of research on rattan', in A. N. Rao and V. Ramanatha Rao (eds) *Rattan: Taxonomy, Ecology, Silviculture, Conservation, Genetic Improvement and Biotechnology*, IPGRI–APO, Serdang, Malaysia, pp15–40

Raven, P. H. (1987) 'The scope of the plant conservation problem world-wide', in D. Bramwell, O. Hamann, V. Heywood and H. Synge (eds) *Botanic Gardens and the World Conservation Strategy*, Academic Press, London, UK, pp19–29

Ray, P. N. (ed) (1994) *Forestry Education in Indira Gandhi National Forest Academy: Present Status and Future Perspectives*, Indian National Science Academy, New Delhi, India, pp154–162

Rebelo, A. G. (1994) 'Iterative selection procedures: Centres of endemism and optimal placement of reserves', in B. J. Huntley (ed) *Botanical Diversity in Southern Africa*, National Botanical Institute, Pretoria, South Africa, pp231–257

Reganold, J. P., Glover, J. D., Andrews, P. K. and Hinman, H. R. (2001) 'Sustainability of three apple production systems', *Nature*, vol 410, pp926–930

Rejmánek, M. (1996) 'Species richness and resistance to invasions', in G. H. Orians, R. Dirzo and J. H. Cushman (eds) *Biodiversity and Ecosystem Processes in Tropical Forests*, Springer-Verlag, Berlin, Germany, pp153–172

Rejmánek, M. (1999) 'Holocene invasions: Finally the resolution the ecologists were waiting for!', *Trends in Ecology and Evolution*, vol 14, pp8–10

Rejmánek, M. (2000) 'Invasive plants: Approaches and predictions', *Austral Ecology*, vol 25, pp497–506

Rejmánek, M. and Brewer, S. W. (2001) 'Vegetative identification of tropical woody plants: State of the art and annotated bibliography', *Biotropica*, vol 33, pp214–228

Rich, T. C. G. (2003) '*Galeopsis segetum* Neck. (Lamiaceae), downy hemp-nettle: Native or introduced in Britain?', *Watsonia*, vol 24, pp1–3

Richard, C. E. (2003) 'Co-management processes to maintain livestock mobility and biodiversity in alpine rangelands of the Tibetan Plateau', in J. Lemons, R. Victor and D. Schaffer (eds) *Conserving Biodiversity in Arid Regions*, Kluwer, Boston, US, pp249–273

Richards, P. and Ruivenkamp, G. (1997) *Seeds and Survival: Crop Genetic Resources in War and Reconstruction in Africa*, International Plant Genetic Resources Institute, Rome, Italy

Richardson, D. M., Pysek, P., Rejmánek, M., Barbour, M. G., Panetta, F D. and West, C. J. (2000) 'Naturalization and invasion of alien plants: Concepts and definitions', *Diversity and Conservation*, vol 6, pp93–107

Richardson, D. M., van Wilgen, B. W., Higgins, S. I., Trinder-Smith, T. H., Cowling, R. M. and McKell, D. H. (1996) 'Current and future threats to plant biodiversity on the Cape Peninsula, South Africa', *Biodiversity and Conservation*, vol 5, pp607–647

Ricketts, T. H. (2004) 'Tropical forest fragments enhance pollinator activity in nearby coffee crops', *Conservation Biology*, vol 18, pp1262–1271

Riley, K. W. (1997) 'Foreward', in A. N. Rao and V. Ramanatha Rao (eds) *Rattan: Taxonomy, Ecology, Silviculture, Conservation, Genetic Improvement and Biotechnology*, IPGRI–APO, Serdang, Malaysia

Robinson, M. D. (2003) 'The importance of native trees in sustaining biodiversity in arid lands', in J. Lemons, R. Victor and D. Schaffer (eds) *Conserving Biodiversity in Arid Regions*, Kluwer, Boston, US, pp395–412

Rodgers, W. A. (1998) 'DNA Bank-Net: A perspective', in R. P Adams and J. E. Adams (eds) *Conservation and Utilization of African Plants*, Missouri Botanical Gardens Press, St Louis, Missouri, US, pp9–24

Rodrigues, A. S. L., Andelman, S. J., Bakarr, M. I., Boltani, L., Brooks, T. M., Cowling, R. M., Fishpool, L. D. C., da Fonseca, G. A. B., Gaston, K. J., Hoffmann, M., Long, J. S., Marquet, P. A., Pilgrim, J. D., Pressey, R. K., Schipper, J., Sechrest, W., Stuart, S. N., Underhill, L. G., Waller, R. W., Watts, M. E. J. and Zie Yan (2004) 'Effectiveness of the global protected area network in representing species diversity', *Nature*, vol 428, pp640–643

Rodwell, J. S., Schaminée, J. H. J., Mucina, L., Pignatti, S., Dring, J. and Moss, D. (2002) *The Diversity of European Vegetation: An Overview of Phytosociological Alliances and Their Relationships to EUNIS Habitats*, Report EC–LNV no 2002/054, Wageningen, The Netherlands

Röling, N. (1988) *Extension Science: Information Systems in Agricultural Development*, Cambridge University Press, Cambridge, UK

Röling, N. G. and Wagemakers, M. A. E. (1998) *Facilitating Sustainable Agriculture: Participatory Learning and Adaptive Management in Times of Environmental Uncertainty*, Cambridge University Press, Cambridge, UK

Ruddle, K. (2000) 'Systems of knowledge: Dialogue, relationships and process', in A. M. Cetto (ed) *Science for the Twenty-First Century: A New Commitment*, UNESCO, Paris, France, pp432–435

Sahai, S. (2004) 'Protection of indigenous knowledge: The Indian experience', Gene Campaign, New Delhi, India, www.genecampaign.org/ikfolio/graham-sahai.doc

Samuelsson, J., Gustafsson, L., and Ingelög, T. (1994) *Dying and Dead Trees: A Review of Their Importance for Biodiversity*, Swedish Threatened Species Unit, Uppsala, Sweden

Schama, S. (1995) *Landscape and Memory*, HarperCollins, London, UK

Schippmann, U. (1995) 'Plant uses and species risk – from horticultural to medicinal plant trade', in J. Newton (ed) *First European Conference on the Conservation of Wild Plants*, Plantlife Hyères, France, pp161–166

Schippmann, U. (2001) *Medicinal Plants Significant Trade Study*, German Federal Agency for Nature Conservation, Bonn, Germany

Schippmann, U., Leaman, D. J. and Cunningham, A. B. (2002) *Impact of Cultivation and Gathering of Medicinal Plants on Biodiversity: Global Trends and Issues*, Inter-Department Working Group on Biology Diversity for Food and Agriculture, FAO, Rome, Italy

Schmitt, S. F. and Maingi, D. R. (2005) 'Certification and woodcarving', in A. B. Cunningham, B. Belcher and B. M. Campbell (eds) *Carving Out a Future: Forests, Livelihoods and the International Woodcarving Trade*, Earthscan, London, UK

Schneider, H., Schuettpelz, E., Pryer, K. M., Cranfill, R., Magallón, S. and Lupia, R. (2004) 'Ferns diversified in the shadow of angiosperms', *Nature*, vol 428, pp553–557

Schrönrogge, K., Walker, P. and Crawley, M. J. (2000) 'Parasitoid and inquiline attack in the galls of four alien, cynipid wasps: Host switches and the effect on parasitoid sex ratios', *Ecological Entomology*, vol 25, pp208–219

Scotland, R. W. and Wortley, A. (2003) 'How many species of seed plants are there?', *Taxon*, vol 52, pp101–104

Shackleton, C. M., Shackleton, S. E., Netshiluvhi, T. R. and Geach, B. (2002). 'Direct-use value of woodland resources in three rural settlements, South Africa', in B. M. Campbell and M. K. Luckert (eds) *Uncovering the Hidden Harvest: Valuation Methods for Woodland and Forest Resources*, Earthscan, London, UK, pp5–7

Shanley, P. and Luz, L. (2003) 'The impacts of forest degradation on medicinal plant use and implications for health care in eastern Amazonia', *BioScience*, vol 53, pp573–584

Sheldon, J. W., Balick, M. J. and Laird, S. A. (1997) 'Medicinal plants: Can utilization and conservation coexist?', *Advances in Economic Botany*, vol 12, pp1–104

Shiva, V. (1996) *Protecting Our Biological and Intellectual Heritage in the Age of Biopiracy*, The Research Foundation for Science, Technology and Natural Resources Policy, New Delhi, India

Siegert, F., Ruecker, G., Hinrichs, A. and Hoffmann, A. A. (2001) 'Increased damage from fires in logged forests during droughts caused by El Niño', *Nature*, vol 414, pp437–440

Sillitoe, P. (2003) 'The gender of crops in the Papua New Guinea highlands', in P. L. Howard (ed) *Women and Plants*, Zed Books, London, UK, pp165–180

Simons, A. J. and Leakey, R. R. B. (2004) 'Tree domestication in agroforestry', *Agroforestry Systems*, vol 61–62, pp167–181

SKEP (Succulent Karoo Ecosystem Programme) (2003) *The Succulent Karoo Ecosystem Programme: 20-Year Strategy*, Fire Escape Design and Publishing, Cape Town, South Africa

Slocum, M., Aide, T. M., Zimmerman, J. K. and Navaro, L. (2000) 'La vegetación leñosa en helechales y bosques de ribera en la Reserva Cienífica Ébano Verde, República Dominicana', *Moscosoa*, vol 11, pp38–56

Smart, J. (1999) 'Plants in peril: How shall we save them?', *Oryx*, vol 33, pp279–280

Smith, M. (2002) 'Opportunities within the school curriculum', in *How Do We Find and Train the Next Generation of Field Botanists?* Presentation at a Joint Meeting of the Botanical Society of the British Isles and the Linnean Society, 4 April 2002, London, UK

Song Yiching and Jiggins, J. (2003) 'Women and maize breeding: The development of new seed systems in a marginal area of south-west China', in P. L. Howard (ed) *Women and Plants*, Zed Books, London, UK, pp273–288

Srivastava, R. (2000) 'Studying the information needs of medicinal plant stakeholders in Europe', *TRAFFIC Dispatches*, vol 15, p5

Stattersfield, A. J. M., Crosby, M. J., Long, A. J. and Wege, D. C. W. (1998) 'Endemic bird areas of the world: Priorities for biodiversity conservation', *BirdLife International*, Cambridge, UK

Stedman-Edwards, P. (1998) *Root Causes of Biodiversity Loss*, WWF, Gland, Switzerland

Steiner, A. M., Ruckenbauer, P. and Goecke, E. (1997) 'Maintenance in genebanks, a case study: Contaminations observed in the Nürnberg oats of 1831', *Genetic Resources and Crop Evolution*, vol 44, pp533–538

Stewart, K. M. (2003) 'The African cherry (*Prunus africana*): From hoe-handles to the international herb market', *Economic Botany*, vol 57, pp559–569

Stiller, J. W. and Hall, B. D. (1997) 'The origin of red algae: Implications for plastid evolution', *Proceedings of the National Academy of Science US*, vol 94, pp4520–4525

Stork, N. E. (1993) 'How many species are there?', *Biodiversity and Conservation*, vol 2, pp215–232

Stuart, S. N. (1981) 'A comparison of the avifaunas of seven East African forest islands', *African Journal of Ecology*, vol 19, pp133–151

Subrat, N. (2002) 'Ayurvedic and herbal products industry: An overview', in *Workshop on Wise Practices and Experiential Learning in the Conservation and Management of Himalayan Medicinal Plants*, Kathmandu, Nepal, 15–20 December 2002, supported by the Ministry of Forest and Soil Conservation, Nepal, the WWF-Nepal Programme, Medicinal and Aromatic Plants Programme in Asia (MAPPA), IDRC, Canada, and the WWF–UNESCO People and Plants Initiative, Kathmandu, Nepal

Synge, H. (2001) *European Plant Conservation Strategy: The Underlying Issues*, Plant Europa Conference, Pruhonice, Czech Republic

Tansley, A. G. (1946) *Our Heritage of Wild Nature: A Plea for Organized Nature*, Conservation Readers Union/Cambridge University Press, London, UK

Teklehaimanot, Z. and Healey, J. (2001) 'Biodiversity conservation in ancient church and monastery yards in Ethiopia', in *Biodiversity Conservation in Ancient Church and Monastery Yards in Ethiopia*, University of Bangor, UK, and Ethiopian Wildlife and Natural History Society, Addis Ababa, Ethiopia, pp2–4

ten Kate, K. and Laird, S. A. (1999) *The Commercial Use of Biodiversity*, Earthscan, London, UK

Tennant, D. J. (2004) 'A re-assessment of montane willow (*Salix L.*, Salicaceae) hybrids in Scotland', *Watsonia*, vol 25, pp65–82

Teshome, A., Fahrig, L., Torrance, J. K., Lambert, J. D., Arnason, T. J. and Baum, B. R. (1999) 'Maintenance of sorghum (*Sorghum bicolor*, Poaceae) landrace diversity by farmers' selection in Ethiopia', *Economic Botany*, vol 53, pp79–88

Tibatemwa-Ekirikubinza (1999) *Women's Violent Crime in Uganda*, Fountain Publishers, Kampala, Uganda

Tilman, D. (1998) 'The greening of the green revolution', *Nature*, vol 396, pp211–212

Tilman, D., Fargione, J., Wolff, B., D'Antonio, C., Dobson, A., Howarth, R., Schindler, D., Schlesinger, W. H., Simberloff, D. and Swackhamer, D. (2001) 'Forecasting agriculturally driven global environmental change', *Science*, vol 292, pp281–284

Timberlake, J. R. and Müller, T. (1994) 'Identifying and describing areas for vegetation conservation in Zimbabwe', in B. R. Huntley (ed) *Botanical Diversity in Southern Africa*, National Botanical Institute, Pretoria, South Africa, pp125–139

Toledo, V. M. (1995) 'New paradigms for a new ethnobotany: Reflections on the case of Mexico', in R. E. Schultes and S. von Reis (eds) *Ethnobotany: Evolution of a Discipline*, Chapman and Hall, London, UK, pp75–88

Trewhella, K. E., Leather, S. R. and Day, K. R. (2000) 'Variation in the suitability of *Pinus contorta* (lodgepole pine) to feeding by three pine defoliators, *Panolia flammea, Neodiprion sertifer* and *Zeiraphera diniana*', *Journal of Applied Entomology*, vol 124, pp11–17

Tsouvalis, J. (2000) *A Critical Geography of Britain's State Forests: An Exploration of Processes of Reality Construction*, Oxford University Press, Oxford, UK

Turner, I. M. and Crolett, R. T. (1996) 'The conservation value of small, isolated fragments of lowland tropical rain forest', *Trends in Ecology and Evolution*, vol 11, pp330–333

Tuxill, H. and Nabhan, G. P. (2001) *People, Plants and Protected Areas*, Earthscan, London

Twaddle, M. (1993) *Kakungulu and the Creation of Uganda*, James Currey, London, UK

Uniyal, R. C. (2000) 'Research for medicinal plants cultivation in India – a reference book', in *Medicinal Plants Stakeholders' Meeting*, TRAFFIC-India, New Delhi, India

Uniyal, S. K., Awasthi, A. and Rawat, G. S. (2002) 'Current status and distribution of commercially exploited medicinal and aromatic plants in upper Gori valley, Kumaon Himalaya, Uttaranchal', *Current Science*, vol 82, pp1246–1252

Valencia, R., Pitman, N., Susna, L.-Y. and Jørgensen, P. M. (eds) (2000) *Libra roja de las plantas endémicas del Ecuador 2000*, Herbario QCA, Pontificia Universidad Católica del Ecuador, Quito, Ecuador

Van den Eynden, V., Cueva, E. and Cabrera, O. (2003) 'Wild foods from southern Ecuador', *Economic Botany*, vol 57, pp576–603

van Opstal, A. (2001) 'The Pan-European Ecological Network and IPAs', lecture at Third European Conference on the Conservation of Wild Plants, Prague, 23–28 June 2001

Van Rammsdonk, L. W. D. (1995) 'The effect of domestication on plant evolution', *Acta Botanica Neerlandica*, vol 44, pp421–438

Van Sway, C. and Warren, M. (eds) (2003) *Prime Butterfly Areas of Europe*, Ministry of Agriculture, Nature Management and Fisheries, The Netherlands

Vandergeest, P. (1996) 'Property rights in protected areas: Obstacles to community involvement as a solution in Thailand', *Environmental Conservation*, vol 23, pp259–268

Veen, P. (2001) 'Natural Grassland Inventory Project', Presentation at Third European Conference on the Conservation of Wild Plants, Prague, 23–28 June

Vogel, J. C., Rumsey, F. J., Schneller, J. J., Barrett, J. A. and Gibby, M. (1999) 'Where are the glacial refugia in Europe? Evidence from pteridophytes', *Biological Journal of the Linnean Society*, vol 66, pp23–37

Wade, R. (1987) 'The management of common property resources: Collective action as an alternative to privatisation or state regulation', *Cambridge Journal of Economics*, vol 11, pp95–106

Walter, K. S. and Gillett, H. J. (eds) (1997) *IUCN Red List of Threatened Plants*, World Conservation Union, Gland, Switzerland

Walters, C. J. (1986) *Adaptive Management of Renewable Resources*, Macmillan, New York, US

Walters, C. J. (ed) (1997) *Adaptive Policy Design: Thinking at Large Spatial Scales*, Springer-Verlag, New York, US

Walters, C. J. and Holling, C. S. (1990) 'Large-scale management experiments and learning by doing', *Ecology*, vol 71, pp2060–2068

Wang Jinxiu, Liu Hongmao, Hu Huabin and Gao Lei (2004) 'Participatory approach for rapid assessment of plant diversity through a folk classification system in a tropical rainforest: Case study in Xishuangbanna, China', *Conservation Biology*, vol 18, pp1139–1142

WCS (Wildlife Conservation Society) (2002) www.peopleandplanet.net

Weber, W. and Vedder, A. (2001) 'Preface', in W. Weber, L. J. T. White, A. Vedder and L. Naughton-Treves (eds) *African Rain Forest Ecology and Conservation*, Yale University Press, New Haven, US, ppix–xii

Welch, D. (2003) 'A reconsideration of the native status of *Linnaea borealis* L. (Caprifoliaceae) in lowland Scotland', *Watsonia*, vol 24, pp427–432

Wells, M., Guggenheim, S., Khan, A., Wardojo, W. and Jepson, P. (1999) *Investing in Biodiversity: A Review of Indonesia's Integrated Conservation and Development Projects*, World Bank, Washington, DC, US

West, H. W. (1972) *Land Policy in Buganda*, Cambridge University Press, Cambridge, UK

Whistler, A. (1998) 'Weeds of Samoa', *Aliens*, vol 7, pp8–9

White, F. (1983) *The Vegetation of Africa: A Descriptive Memoir to Accompany the UNESCO/AETFAT/UNSO Vegetation Map of Africa*, UNESCO, Paris, France

White, L. J. T. (2001) 'The African rain forest', in W. Weber, L. J. T. White, A. Vedder and L. Naughton-Treves (eds) *African Rain Forest Ecology and Conservation*, Yale University Press, New Haven, pp3–29

WHO (World Health Organization) (2003) *Guidelines on Good Agricultural and Collection Practices for Medicinal Plants*, WHO, Geneva, Switzerland

WHO, IUCN (World Conservation Union) and WWF (World Wide Fund for Nature) (1993) *Guidelines on the Conservation of Medicinal Plants*, IUCN, Gland, Switzerland, and Cambridge, UK

Wightman, A. (1996) *Who Owns Scotland*, Canongate, Edinburgh, UK

Wild, R. G. and Mutebi, J. (1996) *Conservation Through Community Use of Plant Resources*, People and Plants Working Paper no 5, Division of Ecological Sciences, UNESCO, Paris, France

Wilkie, D. S. and Laporte, N. (2001) 'Forest area and deforestation in central Africa', in W. Weber, L. J. T. White, A. Vedder and L. Naughton-Treves (eds) *African Rain Forest Ecology and Conservation*, Yale University Press, New Haven, US, pp119–139

Williams, M. (1997) 'Ecology, imperialism and deforestation', in T. Griffiths and L. Robin (eds) *Ecology and Empire: Environmental History of Settler Societies*, University of Natal Press, Pietermaritzburg, South Africa, pp169–184

Wilshusen, P. R., Brechin, S. R., Fortwangler, C. L. and West, P. C. (2002) 'Reinventing a square wheel: Critique of a resurgent "protection paradigm" in international biodiversity conservation', *Society and Natural Resources*, vol 15, pp41–64

Wilson, M. (2003) 'Exchange, patriarchy and status: Women's homegardens in Bangladesh', in P. L. Howard (eds) *Women and Plants*, Zed Books, London, UK, pp211–225

Wong, J. L. G., Thornber, K. and Baker, N. (2001) *Resource Assessment of Non-Wood Forest Products*, FAO, Rome, Italy

Woods, P. (1989) 'Effects of logging, drought and fire on structure and composition of tropical forests in Sabah, Malaysia', *Biotropica*, vol 21, pp290–298

Woods, R. (2002) 'Botanising on the lawn', *Plantlife*, Autumn, p17

Woodward, F. I., Lomas, M. R. and Kelly, C. K. (2004) 'Global climate and the distribution of plant biomes', *Philosophical Transactions of the Royal Society* B, Biological Sciences, vol 359, pp1465–1476

Worah, S., Svendsen, D. S. and Ongleo, C. (1999) *Integrated Conservation and Development: A Trainer's Manual*, WWF, Godalming, UK

Wright, S. J. and Calderon, O. (1995) 'Phylogenetic patterns among tropical flowering phenologies', *Journal of Ecology*, vol 83, pp937–948

WWF (World Wide Fund for Nature) (1997) *The Year the World Caught Fire*, WWF, Gland, Switzerland

WWF (2004) *Ecoregions*, www.worldwildlife.org/science/ecoregions.cfm

WWF and IUCN (World Conservation Union) (1994–1997) *Centres of Plant Diversity: A Guide and Strategy for Their Conservation*, IUCN Publications Unit, Cambridge, UK

Wyse Jackson, P. S. and Sutherland, L. A. (2000) *International Agenda for Botanic Gardens in Conservation*, Botanic Gardens Conservation International, Kew, UK

Xiao Pei-Gen and Peng Yong (1998) 'Ethnopharmacology and research on medicinal plants in China', in H. D. V. Prendergast, N. L. Etkin, D. R. Harris and P. J. Houghton (eds) *Plants for Food and Medicine*, Proceedings of the Joint Conference of the Society for Economic Botany and the International Society for Ethnopharmacology, London, 1–6 July 1996, Royal Botanic Gardens, Kew, UK, pp31–39

Yaa, N.-B. (2001) 'Indigenous versus introduced biodiversity conservation strategies', in W. Weber, L. J. T. White, A. Vedder and L. Naughton-Treves (eds) *African Rain Forest Ecology and Conservation*, Yale University Press, New Haven, US, pp385–394

Yajuan J. Liu and Hall, B. D. (2004) 'Body plan evaluation of ascomycetes, as inferred from an RNA polymerase II phylogeny', *Proceedings of the National Academy of Sciences*, vol 101, pp4507–4512

Yan Zhi-Jian (2004) 'General situation and development of medicinal orchids in Guizhou of China', Paper presented to International Workshop for Revision of Guidelines for the Conservation of Medicinal Plants, Kunming, China

Yin-Long, Yangrae Cho, Cox, C. and Palmer, J. D. (1998) 'The gain of three mitochondrial introns identifies liverworts as the earliest land plants', *Nature*, vol 394, pp671–674

Yin-Long Qlu, Jungho Lee, Bernasconi-Quadroni, F., Soltis, D. E., Zanis, M., Zimmer, E. A., Zhiduan Chen, Savolainen, V. and Chase, M. W. (1999) 'The earliest angiosperms: Evidence from mitochondrial, plastid and nuclear genomes', *Nature*, vol 402, pp404–407

Young, F. (2002) *Benmore Botanic Garden*, Edinburgh Botanic Garden, Edinburgh, UK

Zerner, C. (1999) *Justice and Conservation*, The Rainforest Alliance, New York, US

Zhang Qitai, Tao Guoda, Gong Xun, Feng Zhizhou, Yang Zenghong and Dao Jianhong (eds) (2003) *Wild Ornamental Fruit Plants from Yunnan, China*, Foreign Languages Press, Beijing, China, p245

Zich, F. and Compton, J. (2001) *The Final Frontier: Towards Sustainable Management of Papua New Guinea's Agarwood Resource*, TRAFFIC Oceania, Australia

Index of Scientific Names of Plant Species, Genera and Families

General Index

Aborigines 86, 93, 109
access and benefit sharing 62–69, 168
acid rain 32–33
adaptive management 14, 185, 199–200
Africa 7, 45, 30, 39, 45, 51, 79–82, 84,
    88, 90, 95–96, 110–112, 120–121,
    125, 134–136, 138, 140, 159,
    171–172, 174, 181, 210, 257, 262
African cherry 23
African Union 37, 45, 210
agreements (project/research) 64,
    67–69, 200–201, 222, 267
agricultural subsidies 22, 39, 90
agriculture 24–26, 33, 88–89, 95–97,
    104–107, 153, 197
aid (development assistance) 43, 168
algae 2, 29, 132–133
alien species 140–142
    see also invasive species
Alliance of Religions and Conservation
    (ARC) 49
alternative resources 147, 168, 192,
    204–205, 263
Amazon 22, 24, 27, 36, 88–89, 94, 120,
    125, 135, 161, 166, 211, 262
amchis 55, 63, 79, 153–154, 212–214
America (North and South) 3, 20–22,
    25, 27, 29, 48, 54, 69, 81, 84,
    88–89, 97, 100, 109–110, 125,
    135–136, 141–142, 159, 165, 181,
    212, 262
ancient woodland 83
Andes 88, 100, 135, 141–142, 164–165
angiosperms 2, 132–133
animal fodder 93, 104–107, 220
animals, wild (interactions with plants)
    5–6, 20–21, 84
applied ethnobotany xvii–xviii, 59,
    215–217
    see also ethnobotany

archaeological sites 181–182
area of occupancy/occurrence (species)
    151, 157–158
Areas of Special Scientific Interest (ASSI)
    178
Australia 3–4, 21, 25–27, 30, 47,
    49–50, 52, 84, 86, 93, 98, 109–110,
    129, 134–135, 137, 165, 180
Australian Network for Plant
    Conservation 47, 52, 246
Ayubia National Park 58–59, 104–107,
    205, 220
ayurveda 54–55, 90, 122, 210, 258

bamboo 94–95, 246
banana 110, 141, 247
bans (wild plant collection) 268–269,
    274
benefit-sharing agreements 65–69, 122,
    168
Bern Convention 45
Bhutan 38, 214, 268
Birdlife International 110, 166
biodiversity prospecting 62–69,
    125–126, 209
biosphere reserves 43, 101–102
Bombay Natural History Society 49
botanic gardens 46–47, 142, 239–242
Botanic Gardens Conservation
    International (BGCI) 41, 47
botanical hotspots 164–166
botanical institutes 46–47
Botanical Society of South Africa 49
Botanical Society of the British Isles 49
boundaries of management
    areas/projects 173, 196, 218, 220
Brazil 22, 24, 27, 52, 164, 244, 252
bromeliads 28, 135
bryophytes 2, 33, 132–133, 159, 178
Buganda 171